W9-AKY-155

PRINCIPLES OF INFORMATION SYSTEMS ANALYSIS AND DESIGN

PRINCIPLES OF INFORMATION SYSTEMS ANALYSIS AND DESIGN

Harlan D. Mills
IBM CORPORATION
BETHESDA, MARYLAND
AND
UNIVERSITY OF MARYLAND
COLLEGE PARK, MARYLAND

Richard C. Linger
IBM CORPORATION
BETHESDA, MARYLAND

Alan R. Hevner
COLLEGE OF BUSINESS AND MANAGEMENT
UNIVERSITY OF MARYLAND
COLLEGE PARK, MARYLAND

1986

ACADEMIC PRESS, INC.
Harcourt Brace Jovanovich, Publishers
Orlando San Diego New York Austin
Boston London Sydney Tokyo Toronto

ACADEMIC PRESS, INC.
Orlando, Florida 32887

United Kingdom Edition published by
ACADEMIC PRESS INC. (LONDON) LTD.
24–28 Oval Road, London NW1 7DX

Library of Congress Cataloging in Publication Data

Mills, Harlan D., Date
Principles of information systems analysis and design.

Includes index.
1. Management information systems. 2. System analysis. 3. System design. I. Linger, Richard C., Date . II. Hevner, Alan R. III. Title.
T58.6.M537 1986 658.4'038 86-10954
ISBN 0–12–497545–3 (alk. paper)

PRINTED IN THE UNITED STATES OF AMERICA

86 87 88 89 9 8 7 6 5 4 3 2 1

To
Lolly
Marie and Dick
Susan

Contents

Chapter 4 The Clear Box Behavior of Information Systems

Chapter 5 The Box Structure of Information Systems

Chapter 6 Information Systems Management

Chapter 7 Syntax Structures in Information Systems

Chapter 8 Data Structures in Information Systems

Preface

Information systems development should be practiced as a systematic business engineering discipline. This business engineering discipline can be based on principles of computer science and software engineering that apply directly to business problems. The primary benefit of this is effective management control over information systems development and evolution.

This book presents a systematic approach to the teaching of information systems development. It is based on successful principles of software engineering and systems engineering, which have been distilled to eliminate extraneous complexities and simplified to bare essentials for information systems development. This approach permits a more explicit study of business processes and information systems than do approaches that dwell more on details of computer systems than on the business processes supported by information systems.

The book presents a box structure methodology for information systems development. This methodology uses just three system structures that can be nested over and over in a hierarchical structure. These system structures provide a way to analyze and design information systems and their subsystems to increasing levels of detail without getting lost in that detail.

The three system structures are called black box, state machine, and clear box structures. They give three views of the same information system or any of its subsystems. The black box gives the external, or user's, view. This view consists of stimuli to the system or subsystem and the responses for all stimulus histories. The state machine gives an intermediate view that defines data stored from stimulus to stimulus; that is, it opens up the system to the extent of making its state data visible. The clear box, as the name suggests, opens up the system one more step in an internal, or designer's, view. This view describes how the data are processed and will refer to smaller black boxes in that description. At this point, the hierarchical top-down description can be repeated for each of these other black boxes, identifying next their state machine descriptions and then their clear box descriptions, which refer to even smaller black boxes, and so on.

The power of these simple system structures lies in their universality. Any information system or subsystem, no matter what subject it deals with, can be described as a black box and a corresponding state machine, which can be expanded into a clear box and new black boxes. Thus, these three structures prescribe a methodology for information systems analysis and design that can be studied and applied in any situation for better management control and visibility.

We teach principles rather than appearances in information systems development. The major innovation is the rigor with which logical principles are derived from mathematical foundations and business processes and then taught for information systems development. We can teach a trick dog the appearances of arithmetic but not its principles. When we teach children the principles of arithmetic, we give them an entirely different power than what we give trick dogs. Children, grown to adults, can apply these principles over and over, almost without thought, while solving the real problems they meet. The principles of box structures can also be applied, over and over, in solving the real problems in information systems development.

Box structure principles of information systems are introduced in a spiral approach. Chapter 1 introduces the methodology and briefly illustrates many of the principles. Chapters 2, 3, and 4 develop the principles of black boxes, state machines, and clear boxes, respectively. Chapter 5 integrates these principles into a box structure methodology. Chapter 6 discusses the use of this methodology in managing information systems development and operations. Chapters 7 and 8 develop syntax and data structuring methods of direct use in box structured information systems development. As indicated, Chapters 1–6 represent a course in box structure methodology for information systems development and management. Chapters 7 and 8 can be used selectively to provide deeper techniques for uses of syntax and data in information systems.

The intended audience of *Principles of Information Systems Analysis and Design* includes university students in information systems, practicing professionals, and managers. Students will learn a systematic methodology for information system analysis and design that can be applied throughout their careers. With the logical principles understood, they will be free to focus on the hardest part of information systems development, namely, the application of these principles in business environments, working with people. Practicing professionals will recognize in box structures a means to express their work products in a more systematic and understandable form. Box structures permit information system professionals to communicate with greater precision and completeness. Managers will find box structures a sound basis for defining work, measuring progress, and communicating with users. In every case, the result will be better manageability and responsiveness of information systems to the needs of the business enterprise.

Acknowledgments

The writing of this book was triggered, in part, by a concern of Mr. Vincent N. Cook, President of the Federal Systems Division of IBM, for the technical vitality of systems engineers in complex information systems development at IBM. This concern led to a concerted effort to simplify and codify central concepts of systems and software engineering that have proven successful in the development of complex real-time control systems, such as are reflected in the curriculum of the IBM Software Engineering Institute.

It is a pleasure to acknowledge the encouragement of Dr. Rudolph Lamone, Dean of the College of Business and Management at the University of Maryland, who saw the need to teach the principles of systems engineering in the Information Systems curriculum.

Authors Mills and Linger appreciate the support of IBM for the research and writing of this material. Author Hevner appreciates the support of thc Univcrsity of Maryland. The authors also appreciate the word processing support of IBM, particularly the help of Evelyn Brown in managing and Susan Gary in coordinating this support.

Finally, the authors acknowledge the patience and suggestions of the several hundred students at the University of Maryland who have used previous versions of this material in their coursework and of the many colleagues at IBM who have repeatedly demonstrated the value of box structures, particularly state machines, in the development of large, complex real-time control systems.

Chapter 1 Information Systems Development

1.1 BUSINESS INFORMATION SYSTEMS

Preview: Businesses need information systems to accomplish organizational objectives. These systems can be categorized as Data Processing Systems, Management Information Systems, or Decision Support Systems. The people involved with information systems development are managers, users, operators, and the developers themselves. The development and use of these systems require solutions for logic problems and for people problems.

1.1.1 Business Systems and Information Systems

Information processing is a common ingredient in all businesses. Whatever else they do, make automobiles, sell real estate, run hotels, or whatever, they all process information.

Each business is a system and has many subsystems which are systems in their own right, for example, marketing, manufacturing, financial, and personnel systems. All of these systems are run on the basis of information, which is essential to organize and coordinate the activities of employees in the pursuit of business objectives. So every business has an information system which is used to help manage its many parts. The information system may be based solely on written and spoken communication, particularly in very small businesses, but more likely will require use of computers to manage the gathering, storing, processing, and distributing of the information itself.

The same information may be used in a business in different ways at different times by different people. For example, the rate of pay for a certain type of work in a business will be used not only in the personnel system to calculate payrolls, but also to determine manufacturing or marketing costs, set product prices, and determine cash flow requirements. Thus, the information system of a business must get the right information to the right people at the right time. To achieve this, information systems must be properly designed and managed, just as for the other systems of a business.

The computer is a tool of business just as is a machine or a truck. Computers can be used for manufacturing (robots), research (simulation), and engineering (scientific computing). But a very important use of computers is in the information system of a business to help in the management of the other systems. Note that the information system of a business owes its existence to the business and not to the computers. That is, information systems exist because businesses need them, not because computers make them possible! But computers can permit efficiencies and capabilities for information systems that are not possible any other way. We call such systems **computer** (based) **information systems** (CIS).

1.1.2 Categories of Information Systems

There are three general categories of computer information systems:

1. Data Processing Systems (DPS). These systems carry out straightforward, voluminous, operational data processing for posting results of day to day (minute to minute) operations. Examples include inventory control, customer accounting, and question (query) answering about the status of operations, such as queries on inventory levels, customer balances, etc.

2. Management Information Systems (MIS). These systems carry out automatic analysis of data, possibly periodic, on demand, or triggered by certain data in the DPS, to generate information useful in managing the business. Examples include monthly sales analysis, special analysis of customer complaints, or analysis of raw material quality data triggered by manufacturing defect data in the DPS.

3. Decision Support Systems (DSS). These systems carry out special analyses which address strategic management decision problems. Examples include long-range planning for factory acquisition or construction or selecting raw material suppliers with the best balance between quality, cost, and dependability of supply.

The principles and procedures needed for developing any of these systems are the same. A clerk in a DPS, a middle manager in an MIS, or a long range planner in a DSS all deal with the system as users. In each case it must help them do their work. The developer must identify the needs of the business, specify the CIS behavior required, and design a system to implement it. In every category, the needs of the business or organization must be understood in order to visualize and create the most appropriate system behavior possible.

1.1.3 People in Information Systems

People in information systems can be divided into four categories, namely, managers, users, operators, and developers. Most people will be involved with information systems in each of these roles sometime during their lifetimes. The principal objectives of the system will vary depending upon the category of personal relationship. Managers look for cost/benefit advantages, employee productivity, and customer satisfaction. System users expect easy-to-use interaction, accuracy, and time savings. Operators need reliable, well-documented systems. The developer's objective, then, is to build an information system that will satisfy all of the above objectives in the most effective and cost-efficient manner and be simple to maintain and modify.

The developers include the information systems analysts and designers. The first problem of the developers is to find out what the managers want in an information system. This first problem is one of communication about an enterprise and its needs and the opportunities that information systems technology has to offer. It is an increasingly vital part of the planning process in every enterprise. The second problem of the developers is to find out what the users can use. This second problem is also one of communication with managers and users about the skills and expectations of users. The third problem involves managers and operators in understanding the skills and expectations of operators. The fourth problem of the developers is the developers themselves, specifically, how to manage their own efforts to create the required and usable information system for the managers, users, and operators.

1.1.4 Problems of Logic and People in Information Systems

The deepest and most persistent problems of information systems are people problems. The people problems that the developers face are difficult at best. But they are made even more difficult or impossible by poorly addressed logic problems in information systems development.

A close analogy can be seen in the operation of a bank. Banks have people problems, in maintaining customer satisfaction and employee motivation. They also have logic problems, such as ensuring the accuracy and integrity of accounting procedures. For example, frequent errors in customer statements, caused by inaccurate accounting procedures, a logic problem, could lead to unnecessary people problems in business operations—irate customers, frustrated employees, and overworked executives trying to hold operations together and patch up customer relations. A wrong diagnosis to treat the symptoms rather than the source of the problem by adding more customer relations personnel would only aggravate and perpetuate the underlying logic problem. Indeed, it is not farfetched to imagine customer relations personnel secretly resenting subsequent efforts to eliminate customer statement errors as a threat to their job security. The lesson here is to get the logic problem of accounting under control. The people problems in banking are hard enough without adding unnecessary ones.

There is a career lesson in this banking illustration, as well. If you want to be a banker, learn about accounting in the university—get it out of the way. Accounting principles and procedures learned in the university will be valid throughout your career. But you will have to work at being a banker—at its people problems—all your life. You'll never learn enough about the people problems, but if you don't learn enough about the logic of accounting early you'll be dealing with unnecessary people problems your whole career.

This simple career lesson in banking applies to information systems analysis and design as well. Even though the people problems are the deepest and most persistent, you should learn how to get the logic problems of information system analysis and design out of the way in your university education. The logic problems are finite and bounded. And the logical principles and procedures of information systems will be valid for your whole career, even though you will be learning more about people all your life.

The objective of this book is to teach you principles of information systems analysis and design that will serve you throughout your career in dealing with logic problems. Only after that do we discuss the people problems, but these discussions only give you a start on the problems you will be learning to solve all your life.

> **Summary:** Any business has many business subsystems, including an information system. Information systems are based on business needs, not computer possibilities. Information systems are used for day to day data processing, for management information, and for decision support. Although people problems are the most difficult, it is important to get logic problems out of the way to avoid unnecessary people problems.

1.2 BOX STRUCTURES OF INFORMATION SYSTEMS

> **Preview:** Box structure principles provide a disciplined means to analyze and design business information systems under good management control. The box structures of black box, state machine, and clear box provide different views of any information system or subsystem. Box structures provide a rigorous form for describing business knowledge. Box structure descriptions can be given in graphic or text forms.

1.2.1 Historical Perspective

The introduction of computer technology into business operations brings the potential for more management control in administrative and analytical phases of business. But the rapid, almost pellmell, introduction of computer technology of the past thirty years has sometimes brought a net loss of real management control because of a necessary dependence on personnel more versed in computers than in business operations. On top of that, the explosive growth of the computer industry has created problems of its own in meeting schedule, cost, and reliability targets in information systems development.

Thirty years ago there was no such thing as the data processing systems, management information systems, and decision support systems that dot the information systems landscape today. Even so, it is sobering to reflect how short thirty years is in terms of intellectual development. When civil engineering was thirty years old, the right triangle was yet to be invented; when accounting was thirty years old, double entry principles were unknown. To be sure, many more people are working on information systems than were working in civil engineering or accounting in their first thirty years. But fundamental ideas and deep simplicities take time. Even with all the excitement and progress, there is still a lot to discover—possibly the right triangle for information systems.

The structured revolution that changed trial and error computer programming to software engineering was triggered by a new concept called structured programming. Structured programming cleared a control flow jungle that had grown unchecked in dealing with more and more complex software problems for twenty years. It replaced that control flow jungle with the astonishing assertion that software of any complexity whatsoever could be designed with just three basic control structures: sequence (begin–end), alternation (if–then–else), and iteration (while–do), which could be nested over and over in a hierarchical structure (the structure of structured programming). The benefits of structured programming to the management of large projects are immediate. The work can be structured and progress measured in top-down development in a direct way. Properly done, when a top-down development is 90% done, there is only 10% left to do (in contrast to projects which at 90% done often required another 90% to complete).

Structured programming has a mathematical foundation that can be used for management advantage. First, a so-called Structure Theorem establishes that any flow chart program can be designed as a structured program. Therefore, a management standard of structured programming is technically sound. Second, a Top-Down Corollary to the Structure Theorem establishes that a structured program can be created in a top-down sequence such that each line can be verified correct by reference to previous lines (and not to lines yet to be created). This means that software can be created correctly as it is developed, without a final and unpredictable stage of trying to make it all work together.

The management benefits begin with standard practices for software development that are based on this mathematical foundation. Software personnel can be uniformly educated to these practices, with improved management visibility into the development process and improved communication between programmers in both the design and inspection phases. As a result, large-scale software projects previously risky or impossible can be completed consistently within schedules and budgets. For example, top-down structured programming has been used extensively in the U.S. space shuttle program; it is safe to say that the shuttle could not be flying (orbiting) today without structured programming.

1.2.2 System Structures

Business information systems development is more than software development. The operations of business involve all kinds of data, stored and processed in all kinds of ways. A simple encyclopedic description of such data and their uses leads to a data flow jungle that is even more tangled and arcane than the control flow jungle. Once again mathematics and engineering have come to the rescue by replacing the data flow jungle with just three basic system structures that can be nested over and over in a hierarchical system structure. These system structures are called **black box, state machine,** and **clear box.**

As with structured programming, there is a mathematical foundation for these system structures that can be used for management advantage. They provide a disciplined way to specify, design, and implement information systems and their subsystems to every level of detail. The data flow becomes a by-product of the methodology and now takes its structure from the system, not as an end in itself.

The management benefits of these box structures begin with standard practices of information systems analysis and design that are based on this mathematical foundation. Information systems personnel can be uniformly educated to these practices with improved management visibility into the systems development process and improved communication between analysts and designers. As a result, it will be possible to develop information systems more reliably and more responsively than ever before.

The three basic system structures are called **box structures.** They provide three views of the same information system or any of its subsystems.

The **black box** gives an external view of a system or subsystem that accepts stimuli, and for each stimulus (S), it produces a response (R) before accepting the next stimulus. A diagram of a black box is shown in Figure 1.2-1. The system of the diagram could be a calculator, a computer system, or even a manual work procedure that accepts stimuli from the environment and produces responses one by one. As the name implies, a black box description of a system omits all details of internal structure and operations and deals solely with the behavior that is visible to its user in terms of stimuli and responses. Any black box response is uniquely determined by the system's stimulus history.

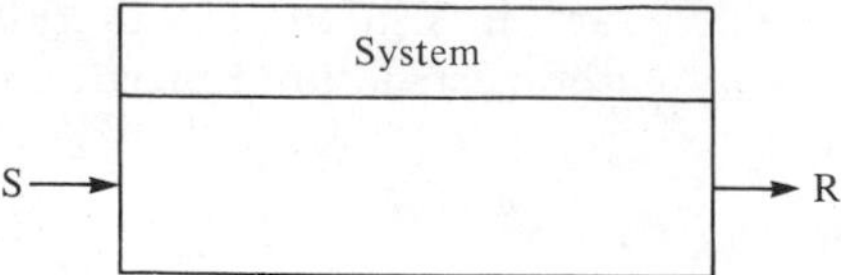

Figure 1.2-1. A Black Box Diagram.

The **state machine** gives an intermediate system view that defines an internal system state, namely the data stored from stimulus to stimulus. It will be established mathematically that every system described by a black box has a state machine description. A state machine diagram is shown in Figure 1.2-2.

The state machine part called Machine is a black box that accepts as its stimulus both the external stimulus and the internal state and produces as a response both the external response and a new internal state which replaces the old state. The role of the state machine is to open up the black box description of a system one step by making its data visible.

The **clear box,** as the name suggests, opens up the state machine description of a system one more step in an internal view that describes the system processing of the stimulus and state (stored data). The processing is described in terms of the three control constructs of structured programming, namely, sequence, alternation, and iteration, and a concurrent structure as shown in Figure 1.2-3. Machine parts M1, M2 are black boxes; each accepts as its stimulus both a stimulus and state and produces as its response both a response and a new state. For example, in the sequence structure, the clear box stimulus is the stimulus to black box M1, whose response becomes the stimulus to M2, whose response is the response of the clear box. Machine part C is a conditional switch that accepts a stimulus and a state, and then transmits that stimulus to one of

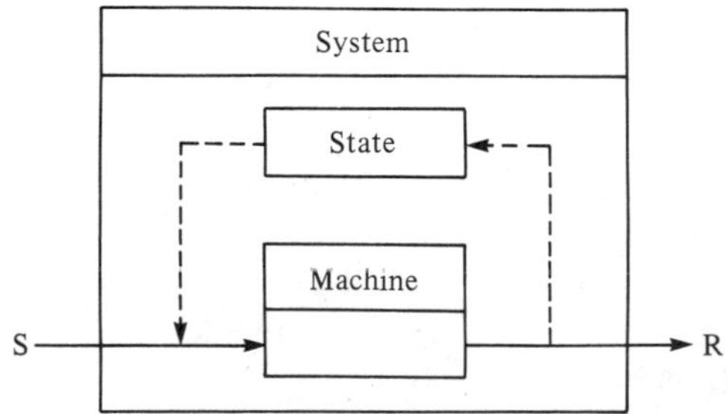

Figure 1.2-2. A State Machine Diagram.

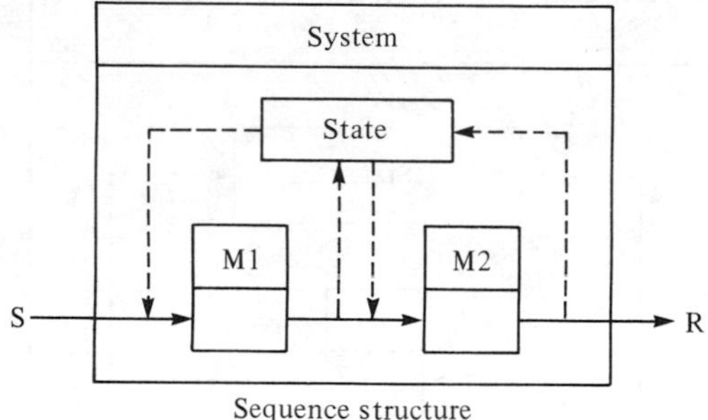

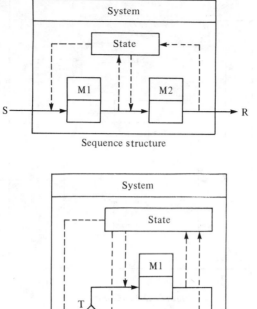

Sequence structure

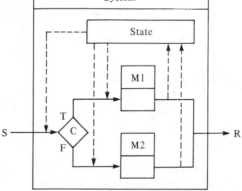

Alternation structure

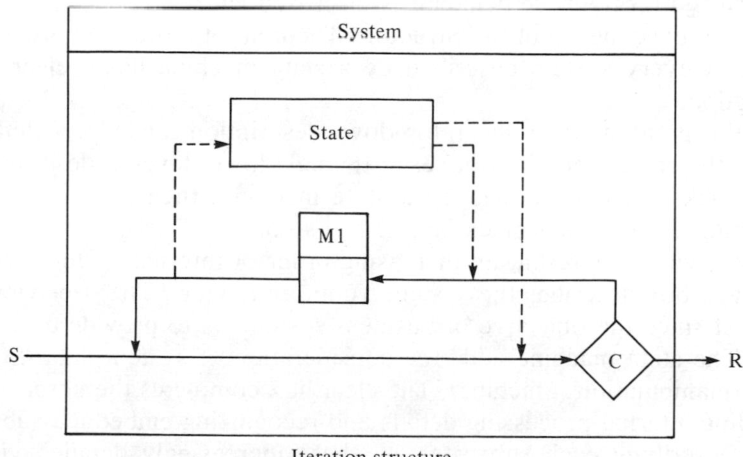

Iteration structure

Figure 1.2-3. Clear Box Diagrams (Mi = Machines).

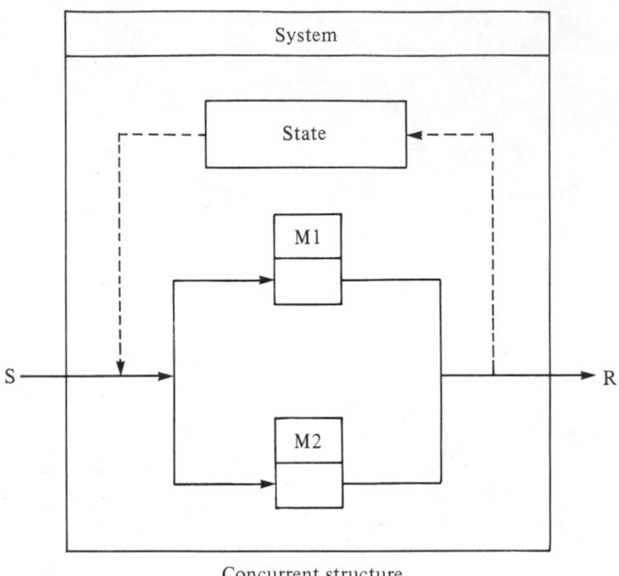

Concurrent structure

Figure 1.2-3. (*Continued*)

two other parts but does not affect the state. For example, in the alternation structure, conditional switch C transmits the stimulus to either M1 or M2, while in the iteration structure, C transmits the stimulus to either M1 or the next part of the next higher structure. The concurrent structure is an advanced form whose behavior is discussed later.

It is a consequence of the Structure Theorem of structured programming that every system described by a state machine has a clear box description.

At this point, a hierarchical, top-down description can be repeated for each of the embedded black boxes at the next lower level of description. Each black box is described by a state machine, then by a clear box containing even smaller black boxes, and so on.

These views represent an increasing order of internal system detail. The black box describes the system from a user view. The user view is foremost since the objective of business systems is to provide user services. The state machine adds the consideration of system data (State) and its manipulation (Machine). The clear box completes the description by adding internal processing details and recognizing embedded subsystems. Describing each subsystem in these increasingly detailed views provides an internal consistency that is essential in developing and managing systems. The data structure must be consistent with the user view, and the processing structure must be consistent with the data structure. System management is helped by the thorough documentation of the mappings between the system views.

1.2.3 Box Structures in Business Operations

Although the concept of a box structured hierarchical system is easy to see, its use in actual business systems requires business knowledge as well as computer knowledge. In fact, the box structures provide a form in which to describe business knowledge in a standard way. The principal value of a black box is that any business information system or subsystem will behave as a black box whether consciously described as such or not. In turn, any black box can be described as a state machine (actually in many ways), and any state machine can be described as a clear box (also in many ways), possibly using other black boxes. In practice, information systems or subsystems often have their own natural descriptions that can be reformulated as box structures.

In illustration, a 12-month running average defines a simple, low-level black box that might be used in sales forecasting. A stimulus of last month's sales of an item would produce a response of the past year's average monthly sales of the item; each month a new sales amount produces a new average of the last 12 months. Figure 1.2-4 shows the running average black box diagram. Using the stimulus history, the black box transition formula for the response R(i) at the end of month i is

$$R(i) = \frac{S(i) + S(i-1) + \cdots + S(i-11)}{12}$$

where for month i, S(i) is this month's sales, S(i − 1) is last month's sales, and so on. Although the stimulus history of the black box may contain

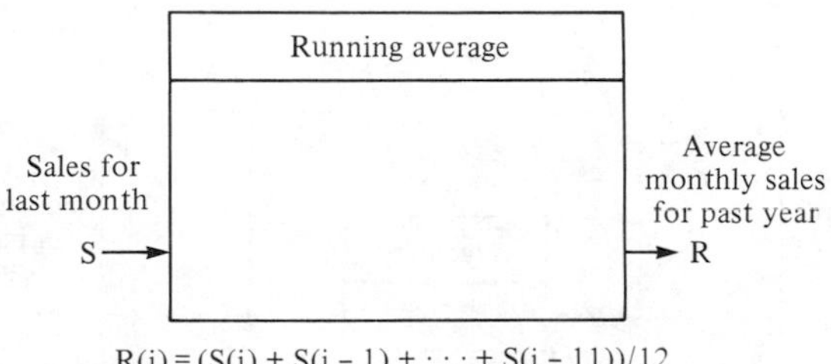

Figure 1.2-4. Running Average Black Box.

years of sales values, only sales from the most recent 12 months are required in the transition formula.

A possible state machine of this black box would identify that the previous 12 monthly sales are to be stored in the state. This state replaces the stimulus history of the black box. The machine, upon receiving the month's sales, would update the state by discarding the oldest sales value and storing the newly input sales value, then calculate the new running average response from state data rather than from the stimulus history. Figure 1.2-5 shows the state machine diagram.

Note a distinction between S1, S2, . . ., which are data recorded in the state and S(i), S(i − 1) which are the monthly sales. The values are the same, but unless S(i), S(i − 1), . . . are recorded as S1, S2, . . ., they will be lost to the state machine which does not access the stimulus history.

A clear box will describe how the state updating process and the averaging process are performed. One possible design is shown in Figure 1.2-6. The Update state and Find average machines are simple enough to include their processing details directly in a sequence structure. In this case, no further black box description is needed because neither Update state nor Find average introduces any new state data.

In this case the assignment operator := in Figure 1.2-6 means to assign the current value of the expression on the right side to the variable on the left side. For example,

S12 := S11

means to assign the value of S11 at month i-1 to the variable S12 at month i, so that for month i

S12(i) = S11(i − 1),

since the current value of S11 was not determined now but in the previous month i-1. When more than one assignment appears in a box, all such

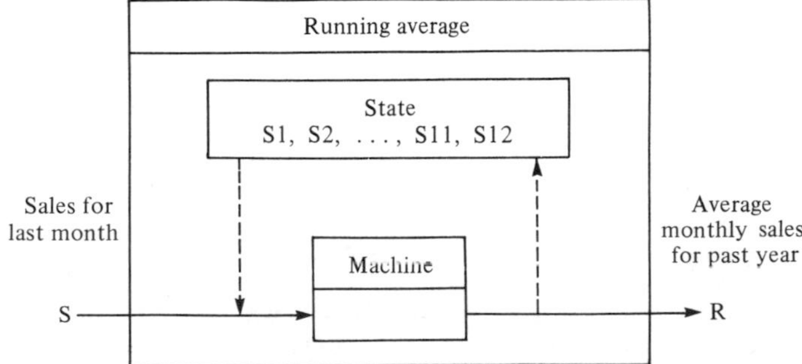

Figure 1.2-5. Running Average State Machine.

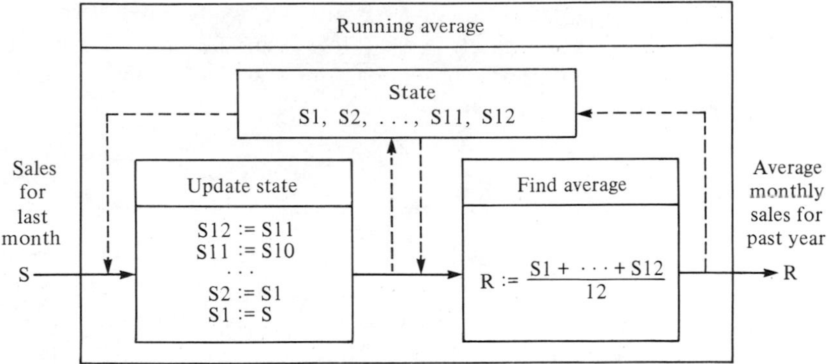

Figure 1.2-6. Running Average Clear Box.

assignments are simultaneous. In the second step, Find average, the values of S1, S2, . . . , S12 have already been updated for month i, so the assignment

$$R := \frac{S1 + \cdots + S12}{12}$$

means that

$$R(i) = \frac{S1(i) + \cdots + S12(i)}{12}$$

The assignment notation allows the subscripts that identify the months to be suppressed for simpler expressions.

Note that many other state machine and clear box designs could have been chosen to implement the running average black box. For example, the state data could be stored as monthly sales values divided by 12. Then, the running average would be found by adding all the state data.

A running average black box is a simple sales forecaster. However, if sales are seasonal or have definite trends, a more suitable black box is required. Such a forecaster will differ in details, but can still be described in a black box/state machine/clear box structure.

If a human forecaster is known to be successful, it will be useful to incorporate that wisdom into a forecasting black box for an entire inventory, e.g., for 10,000 items, which are beyond the ability of the human forecaster to deal with one by one. In this case, the human forecaster may not be able to describe a black box behavior directly. Instead, the description may come out as a mental process of recollections and calculations that involve both state machine and clear box behavior. The box structure discipline gives a systematic basis for interviewing such a human forecaster and converting that human wisdom into systematic form. The result will be a forecasting black box/state machine/clear box structure that can be analyzed as part of a larger system, e.g., an inventory control system with its own box structure.

1.2.4 Box Structure Descriptions

Box structure descriptions will be so useful that we will express them in two forms, called Box Description Language (BDL) and Box Description Graphics (BDG). We have already seen Box Description Graphics for black boxes, state machines, and clear boxes in the previous definitions and examples. BDG consists of standard diagrams in which descriptive text can be embedded, such as the assignment statements and state variables in the Running average clear box. In more complex descriptions, the text may be more general to describe data or operations in English phrases. For example, Figure 1.2-7 illustrates a clear box that might describe a human forecaster's approach for a seasonal product in which each English phrase is expanded separately, as in Table 1.2-1.

Box Description Graphics will be especially useful in information systems analysis for recording current manual procedures and verifying their correctness with managers, users, and operators. The descriptions are readily understood by others outside the development group and help in precise communication about current or desired procedures.

The other form of expression, Box Description Language (BDL), is in text to serve as a formal description language. BDL describes no more

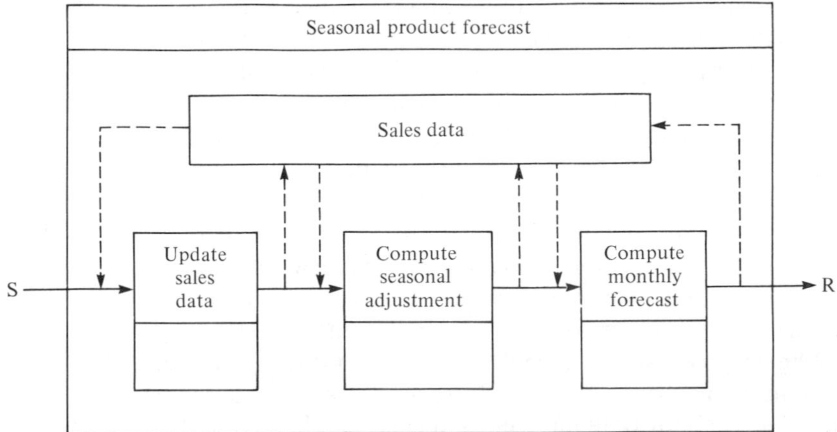

Figure 1.2-7. Seasonal Product Forecast Clear Box.

Table 1.2-1

Seasonal Product Forecast Term Definitions

Term	Definition
Sales Data	Past five years of monthly sales
Update Sales Data	Delete oldest sales and add newest sales to state
Compute Seasonal Adjustment	Divide total of sales this month for the past five years by total sales for the past five years
Compute Monthly Forecast	Multiply seasonal adjustment by total sales for past year

nor less than BDG, but is more concise and rigorous. Box Description Language is most useful in design for recording the evolving box structure of an information system. The various box structures and control structures are defined by use of keywords in a typographic format. BDL will be developed in the next three chapters.

Summary: Any information system or subsystem can be described in terms of a black box, a state machine, and a clear box. The translations between these descriptions provide insights into the system structure.

1.3 THE U.S. NAVY SUPPLY SYSTEM REORDER POLICY

Preview: A real life case study of the U.S. Navy Supply System Reorder Policy demonstrates the use of box structures for information systems analysis.

The creation of clear box descriptions out of existing business processes and their conversions into state machine and black box descriptions can be useful directly. For example, in the middle fifties, an analysis of the U. S. Navy multiechelon supply system led to a radical revision and improvement in inventory control. The basis for this analysis was the conversion of a clear box description of inventory reordering into a state machine, then into a black box description. At the time, the current Navy Supply System reorder policy, called the "k months of supply" policy, seemed sensible enough. It called for maintaining some factor k times an average month's demand of an item either in inventory or on order. The factor k was chosen by the inventory manager to reflect the length of the pipeline, the variability of demand, and the consequences of outage for the item. This k varied from item to item, say from anchors to socks, but once chosen, the rest of the calculation of each month's reorder was simple and automatic. The average demand was calculated by a 12-month running average, so the effects of an unusual month would seem to be averaged out. For example, if the manufacturing cycle for making a certain size anchor is 9 months, a variation of 3 month's demand could be expected and the consequences of outage indicate another 3 month's safety factor, then k would be 15 (9 + 3 + 3) months.

1.3.1 The Clear Box Formulation

The clear box of the reorder policy can be formulated directly from the business process. The clear box description of the k months of supply policy for a particular item has as state data the value of k for the item, the current inventory (including items on order), and the past 12 months of demands. With the stimulus of last month's demand, the new state is obtained by discarding the oldest demand, retaining the current one, and subtracting it from last month's inventory to get current inventory. Next, the running average of the past 12 months is computed, multiplied by k, and then the current inventory is subtracted to get the reorder value. Finally, inventory (which includes items on order) is increased by adding the reorder just calculated. This clear box is depicted next in Figure 1.3-1,

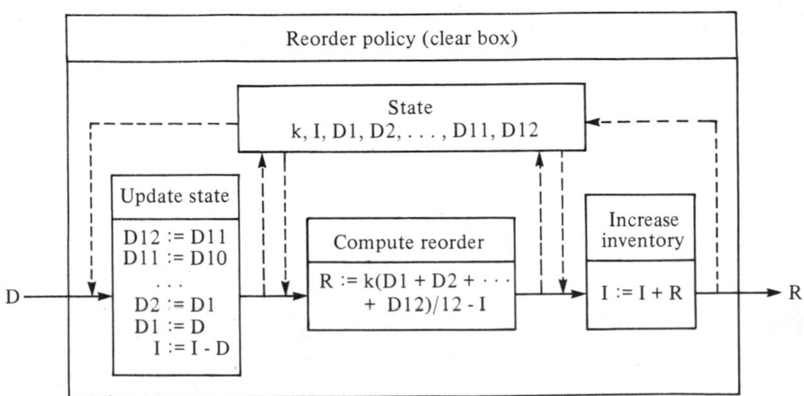

Figure 1.3-1. Reorder Policy Clear Box.

using variables k, I (for inventory), D1, D2, . . . , D11, and D12 for the past 12 months of demands; D is the current demand and R the resulting reorder.

This clear box description represents an actual business process developed on a pragmatic basis that seems to make a lot of sense. Once formulated, however, it can be converted rigorously into a state machine and then into a black box for further understanding and insight.

1.3.2 The State Machine Derivation

The state machine of the reorder policy can be determined by replacing the clear box sequence structure with a sequence-free state machine transition. The transition can be determined by finding single expressions for the response R and each state variable I, D1, D2, . . . , D12 in terms of the stimulus D and the last values of the state variables. On examination of the clear box, it can be seen that the new values of D1, D2, . . . , D12 are given by the Update state part because that is the only place they are updated. The expression for I can be determined from the two parts in which I is updated. In this case, D is subtracted from the last state value of I in Update state, then R is added to this intermediate value of I − D, so the new state value for I is

$$I := I - D + R.$$

However, R must be worked out before I is known in terms of the stimulus and old state. In this case, R is updated only in Compute reorder in an expression that contains D1, D2, . . . , D12, and I. But all these variables were just updated in Update state, which replaces D1 by D, D2 by D1, . . . , D12 by D11, and I by I − D. Therefore, the expression

$$R := k(D1 + D2 + \cdots + D12)/12 - I$$

in the intermediate state data becomes

$$R := k(D + D1 + \cdots + D11)/12 - (I - D)$$

in terms of the old state data. Now, I can be finally worked out, from

$$I := I - D + R$$

to

$$I := I - D + k(D + D1 + \cdots + D11)/12 - (I - D)$$

and the last term (I − D) cancels the first two terms, so I is simply

$$I := k(D + D1 + \cdots + D11)/12$$

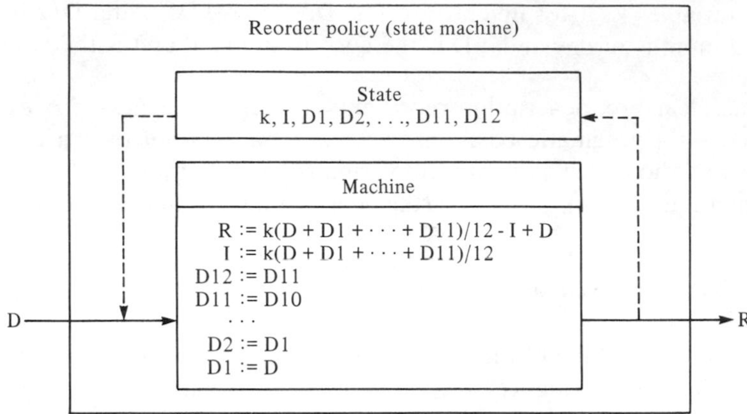

Figure 1.3-2. Reorder Policy State Machine.

At first glance, it may seem surprising that I is just k times the average of the last 12 months of demands, but on second thought, that is just what the k months of supply reorder policy should produce. The state machine so derived above is depicted in Figure 1.3-2.

In this case, there are no real surprises in the Reorder policy state machine. But, it has been distilled down one step by removing the sequential dependencies of the clear box. Note, however, that the so-called "material balance" equation—that new inventory should equal old inventory plus additions minus deletions—is automatically accounted for in this state machine; it is not a required inspiration of an analyst to remember or account for it. One simplifying action can be observed. The state variable D12 and the assignment D12 := D11 can be eliminated since D12 is not used in the assignments for R or I.

1.3.3 The Black Box Derivation

With sequential dependencies of the clear box eliminated to get the state machine, the next step is to eliminate the state dependencies of the state machine to get the black box. In doing so, it will be necessary to introduce previous demands into the single expression for the response. Let D(m) be the demand for month m, the state data for the state machine that accepts stimulus D(m) be I(m − 1), D1(m − 1), D2(m − 1), . . . , D11(m − 1), and the response to this stimulus be R(m). Now, the new state that will be updated from stimulus D(m) will be I(m), D1(m), D2(m), . . . , D11(m) for next month. Then, an inspection of the Reorder policy state machine of Figure 1.3-2 shows that the response and new state values will be as follows:

$$R(m) = k(D(m) + D1(m-1) + \cdots + D11(m-1))/12 - I(m-1) + D(m)$$

$$I(m) = k(D(m) + D1(m-1) + \cdots + D11(m-1))/12$$

$$D11(m) = D10(m-1)$$

$$\ldots$$

$$D2(m) = D1(m-1)$$

$$D1(m) = D(m)$$

Note that these are equations (=), rather than assignments (:=); the month indexes make this possible and correct. In particular, these equations hold for m replaced throughout any equations by another expression for m. For example, these values could be computed on a spreadsheet, with headings for the stimulus, response and state data and initial values for the state; then a column of stimuli values could produce the rest of the values of the spreadsheet automatically. More concretely, given initial values for the state

$$k, I(0), D1(0), D2(0), \ldots, D11(0)$$

in the first row of the spreadsheet and an input column of values under D, referred to as D(1), D(2), . . ., D(m); the spreadsheet process will compute first R(1), then I(1), D1(1), D2(1), . . ., and D11(1). The second demand, D(2), would produce the second response, R(2), and the state for the second iteration; and so on. Of course, all of this processing is done so rapidly that the step-by-step calculation may not be noticed by the spreadsheet user. But, while some intuition could be obtained by trying various columns of stimuli, we will see, in this particular case, that a symbolic mathematical analysis of these equations will lead to a major revelation.

In order to carry out the elimination of state data from this set of equations, it turns out to be convenient first to express D1, . . ., D11 in terms of demands D. Since

$$D1(m) = D(m)$$

is an equation, replace m by m − 1 on both sides to get

$$D1(m-1) = D(m-1)$$

Next,

$$D2(m-1) = D1(m-2) = D(m-2)$$

$$D3(m-1) = D2(m-2) = D1(m-3) = D(m-3)$$

$$\ldots$$

$$D11(m-1) = D(m-11)$$

as can be expected with a little thought. Now, both R(m) and I(m) can be expressed in terms of demands D instead of state data D1, . . ., D11, but it will be convenient, also, to substitute the expression for $I(m-1)$ (obtained by replacing m by $m-1$ throughout the equation for I(m)) in R(m) to obtain

$$\begin{aligned} R(m) = {} & k(D(m) + D(m-1) + \cdots + D(m-11))/12 \\ & - k(D(m-1) + D(m-2) + \cdots + D(m-12))/12 \\ & + D(m) \end{aligned}$$

Now, the surprise is that 11 terms of the first line of the right side are exactly the same as 11 terms of the second line, but with opposite signs—they cancel out! Therefore, R(m) reduces to

$$R(m) = (kD(m) - kD(m-12))/12 + D(m)$$

which simplifies to

$$R(m) = (1 + k/12)D(m) - (k/12)D(m-12)$$

That is, the Reorder policy black box is given by a weighted combination of exactly two demands as shown in Figure 1.3-3.

The surprise is that R(m) depends on only two demands D(m) and D(m-12), a year apart, even though a running average of these demands was used in defining R(m) in its business process and clear box description. It just happens that the interactions of the material balance and the reorder policy cancels out the effect of all the intermediate demands. These interactions and cancellations would also be taking place, over and over, in spreadsheet calculations, but the chances of discovering such a pattern would be very remote. As evidence, this reorder policy had been in use by many organizations over many decades without any hint that such a pattern existed. That is, a lot of human thought and observation of results did not even lead to a suspicion of this pattern!

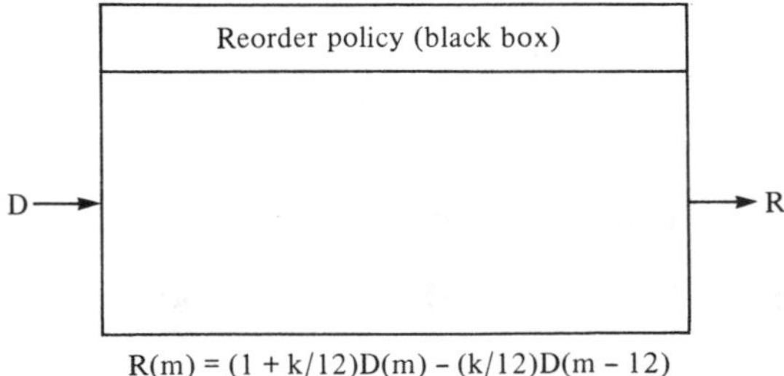

Figure 1.3-3. Reorder Policy Black Box.

1.3.4 Analysis of the Reorder Policy

Even though the form of the Reorder policy black box is a surprise, is that bad? The reorder policy is used in a multiechelon hierarchy from many small supply points at the bottom up through a few large ones (ultimately a few suppliers, possibly only one) at the top. The objective of the reorder policy, beyond providing supplies, is to smooth or dampen the demand variability necessarily expected at its bottom to get a more level aggregate of demands higher up in each echelon, so that the ordering to outside suppliers at the top is as level as possible. The effort of such smoothing through several echelons is multiplicative and can be very effective. For example, if each echelon reduced the demand variability by a factor of two, then the effect through two echelons would reduce the variability by a factor of four and through three echelons by a factor of 8 over the variability at the bottom. In turn, steady orders on the outside suppliers can mean lower costs per unit because of the economics of stable production. That is, if the k months of supply policy, used throughout the multiechelon system smoothed demand variability at each reorder point, it could effect the economics of supply significantly.

Now that the black box of the reorder policy has been derived, it is possible to analyze the smoothing of reorders from demands. The reorder R has the form (simplifying notation)

$$R = (1 + k/12)D - (k/12)D'$$

where D is last month's demand and D′ is the demand a year ago. First of all, if demand is constant, say D0, then

$$\begin{aligned} R &= (1 + k/12)D0 - (k/12)D0 \\ &= (1 + k/12 - k/12)D0 \\ &= D0 \end{aligned}$$

so reorders will exactly match demands, a good thing because inventory will be completely stable. Now, consider the variability of demands D and D′. If D or D′ are unusually high or low, the other may compensate or may not.

In order to develop a concrete numerical illustration, suppose k = 12, so R has the especially simple form

$$R = (1 + 12/12)D - (12/12)D' = 2D - D'$$

Suppose that D and D′ average 100 units, but are 75, 100, and 125, each with an independent probability 1/3. Then there are 9 equally likely cases for (D, D′) values each with probability 1/9. For example, if D = 75, D′ = 125, then

$R = 2(75) - 125 = 25$

When these cases are listed, the values of R are given in Table 1.3-1. Surprisingly, Table 1.3-1 shows demands D and D′ vary only at most 25 from their average value 100, but the reorder R varies up to 75 from its average value of 100. In fact, Table 1.3-1 shows that the reorder policy does not dampen the variability of demands at all; it amplifies them—in this case up to a factor of 3! A more extensive statistical analysis verifies this illustration. The standard deviation of R turns out to be $\sqrt{5}$ (=2.236 . . .) times that of the standard deviation of D and D′. That is, the k months of supply policy is an inadvertent demand variability amplifier in the multiechelon supply system. Just as dampening is multiplicative so is amplification. Through 3 levels, rather than reducing variability by a factor of 8, this reorder policy in fact increases variability by a factor of $(\sqrt{5})(\sqrt{5})(\sqrt{5}) = 11.18 \ldots$!

This clear box to black box analysis showed that most of the variability of inventory levels and reorders in the upper echelons of the Navy supply system was self-induced by a seemingly sensible reordering policy. Once the problem was revealed, it was possible to devise a new kind of reordering policy, called an exponential smoothing policy, that reduced the variability of demands up the echelons rather than amplifying them.

It may seem a surprise in a book on information systems development that the first major example does not even depend on a computer! There is a good reason. The example is about a business process and its analysis. It would be possible to automate the k months of supply policy in an information system. The best design techniques could be used to store the data and process it. The best documentation techniques could be used to make the system understandable to inventory managers. The best human factors could be employed for entering the data for k and demands. But it would all be wrong—not wrong in the implementation, but wrong in information systems analysis and design. This example illustrates the

Table 1.3-1

Values for R

		D′: 75	D′: 100	D′: 125
D:	75	75	50	25
	100	125	100	75
	125	175	150	125

important truth that the reason for information processing is the business not computers. So every part of an information system must begin with a sound analysis of the business process. Only then do computers come into the picture.

Summary: This case study illustrates insights gained through box structure analysis of existing information systems. Such analysis may reveal unsuspected behavior and lead to better information systems designs. Sound information systems begin with sound analysis of business processes.

1.4 MANAGING INFORMATION SYSTEMS DEVELOPMENT

Preview: Box structure hierarchies provide effective means for analysis, design, and management in information systems development. Black box replacement and state migration are important techniques in developing box structure hierarchies. Analysis and design libraries are repositories for evolving box structures. The concepts of box structure derivation and expansion are precisely defined in this methodology. The system development process defines activities of investigation, specification, and implementation that are scheduled in a development plan. Providing information systems integrity requires consideration of many system issues inherent in an operational system.

1.4.1 Box Structure Hierarchies

The low level examples of a running average and the inventory reorder policy illustrate the concepts but not the scope of box structures. Any business information system behaves as a black box for its users. They enter data (stimuli) and receive data (responses). Data entry may be by keystroke, by punched cards, even by automatic sensors such as optical scanners. Data output may be on computer displays, hard copy, even machine readable media. For example, an airline reservation clerk uses the reservation system as a black box. But inside is a gigantic state machine (the state is the data of the entire system) and a corresponding clear box (the system state and top level programs of the system).

A database system such as IMS behaves as a black box, with application programs in COBOL or PL/I as its users. The state machine can be visualized in storage and retrieval terms, while the clear box will be involved with storage and retrieval computation. In this case, the information system using the database system as a black box component will itself behave as a black box for its human users.

That is, business information systems and their subsystems all exhibit black box behavior, and thereby admit description by black box/state machine/clear box structures. As a result, identical structures and methods of reasoning can be used during information systems analysis and design in a hierarchy of smaller and smaller subsystems, as shown in Figure 1.4-1.

A box structure hierarchy itself provides an effective means of management control in developing large, complex information systems. By identifying black box subsystems in higher levels of the system, only a manageable amount of state data and processing needs to be handled within each box structure.

Each subsystem becomes a well-defined, independent module in the overall system. Although the progression from black box to state machine to clear box at any point in the hierarchy may appear to be a triplication of effort, this is not the case. Each subsystem should be initially described in its most natural form, with the other forms determined as necessary for analysis and design. Two important concepts in developing a box structure hierarchy are **black box replacement** at any point of the hierarchy and **state migration** between points of the hierarchy.

The concept of **black box replacement** is key in system development for the management flexibility it provides. A black box is a unit of descrip-

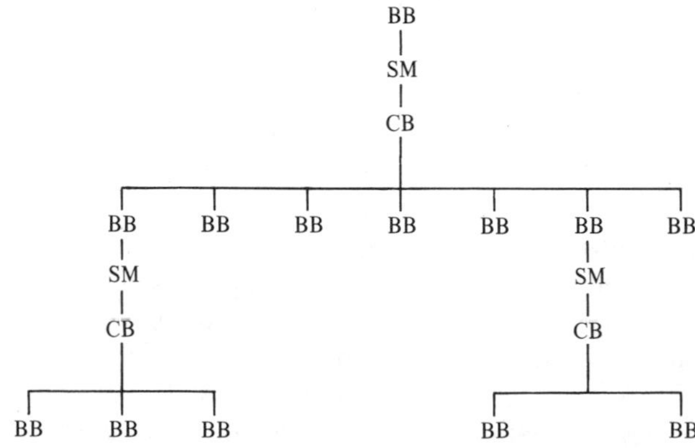

Figure 1.4-1. A Box Structure Hierarchy.

tion that can be isolated and is independent of its surroundings in a system. In particular, a black box can be redesigned as many different state machines and clear boxes. As long as the black box behavior of these state machines and clear boxes is identical to the original, the rest of the system will operate exactly as before. Such black box replacement may be required or desirable for purposes of better performance, changing hardware, or even changing from manual to automatic operations.

State migration in an evolving box structure hierarchy is a powerful design technique. It permits placement of state data at the most effective level for its use. Downward migration of state data is possible whenever new black boxes are identified and used in a higher level clear box. Any state data used solely within one of the new black boxes can be migrated to the state machine expansion of that black box at the next lower level of the box structure hierarchy. Thc isolation of state data through state migration in the system hierarchy provides important criteria for the design of database systems and file systems. Upward migration is desirable when duplicate state data is updated in identical ways in several places in the hierarchy. This data can be migrated up to the closest common parent subsystem for consistent update at one location.

Box structure concepts provide a solid basis for management and control of all development activities. New information, better ideas, and even setbacks can be expected throughout information systems development. The box structure hierarchy provides a framework for orderly control of the development process, rather than the chaos that such new information, better ideas, and setbacks can generate in a less disciplined development. Black box replacement and state migration provide creative flexibility during system development by allowing improvements in the design without losing its integrity.

1.4.2 Box Structure Derivation and Expansion

The box structure of information systems leads to a precise definition of the tasks of **derivation** and **expansion,** as shown in Figure 1.4-2, using a sequence clear box for illustration (with alternation, iteration, and concurrent clear boxes possible).

It is a derivation to deduce a black box from a state machine or to deduce a state machine from a clear box, while it is an expansion to induce a state machine from a black box or to induce a clear box from a state machine. That is, a black box derivation from a state machine produces a **state-free** description, and a state machine derivation from a clear box produces a **procedure-free** description. Conversely, a state machine

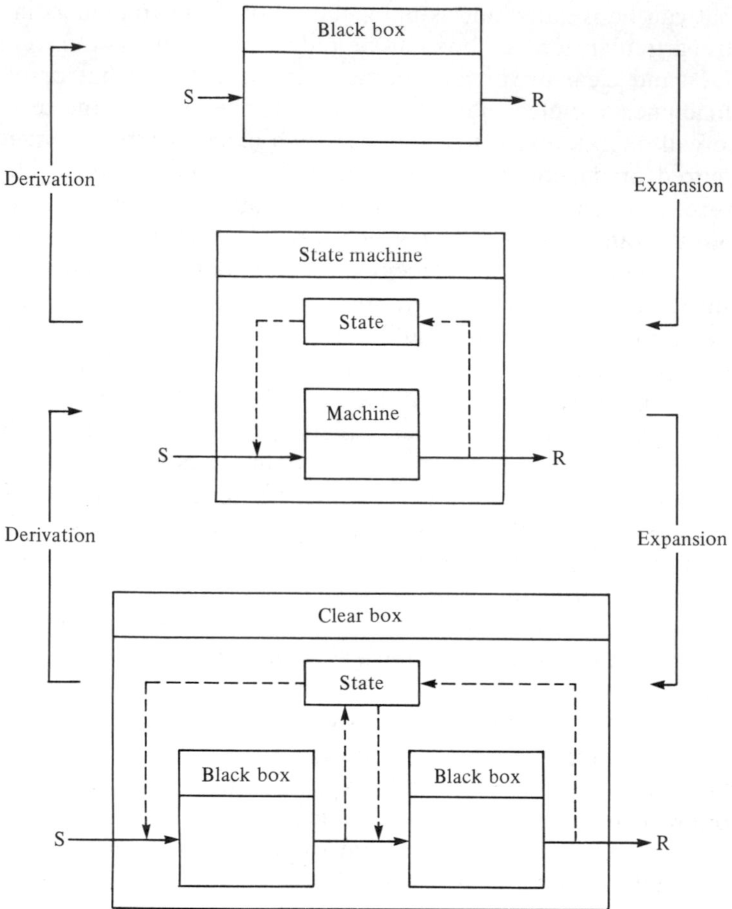

Figure 1.4-2. Box Structure Derivation and Expansion.

expansion of a black box produces a **state-defined** description, and a clear box expansion of a state machine produces a **procedure-defined** description. The expansion step does not produce a unique product because there are many state machines that behave like a given black box, and many clear boxes that behave like a given state machine. The derivation step does produce a unique product because there is only one black box that behaves like a given state machine and only one state machine that behaves like a given clear box. Throughout this text, we will present many examples of derivation and expansion steps.

These definitions of derivation and expansion allow work assignments and reporting to be precise and comprehensive in managing information systems development. Each box structure derivation or expansion step represents a discrete unit of work, which altogether create the analysis and design of an entire system. The analysis activities of manual procedure reviews and interviews fit directly into these definitions. A person interviewed will describe procedures that the analyst will formulate as a clear box and then convert by derivation into state machine and black box terms. The design will then proceed from the derived black box by expansion back through a state machine and clear box better suited for automatic processing.

Information system development may take months or years and require from a few to a few dozen, even a few hundred, people. Each of these people are discovering new facts about the business or the system, identifying new problems and old problems, finding solutions to those problems, making logical decisions about data storage and processing, and so on, every day. Even a small information system involves a large amount of logical structure and detail in its development. It is imperative to keep all this structure and detail organized and accessible for the developers in the conduct of the work.

The hierarchical box structure of black boxes, state machines, and clear boxes is designed explicitly to keep the details of derivation and expansion accessible during information system development. But there still must be a physical medium for recording this structure and its details. For this reason, this methodology introduces two systematic documentation structures, an **analysis library** and a **design library.** The analysis library records findings about the business and its needs for the information system in question, and is created in terms understandable to users in the business. The design library records the logical inventions and solutions the developers have discovered which address the needs of the business in a potential information system and uses more a precise and concise design language understood by the developers. Both libraries are organized in the same way, in the box structure of the information system under development. The developers understand and create both libraries, using the analysis library to interface with management, users, and operators in the language of the business and the design library to ensure the completeness and consistency of the information system in more concise language.

For example, the results of management or user interviews will appear first in the analysis library and will be confirmed in that form with the management or users. Such interviews may not cover unusual cases in computer operations that the users never see, such as how files are protected during an electrical power outage. But as those results are translated into the design library, additional technical considerations may arise, as, in this case, how power outages are to be handled. Once problems and solutions are recorded in the design library, their results may be fed back to the analysis library by subsequent discussion with management or operators, in this case to decide whether to provide for emergency power facilities in computer operations.

1.4.3 The System Development Process

One of the obvious appearances in information systems is the **life cycle.** It is certainly apparent that information systems go through various stages of conception, specification, design, implementation, operation, maintenance, modification, and so on. But although these terms are suggestive, real information systems do not pass through these stages in any simple or straightforward way.

If information systems were developed for their own intrinsic worth by people with infinite knowledge and intelligence, and given unlimited time and budgets, such an information system life cycle might be possible and sensible. But, as already discussed, information systems are developed for business purposes with limited time and budgets by real people, often under conditions that are far from ideal because of business pressures.

If a competitive hotel chain announces a new reservation system, the business needs a quick response with whatever system that can be put into operation, not a system developed to an orderly timetable that arrives too late to save the business. If a banking law changes and more immediate financial information can save interest rates, every day spent in a fixed development cycle is money lost.

Faced with such business pressures, it is easy to fall into a harum-scarum, disorderly mode of operation that generates random activity but no real progress. What are needed in information systems development in real business environments are management principles to balance urgent business needs with requirements for systematic work. Such principles are not new in business and management. They involve a spectrum of short range to long range planning. Long range plans deal with fundamental business objectives and trends; short range plans deal with near term needs and account for current conditions.

In information systems development these principles are embodied in the definition of a set of limited, time phased **activities** to decompose and manage the various kinds of work required, and a **development plan** that defines and schedules the specific activities needed to address a specific problem. The development plan represents long range planning for information system development and the activity plans represent short range planning. As each activity is completed, the entire development plan should be updated to account for the current situation.

Although the activities of a development plan are always specific to a particular system's development problem, they can be categorized into three general classes, **investigation, specification,** and **implementation.** An investigation is a fact finding, exploratory study, usually to assess the feasibility of an information system. A specification is more focused to define a specific information system and its benefits to the business. An implementation converts a specification into an operational system. To summarize:

> The **system development process** is defined by a development plan that specifies a time phased set of activities to address business needs. The development plan should be updated at the completion of each activity to account for progress made, lessons learned, and changing needs of the business.

System development requires focused, creative work carried out with strict discipline. It requires mental inspiration and mental perspiration in the usual ratio of 5% inspiration to 95% perspiration. This need for both creativity and discipline calls for a management process to define short term and long term objectives, measure progress, introduce midcourse corrections, and ensure completion and success in systems development.

Box structures provide continuity of form for managing the systems development process. They can be used extensively and continuously in the three activities of development:

Investigation. Do the developers understand the problem? They can demonstrate they do by describing current operations, manual or automated. Formulated in box structures, these descriptions should be verified with the users and management, along with a preliminary estimate of the costs and benefits of a new system.

Specification. Do the developers have a solution to the problem? They can demonstrate they do by describing a possible information system to improve current operations, with a comprehensive treatment of the inputs, outputs, storage, and processing proposed. Formulated in box structures, the proposed information system should be augmented with benefits and firm cost and schedule estimates.

Implementation. Can the developers make good on their proposed solution? The box structured specification is the right foundation for a box structured implementation in a top down hierarchical development of the information system to meet specifications within budgets and schedules.

Hierarchical box structures provide a natural framework for cost and schedule control. Once analysis is completed, the initial design task is to develop a top level black box, state machine, and clear box. The clear box will make use of black boxes at the next level of refinement, for which the design process will repeat. The top level design effectively partitions the original design problem into a structure of component problems, each of which can be dealt with independently using the same box structure methodology. Each new black box in the structure represents a new top, which must in turn be elaborated into a box structure hierarchy of its own. Since each new black box is smaller and simpler than those above it in the hierarchy, eventually black boxes will be reached which do not introduce new black boxes, and the design will be complete. That is, box structures permit a rigorous design-to-cost management process, in the stepwise allocation and consumption of project resources. Every new black box in the evolving hierarchy represents a new subproject to be designed to cost given the resources available.

Each activity step, regardless of its type, can be viewed as going through three stages, as shown in Figure 1.4-3:

(a) **Planning.** Planning involves a proposal detailing the objectives and statement of work for that step and defining resources and schedules required for completion. The proposal must be reviewed and accepted by the appropriate individuals.

(b) **Performance.** The tasks involved in the activity step are performed. For example, an investigation task may require interviewing significant system users; a specification task may be to design a subsystem from black box to state machine to clear box description; and an implementation task might be to program the resulting clear box specification.

(c) **Evaluation.** Within each step, an evaluation of the results must be performed and the development plan updated. The type of evaluation will vary based upon the tasks performed. Forms of evaluation include management reviews, design verification, system testing, etc.

Just as box structures have their hierarchies, so do these activity steps. A major task may define a schedule of smaller tasks, each with their own planning, performance, and evaluation stages. These hierarchies will

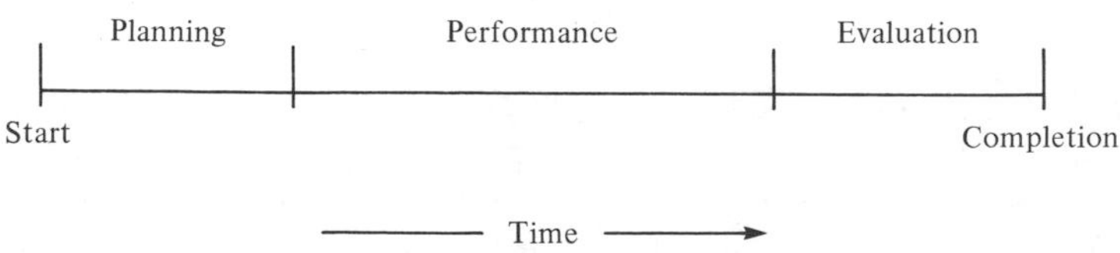

Figure 1.4-3. Stages of an Activity Step.

correspond closely with management hierarchies in large information systems development. An upper level manager may not be fully aware of detailed activities planned, performed, and evaluated under lower level managers to meet upper level planning, performance, and evaluation steps.

It is convenient to visualize the effect of the system development process in a specific situation as a sequence of feedback driven development activity steps, as shown in Figure 1.4-4 in the form of a **system development spiral.**

In this system development spiral, each loop of the spiral is a distinct activity step with its three stages of planning, performance, and evaluation. An approval by management, based on the development plan, is shown preceding each activity step and at the completion of the development. This particular spiral shows one pattern of activity steps, namely,

Investigation
Specification
Investigation
Implementation

Possibly the second investigation step was to confirm a cost/benefit analysis produced by the specification step.

An ideal pattern is

Investigation
Specification
Implementation

but, in fact, this turns out to be too simple for most situations. In the absence of business opportunities or pressures, it can be followed for straightforward developments. The problem is that most information systems are developed just because there are business opportunities or pressures. As noted, if a competitive hotel chain announces a new reservation system, the business needs an implementation as soon as possible, not an investigation. If an advanced system is being developed, the implementation may need to be carried out incrementally, with investigation and specification steps interspersed, for example, in an intended pattern:

Investigation
Specification
Implementation
Investigation
Specification
Implementation
...

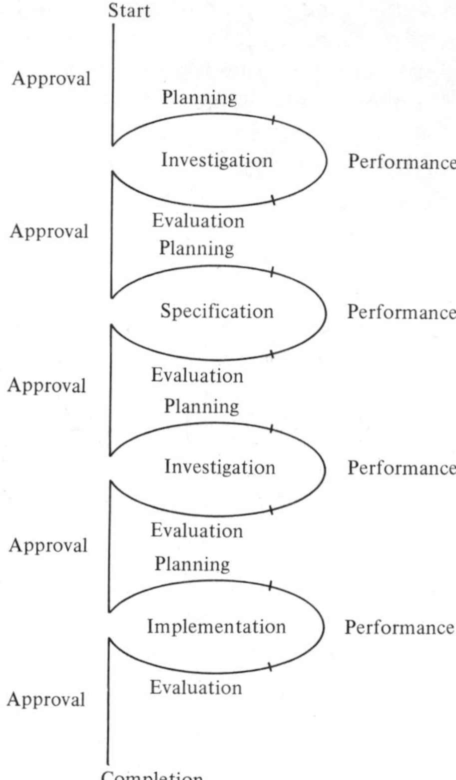

Figure 1.4-4. A System Development Spiral.

Even though this pattern is intended, extra steps of investigation or specification may be required to meet unexpected problems that might arise during the development.

The system development process can itself be described in two coupled black boxes that show the interactions between the system development group and the environment in the business as shown in Figure 1.4-5. Each interaction consists of a single transmission of information between the business environment and the system development group. For example, a proposed information system specification with its benefits, cost, and schedule estimate is simultaneously a response from system development (to prior stimuli) and a stimulus to the business environment. That stimulus to the business environment may produce a response to approve the implementation of the proposed system. This approval response in turn becomes a stimulus to system development and so on. This interac-

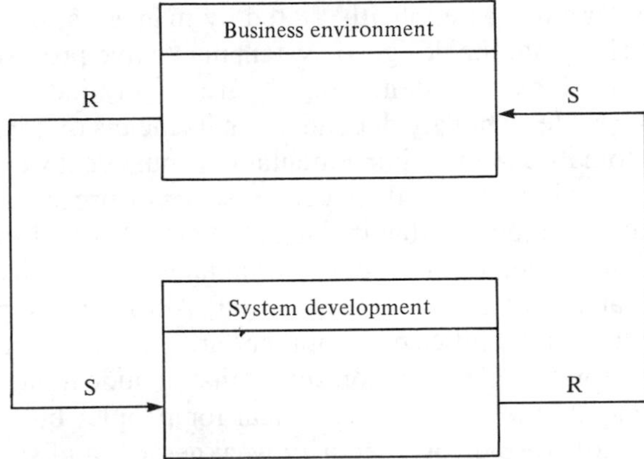

Figure 1.4-5. The System Development Black Box and Its Business Environment.

tion goes on during analysis and design at lower levels, right down to individual conversations between developers and managers, users, or operators. For example, an interview about sales forecasting methods in the business would take the form of a lengthy series of stimulus/response messages between a developer and a forecaster. Each question or answer will be itself a response for one person and a stimulus for the other.

Figure 1.4-5 represents the constant interaction between the system development process and the business environment in which it exists. The system environment includes the managers, users, and operators of the system. The transitions of the System development black box can be organized into activity steps—an investigation step, a specification step, or an implementation step. The type of step taken depends upon the development plan, which should account for results of the previous step and the feedback received from the business environment.

1.4.4 Information Systems Integrity

Box structure descriptions of systems are conceptual representatives of real systems, manual or automatic, that process information. Such box structures describe intended data processing and storage, but between these intentions and actual implementations there may be many system issues. In simplest terms, information systems integrity is the property of the system fulfilling its function, while handling all of the system issues inherent in its implementation. For example, systems are expected to be correct, secure, reliable, and capable of handling their applications. These requirements may not be explicitly stated by managers, users, or operators, but it is clear that the designed system must have provisions for such properties. Questions of system integrity are largely independent of the function of the system, but are dependent on its means of implementation, manual or automatic. Manual implementations must deal with the fallibilities of people, beginning with their very absence or presence (so back-up personnel may be required), that include limited ability and speed in doing arithmetic, limited memory capability for detailed facts, lapses in performance from fatigue or boredom, and so on. Automatic implementation must deal with the fallibilities of computer hardware and software, beginning with their total lack of common sense, that include limited processing and storage capabilities (much larger than for people, but still limited), hardware and software errors, security weaknesses, and so on.

Even though manual and automatic implementations of information systems deal with quite different fallibilities, the questions of integrity can be divided into categories that are common to manual and automatic implementations because they are properties of systems, not of implementations. Six such categories are:

Security
Operability
Capability
Correctness
Auditability
Reliability

The means for achieving integrity in these categories vary between manual and automatic implementations, as indicated in Table 1.4-1. System integrity begins with the recognition of potential system malfunctions in analysis and design, the earlier recognized, the better. The levels of integrity necessary should be identified during analysis and the means for achieving it determined during design. Integrity has its own costs and benefits, so cost/benefit trade-offs are required during analysis and design for integrity as well as for the functions of a system. For example, security features such as data encryption are expensive in terms of software and system performance. A high level of security requirement should exist before encryption is considered.

The example in the preceding section of protecting files from electrical power failures illustrates analysis and design by integrity questions. It also illustrates that analysis and design is usually an iterative process. Some questions of integrity may surface only when implementation issues are faced in design. When they do, the analysis activity should be resumed to ensure their proper treatment. Of course, it is in the best interest

Table 1.4-1

Examples of Means for Achieving Information Systems Integrity

	Manual	Automatic
Security	Personnel Access, Legal Agreements	Passwords, Secure Systems
Operability	Procedures Manuals, Training Programs	Operator Manuals, Control Consoles
Capability	Adequate Staffing, Temporary Help	Adequate Resources, Archive and Restoration Procedures
Correctness	Double-checking, Reviews	Domain Checking, Consistency
Auditability	Accounting Records and Practices	Program Inspection, Logs, Audit Reports
Reliability	Redundancies, Cross-Checks	Checkpoint, Restart Procedures

to identify such questions as early as possible, to minimize the iterations and backtracking required to deal with them.

Summary: The system development process is a management paradigm for defining and scheduling work in investigation, specification, and implementation activities. The concepts and principles of the box structure methodology provide a comprehensive and rigorous way to manage and control information systems development. The system development spiral reflects the actual way work unfolds in development. Integrity requirements must be addressed together with the functional requirements of an information system.

EXERCISES

1. Give examples of a data processing system, a management information system, and a decision support system. Identify the managers, users, operators, and developers of each system.
2. A system accepts hourly temperature readings as stimuli. The response is the highest and lowest readings in the past 24 hours. Describe this system as a black box, a state machine, and a clear box.

3. A system receives a stimulus that has two possible values, YES and NO. The response of the system is the current count of YES's received and the count of NO's received. Describe this system as a black box, a state machine, and a clear box.
4. Modify exercise 3 to include a CLEAR stimulus value. Upon receiving CLEAR, the system sets the YES and NO counts to zero. Describe this system as a black box, a state machine, and a clear box.
5. Briefly define the box structure concepts of black box replacement and state migration. Give an example of each.
6. State migration may be performed upward or downward in a box structure hierarchy. Discuss the advantages and disadvantages of a centralized state at a high level versus decentralized states at lower levels.
7. Distinguish between the concepts of analysis and design in the box structure methodology. How are the analysis and design libraries used?
8. A company has found from experience that the best forecast of this month's sales is a weighted average of the last month's sales and the average monthly sales for the past year. The weight given to last month's sales is a fraction x and the weight for the past year's average is (1-x). Give a clear box description of this sales forecasting method and derive the state machine and black box descriptions via analysis.
9. List and discuss the types of activities performed by the system development team during information systems development.
10. Why is the system development spiral a convenient way to describe a system's development?
11. List and describe the categories of system integrity. Find examples of systems that have significant requirements for each type of integrity.
12. Propose and discuss an appropriate system development spiral for each of the following system examples. Detail the activity step represented by each loop in the spiral.
    (a) A company wants to go from a manual payroll procedure to a computer automated payroll. They have no computer software or hardware.
    (b) You want to develop a calendar/appointment system for your personal computer. You don't know whether to buy a software package or code the system yourself.

(c) An organization's computerized tax preparation system has become obsolete because of a complete overhaul of the tax laws.

(d) A computerized mailing system must be changed to accommodate 9 digit zip codes. Note that this simple system change will result in a system development spiral of its own.

Chapter 2 | The Black Box Behavior of Information Systems

2.1 BLACK BOX BEHAVIOR

> **Preview:** Any information system operates in a consistent way based on its history of use. This history of use can be described as its black box behavior.

We encounter information systems every day that do useful things for us. We can learn **what** these systems do for us without knowing exactly **how** they do it. We use information systems directly when we make airline or hotel reservations and indirectly in automatic teller machines or daily work. On a smaller scale, we can learn to use a hand calculator to do arithmetic and a personal computer to do word processing without understanding the internal operations of their programs and circuits.

However, in order to put our trust in such a system, we must be convinced that the system operates in a consistent way based on its history of use. That is, the system cannot be capricious or produce different results one time or another with the same history. For example, we expect a hand calculator to give the same result (correct answer) every time we give it the same problem in a history of keystrokes.

In each case, we can treat the system we are using—the calculator or the computer—as a **black box,** such that each time we give it a **stimulus,** it gives us a **response.** When a black box accepts a stimulus, it will return a response before it will accept another stimulus. In the case of calculators and computers, a stimulus is a key or a button that we press, one after another, and the response is a display of some kind, often on a video screen.

A diagram of a black box is shown in Figure 2.1-1. The box labeled "System" could be a calculator or a computer that accepts a stimulus from and then gives a response to the user, who may then enter another stimulus, etc. As the name implies, a black box description of a system intentionally omits all details of internal structure and operations and instead deals solely with behavior that is visible to its user in terms of stimuli and responses.

> **Definition. Black Box:** A black box is a mechanism that accepts stimuli and for each stimulus, produces a response before accepting another stimulus; furthermore, each response is uniquely determined by the history of stimuli accepted by the black box.

The examples of calculators, computers, and information systems involve more than providing a response to every stimulus. They involve the predictable use of data entered by previous stimuli and possibly computations with such data in producing that response. For example, a roulette wheel will produce a response (a number) with every stimulus (a spin), but based on no information about previous stimuli or numbers that may have already turned up. So a roulette wheel, even an electronic one, would not be considered a black box. Similarly, an electronic device that produces the temperature at the push of a button would not be regarded as a black box.

These examples illustrate that the definition of black box behavior should not depend on how a system is constructed—electronic, mechanical, or whatever. Instead it should depend only on the stimulus, response properties of the system.

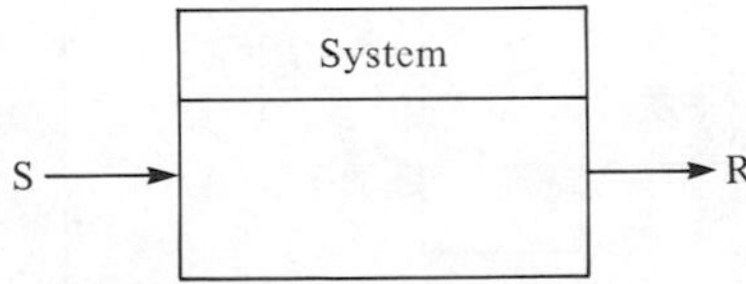

Figure 2.1-1. A Black Box Diagram.

2.1.1 Discovering Black Box Behavior

Calculators and computers usually have instruction books that go along with them and explain their black box behavior. However, in information systems development, we often encounter black box behavior that is not well explained, usually because certain business operations and practices have evolved unconsciously without explicit logical design. For example, a valuable employee may be able to exercise consistent judgements without being able to explain them. In such a case, the only means available to understand a system as a black box is to observe the responses associated with various stimuli and seek to understand how they are related.

In illustration, imagine the simple hand-held device shown in Figure 2.1-2. It has keys labeled 0 through 9 for accepting stimuli from its users, and a two-digit display for producing its responses. Although the device resembles a hand calculator, its function is not obvious. But we can study its function by experimenting and observing relationships between the responses the device produces and the stimuli that cause them.

A **stimulus-response** table is a convenient means of recording black box behavior. Imagine that the following table of key stimuli and display responses was produced by using the device in Figure 2.1-2:

Stimulus	Response
3	3
6	9
1	7
9	10
6	15

Each row of the table represents an action of the black box in accepting a stimulus provided by its user and then returning a response. This

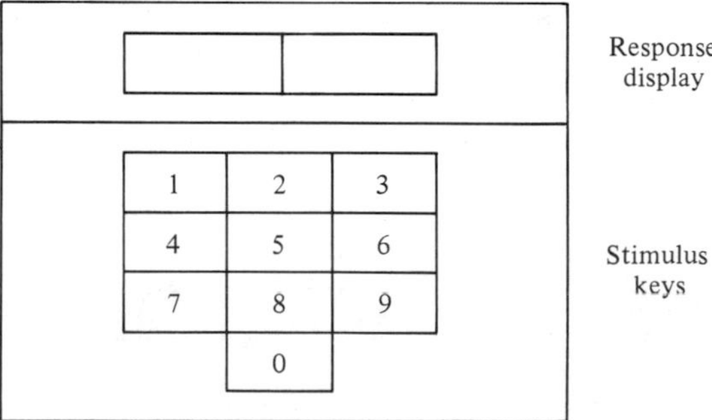

Figure 2.1-2. A Hand-Held Device That Accepts Stimuli and Produces Responses.

action of a black box in converting from a stimulus to a response is called a **black box transition.** The sequence of individual transitions from stimulus to response in the black box table is $3 \rightarrow 3$, $6 \rightarrow 9$, $1 \rightarrow 7$, $9 \rightarrow 10$, and $6 \rightarrow 15$, where each arrow represents a transition.

Definition: Black Box Transition: A black box transition is an ordered pair whose first member is a stimulus and whose second member is the response to that stimulus of the black box.

2.1.2 Stimulus Histories

The transitions of a black box from stimulus to response are not necessarily unique. For example, the stimulus of 6 in the table evokes a response of 9 in one transition and 15 in another. But to be useful, a system described as a black box must provide consistent, repeatable behavior to its users. How is this consistency achieved? The answer is that black box behavior may depend on more than the current transition of stimulus to response; it may also depend on the history of transitions from stimuli to responses.

Therefore, we consider a new table that records **stimulus history** and **response** on each row, rather than stimulus and response alone:

Stimulus History	Response
3	3
3 6	9
3 6 1	7
3 6 1 9	10
3 6 1 9 6	15

In each line of the new table, the current stimulus is the last number, reading left to right in the stimulus history. Note that the current stimulus becomes the previous stimulus on the next line of the table, as the stimulus history is built up by successive transitions.

This table emphasizes the dependency of unique responses on stimulus history. Thus, the table shows that a stimulus of 6 with a stimulus history of 3 6 evokes a response of 9; an identical stimulus of 6 with a different stimulus history of 3 6 1 9 6 evokes a response of 15.

What explanation can we offer for the behavior shown in the stimulus history—response table? In the first transition, the stimulus is replicated as a response, but the subsequent stimuli evoked different responses. Since the response of a black box depends on its stimulus history, a useful analysis strategy is to look for relations between responses and the history that evoked them. With a little thought, it is apparent that every response except the first one can be explained as the sum of the current and previous stimuli:

response := current stimulus + previous stimulus

And if the first response is regarded as the sum of 3 and 0, then this assignment explains every transition in the table. Thus, an apparent black box behavior of our device is to compute and display pairwise sums of current and previous stimuli entered by its user. Add2 seems an appropriate name for this black box that apparently represents the behavior of the device.

That is, when the history of transitions is taken into account, we have found an explanation for the behavior of the device that is consistent and predictable. More transitions may confirm or refute this explanation, but we have an explanation for this data.

We take this history of transitions as the means of defining black box behavior. That is, given any stimulus history a black box will produce a predictable response history. Note that roulette wheels and thermometers do not satisfy this property, whereas calculators and computers do.

Does the description of black box behavior require its response history as well as its stimulus history? No, because each response in its response history simply depends on prior stimuli in its stimulus history. Thus, although a response may appear to depend on prior responses, those responses can always be traced back to prior stimuli. So the stimulus history of a black box is itself sufficient to guarantee unique responses.

Black box behavior is an extremely important property for an information system to possess, and in fact, virtually all information systems do indeed exhibit black box behavior. Such systems can be described as black boxes with no discussion of their internal operations required.

Complex Stimuli and Responses

Although the examples so far have used a single number as a stimulus or response for a black box, **complex stimuli** or **complex responses** are possible for other black boxes. For example, a batch computer run may use an entire file of data as a stimulus for black box behavior; the response from this transition may be an entire report.

As a smaller example, consider a majority voting system where several persons, say five, may vote on issues, one at a time. This system can be considered a black box with a five part stimulus. On each issue, the black box will not accept more than one vote from one person. That is, just as defined, the black box will not accept a new stimulus (or part of a stimulus) until it has produced a response. In this case, all five persons must vote with a stimulus to make up a complex stimulus for the voting system. Then, the result can be broadcast to the voters as a complex response and the next issue taken up on the next transition.

2.1.3 Black Box Initial Conditions

Now that we have defined some principles of black box behavior, we are ready to continue our study of the device in Figure 2.1-2 to confirm our explanation of its black box behavior. Now suppose we decide to replicate the previous experiment with the Add2 hypothesis in order to check results. To our surprise, the second experiment leads to the following table:

Stimulus History	Response
3	9
3 6	9
3 6 1	7
3 6 1 9	10
3 6 1 9 6	15

This table is identical to the previous table except for the first transition, $3 \rightarrow 9$, where $3 \rightarrow 3$ was expected. We have provided the same stimulus history, but the responses are not identical! What has happened?

Our current explanation may be inadequate, but another answer lies in our assumptions about the **initial condition** of the black box. In the first experiment, we assumed the initial condition, that is, the initial "previous stimulus," was 0, so that $3 + 0$ produced a response of 3. But in our second experiment, this assumption did not hold.

In order to get a response of 9 for the first transition, we need a "previous stimulus" of 6 to add to the current stimulus of 3.

But recall that 6 was the final stimulus in the first experiment! Thus, the black box simply retained the final stimulus from the first experiment, just as it had all the others, in turn, to use as the previous stimulus for the first transition of the second experiment. Thus, the Add2 hypothesis is still valid. Note that the term initial condition refers to the experiment, not the black box.

> **Fundamental Principle:** A black box will produce identical responses for identical stimulus histories only when it starts from identical initial conditions.

In effect, the term initial condition means no more (or less) than the (unstated) previous stimulus history. The complete stimulus history for this second experiment is the combined history of both experiments. In this case, the only thing that matters about the previous stimulus history for the black box is the previous stimulus, so the initial condition depends only on that previous stimulus. In general, however, no two distinct stimulus histories in the same device can have identical complete previous stimulus histories, because one of these previous stimulus histories will be necessarily part of the other one. Therefore, identical initial conditions can only be defined when the response is determined by a finite history of stimuli, as in this example.

Alternatively, the inital condition of this device can be set to a known value before use—for example, to 0—by entering a 0 and ignoring the response:

Stimulus History	Response	
? 0	0 + ?	} Initialization
0 3	3	
0 3 6	9	
0 3 6 1	7	
0 3 6 1 9	10	
0 3 6 1 9 6	15	

The "?" symbol in the table represents an unknown initial condition. After initialization, the 0 becomes the "previous stimulus," that is, the initial condition, for any subsequent use, that will evoke black box behavior that is repeatable from the same initial condition.

2.1.4 Finite Black Boxes

If any response of a black box depends on at most a number m of immediately preceeding stimuli, for some finite number m, it is called a **finite black box.** If k is the smallest possible number for m, it is called a finite black box of order k. For example, Add2 is a **finite black box of order** 2.

A small set of examples illustrates these ideas (stimuli are always digits in these examples)

1. **Echo: response := stimulus**
   Echo is a finite black box of order 1.
2. **Previous: response := previous stimulus**
   Previous is a finite black box of order 2.
3. **Constant: response := constant**
   Constant is a finite black box of order 0.

4. **OddEven: if stimulus is odd digit, response := stimulus**
   if stimulus is even digit, response := previous stimulus
   OddEven is a finite black box of order 2.
5. **First: response := first stimulus**
   First is not a finite black box.
6. **Max: response := maximum of all previous stimuli**
   Max is not a finite black box.

We have discussed black box concepts so far in terms of a very simple device. But as noted earlier, any information system whatsoever exhibits black box behavior in its operation. This is because any information system simply accepts stimuli from its users and returns responses to them, based on stimulus history, and this black box behavior can be explained without discussion of internal system structure and operations.

2.1.5 Black Boxes in Business Operations

At first glance a black box such as Add2 may seem simple and remote from business operations, but that is not the case. A running average of sales is frequently used in inventory control and ordering policies; for example, a black box called RA12 can be defined to return the running average of the past 12 months of sales of an item. Its stimulus is this month's sales, its response is the running average of the past 12 months of sales. In this light, Add2 can be seen to return a running total of 2 stimuli and to illustrate a variety of black boxes such as RA12 that abound in business operations.

In a store with thousands of items, there will be thousands of such inventory policy (IP) black boxes, one for each item. The inventory manager may not use or know the term black box, but the inventory policy defines a black box for each item.

As defined, RA12 is a finite black box of order 12. But a sales average need not be defined by a finite black box. For example, a sales forecasting method that uses a weighted average of all previous sales, but weights more recent sales more heavily, is given by a black box called WA, which returns 1/2 the current sales, 1/4 the last sales, 1/8 the last sales before that, and so on ($1/2 + 1/4 + 1/8 + \cdots = 1$). The black box WA is not finite.

Another important statistic in business operations is peak activity, for example, peak demand of electricity over a 24 hour period. In this case a black box called Max24 can be defined that returns the maximum demand of the past 24 hours. A simpler black box Max2, that returns the maximum of the past two stimuli, can be seen to illustrate a variety of black boxes such as Max24 that also abound in business operations.

Any experiment with a finite black box of order k can be initialized by a stimulus history of length k − 1. That is, the future behavior of the black box will be completely determined by such an initializing stimulus history and the stimulus history from then on. For Add2, k − 1 happens to be 1, so the previous stimulus is a sufficient initial condition.

Summary: Black box behavior is given by stimulus histories, and is explained by a relationship between stimulus history and response. The stimulus history must also account for the initial condition at the start of the history. The behavior of a finite black box can be explained by finite stimulus histories.

2.2 THE BLACK BOX BEHAVIOR OF A HAND CALCULATOR

Preview: A hand calculator black box accepts a history of stimuli known to the user. Each stimulus invokes a transition and a response, leading to a final response unknown to the user. Every response depends on the current stimulus history and the initial condition.

2.2.1 Finding a Sum with a Hand Calculator

As surprising as it may seem, a simple hand calculator can serve to illustrate most of the logical principles and procedures of information systems analysis and design. At first glance this may seem impossible. What about databases, terminals, and other complex aspects of information systems? The answer has two parts. First, a hand calculator can be used to explain the principles and procedures of information systems analysis and design, not to explain information systems per se. It is these principles and procedures, applied over and over, that must be used to deal with information systems. Second, as already noted, the logic problems are the small part of the total problems of information systems analysis and design. But getting these logic problems out of the way allows you to deal with the people problems without unnecessary distractions.

Each key on a hand calculator represents a point of entry whose depression creates a stimulus. Each stimulus invokes a transition of the hand calculator black box. Responses from these transitions are shown as numbers on the display of the hand calculator.

To illustrate, consider the following sequence of stimuli and responses to find the sum 14 + 43:

Stimulus	Response
C	0
1	1
4	14
+	14
4	4
3	43
=	57

The stimuli are the successive key entries made to find the sum. The left-hand column shows the key entries made by the user. The right-hand column shows the responses contained in the calculator's display following each key entry.

Entries begin with the depression of the C (Clear) key, which produces a display of 0. The user then depresses the keys 1 and 4. After the 1 key the display shows the value 1, and after the 4 key the display shows the value 14. Next, the + key is depressed, and the display retains the value 14. That is, there is no new response resulting from the + input.

Now the user depresses the 4 key and the display shows a 4. Continuing, when the 3 key is depressed, the display presents the value 43. Finally, depressing the = key leads to the display of the sum of 14 + 43, namely, 57.

This example illustrates a fundamental property of black box behavior already discussed: the same stimulus can produce different responses at different times. When the first 4 stimulus was entered the response was 14, but when the second 4 stimulus was entered the response was 4. So it is clear that the response produced by a black box depends on more than the current stimulus alone. In fact, the history of stimuli to a black box at the time a new stimulus is received determines the response. Because of the stimulus history in our example, the calculator treated the first 4 as part of a number being built, digit by digit, and that 4 was displayed in sequence following the 1. But when the second 4 was entered, the calculator treated it as the first digit of a new number.

2.2.2 Stimulus History in Black Box Behavior

As discussed previously, the response of any black box is uniquely determined by the history of previous stimuli it has received. But every new stimulus, once processed by a black box transition, itself becomes the most recent addition to the stimulus history.

Table 2.2-1

Accumulating Stimulus History through Black Box Transitions

(stimulus, stimulus history)		→	(response, new stimulus history)	
(C ,	*)	→	(0 ,	C)
(1 ,	C)	→	(1 ,	C1)
(4 ,	C1)	→	(14 ,	C14)
(+ ,	C14)	→	(14 ,	C14+)
(4 ,	C14+)	→	(4 ,	C14+4)
(3 ,	C14+4)	→	(43 ,	C14+43)
(= ,	C14+43)	→	(57 ,	C14+43=)

* The clear key makes previous history irrelevant.

Thus, black box behavior can also be defined as follows,

(stimulus, stimulus history) → (response, new stimulus history)

where the arrow represents a black box transition. That is, a black box takes in a stimulus and, depending on its stimulus history, produces a particular response, and then has a new stimulus history that will influence its response to the next stimulus. So we can summarize the black box behavior of our hand calculator example as depicted in Table 2.2-1.

Because of the dependency on stimulus history, even slight variations in a stimulus sequence can produce totally different responses from a black box.

For example, suppose the stimulus sequence was changed so that the second 4 stimulus was entered immediately after the first 4 stimulus. The display would then have shown the value 144, rather than 14. So it appears that the transition invoked by the + key in the original sequence ended the entry of digits for the first number and conditioned the black box to accept the digits for a second number beginning with the next stimulus. After both the first 4 entry and the + entry, the response was to display the value 14. Although there was no change in response after the + was entered, we can conclude, based on what followed, that the + entry had a dramatic effect on the subsequent behavior of the black box.

2.2.3 The Clear Key Makes History Irrelevant

The hand calculator exhibits black box behavior, but not finite black box behavior because there is no limit to the size of the stimulus history that can affect the response. Nevertheless, the hand calculator has another facility which can be used to guarantee identical **initial conditions** for two distinct stimulus histories. It is the C key, as we discuss next.

That is, nonfinite black boxes may (or may not) have facilities to establish standard initial conditions at the beginning of stimulus histories. The C key does just that for the hand calculator.

Suppose a careless user, trying to find the sum of 14 + 43, neglected to start with the C key and to look at the display before depressing the keys 1 and 4, and so on. Suppose, further, that the previous history of stimuli just happened to be C 31 +. Then we know that the additional history 14 + 43 = will not find the sum 14 + 43 = 57, but rather the sum 31 + 14 + 43 = 88. However, our careless user might well assume that the sum of 14 + 43 is indeed 88 because of a trust in the hand calculator. In this case, there is an apparent history and a real history for this calculation:

Stimulus	Response		
C	0		Real History
3	3		Real History
1	31		Real History
+	31		Real History
1	1	Apparent History	Real History
4	14	Apparent History	Real History
+	45	Apparent History	Real History
4	4	Apparent History	Real History
3	43	Apparent History	Real History
=	88	Apparent History	Real History

Once the user neglected to use the C key, there is only one opportunity to observe that something is amiss, namely, in the response to the + key, which the user should know would be 14, not 45. But it is just this response that the user might not notice because 14 is expected. In every other stimulus, a digit has been entered, so there is good reason to look at the display to verify its correct entry. But the user can't verify the correct entry of +, instead of − or *, say, by looking at the display.

However, if the user remembers to start with the C key, the previous history is irrelevant, and a correct calculation can be carried out:

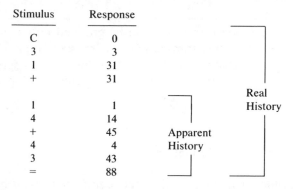

Stimulus	Response
C	0
3	3
1	31
+	31
C	0
1	1
4	14
+	14
4	4
3	43
=	57

Summary: A simple hand calculator illustrates the logical principles of a black box. It is a black box in which each keystroke is a stimulus and each display following a keystroke is a response. The clear key establishes a standard initial condition by eliminating the effect of previous stimulus history.

2.3 BLACK BOX TRANSITIONS AND TRANSACTIONS

Preview: Black box transitions can be grouped into sequences that represent black box transactions. A transaction provides in its last response new information for the user, while previous responses in a transaction provide user confirmation that stimuli are being received correctly by the black box.

2.3.1 Known and New Information

There is a fundamental distinction between two types of information found in the stimuli and responses of a black box. Some responses are known to the user in advance, while others are not. In the hand calculator example, the entry stimuli C 14 + 43 = were known to the user in advance, and the corresponding responses of 0, 1, 14, 14, 4, 43 in the display simply verified to the user that the correct digits had been entered. That is, these responses replicated known information. But the response of 57 in the display was different. It represented new information, not known in advance by the user. It was information that was not keyed in, but rather was generated by the black box as an answer to the question that the user posed. So black boxes are capable of creating new information out of old; in fact, that is the reason for their existence!

Note, however, that not all known information was replicated in the black box display. For example, when the + was entered, no new output was provided—that is, the black box did not explicitly inform the user that a + stimulus had been received. So the user was forced to assume that the addition function had been properly recorded within the black box. This assumption could only be based on concentration in making key entries, so as to know that, say, the − (minus) key had not been inadvertently depressed when the + key was intended.

In the example, the hand calculator black box required another assumption from the user. When the value 57 was displayed following entry of the = stimulus, the user assumed that 57 was indeed the correct answer. This seems an easy assumption to make because we have learned to trust the arithmetic capabilities of hand calculators. But it is not necessarily a valid one, since the answer could be affected by forgetting to start with the C key, by a low battery, or even by an intermittent component failure in the adder circuitry.

2.3.2 Transitions and Transactions

Now consider the overall meaning of the hand calculator stimuli and responses. In effect, the sequence of entries serves to ask the question, "What is the sum of 14 + 43?" Although the user took note of successive responses from each transition to help verify the correctness of key entries, it was only in the final response that the user received the answer to the question posed. So from the user's viewpoint, it is the entire stimulus sequence that produces the final response of interest. Once the user has the answer embodied in the final response, the intermediate responses, useful at the time of their display, are of no further value and can be forgotten. This leads to the following definitions:

> **Definitions. Black Box Transactions, Input, and Output:** A black box transaction is a sequence of black box transitions in which all responses, but the last, are predictable by the user. The last response is not predictable. The entire sequence of stimuli is called an input, the last response is called an output.

That is, a black box transaction is a sequence of one or more black box transitions that produces a response required, but unknown, by the user. A black box input is a sequence of stimuli that defines a transaction. A black box output is the final response of a transaction. Just as a black box transition produces a single response from a single stimulus, so too a transaction produces a single output from a single input. In effect, a transaction is an abstract description of black box behavior.

For example, the add transaction in the hand calculator problem is:

Input
C 14 + 43 =

Output
57

Note that this transaction says nothing about the black box itself or how it operates, and the desired computation could as easily be carried out by a human as by a hand calculator.

The input of the add transaction corresponds to many possible stimulus sequences.

For example, the sequence

Stimulus	Response
C	0
4	4
3	43
+	43
1	1
4	14
=	57

which reverses the order of number entry, would work just as well, as would the sequence

Stimulus	Response
C	0
1	1
4	14
+	14
3	3
4	34
CE	0
4	4
3	43
=	57

which contains a digit sequencing error that was corrected by depressing the CE (Clear Entry) key and entering the proper sequence before continuing with the problem. It is clear from this last example that much behavior of black boxes is directed to assisting their users in carrying out transactions. In fact, all of the responses of the hand calculator black box except the last one exist just for this purpose. Thus, the following black box behavior

Stimulus	Response
C	0
1	0
4	0
+	0
4	0
3	0
=	57

with no useful intermediate responses at all, is theoretically sufficient, given a very careful user!

We observe that the add transaction specified an operation that can be carried out on any two input numbers, not just 14 and 43. So it is easy to generalize the transaction to cover a wide range of desired black box behavior.

It is also important to note that the definition of transactions for a black box is based on what the user knows at each response, and depends on how the work of the user is perceived and organized. So the definition of transactions is very much for the benefit and use of humans, rather than for the black boxes that carry them out. In fact, a black box simply performs transitions as directed, one after another in a mechanical fashion, without ever knowing that it is performing the transaction that its user has in mind.

In this case the same input will always produce the same output. But for more complex black boxes the same input can produce different outputs. For example, a black box input to request a bank balance will produce as output the current balance, which will change from time to time. As with black box stimuli and responses, the key to this discrepancy is in input histories. If the history of deposits and withdrawals is taken into account the outputs can be explained in turns of the input histories.

Summary: A hand calculator black box accepts an input known to the user, which invokes a transaction and produces an output unknown to the user. The output depends on the input history and the initial condition.

2.4 ANY INFORMATION SYSTEM EXHIBITS BLACK BOX BEHAVIOR

Preview: The behavior of any system of people and/or machines whose responses depend on initial conditions and stimulus histories can be described as a black box.

2.4.1 A Personal Computer Exhibits Black Box Behavior

A personal computer provides another, more complex example of black box behavior. In this case, the input devices are the keyboard keys and the output devices include the video screen and printer. As with the hand calculator, these output devices can replicate information known to the user, for example, in displaying responses to keystroke stimuli as they are entered, and can also present new information, for example, in displaying the results of a calculation or a spreadsheet analysis.

Much of the utility of a personal computer arises from its removable storage media, in diskettes or tape cassettes, which can be used to configure its black box behavior to suit particular user needs of the moment. Thus, a personal computer exhibits the black box behavior of a text editor when a word processing diskette is inserted, and the black box behavior of a spreadsheet analyzer when a spreadsheet diskette is inserted. In fact, a personal computer can be programmed to simulate the behavior of a black box. For example, a personal computer can easily be programmed to simulate the black box behavior of the simple hand calculator discussed above, so that a user, suitably isolated from physical clues, could not tell whether a hand calculator or a personal computer was solving arithmetic problems!

We can illustrate the black box behavior of a personal computer through a text editing example—namely, to enter the phrase "regional sales." Consider the sequence of stimuli and responses in Figure 2.4-1 for a personal computer with a text editor diskette inserted. Each keystroke stimulus results in a new display screen response as shown. Successive screens are numbered in the figure for ease of reference.

Each stimulus of a keyboard character at screens 1–8 and 10–14 produces a similar black box transition,

Screen	Stimulus	Response
1	r	r_
2	e	re_
3	g	reg_
4	i	regi_
5	o	regio_
6	n	region_
7	a	regiona_
8	l	regional_
9	<sb>	regional _
10	s	regional s_
11	a	regional sa_
12	l	regional sal_
13	e	regional sale_
14	s	regional sales_

<sb> means "spacebar"

Figure 2.4-1. Text Editor Black Box Behavior of a Personal Computer.

> Display the stimulus character at the cursor position and move the cursor one position to the right,

while the space bar stimulus at screen 9 produces the transition,

> Display a blank character at the cursor position and move the cursor one position to the right.

Now consider the sequence of stimuli and responses that also enter the phrase "regional sales," shown in Figure 2.4-2. The sequence is more complex and reflects use of the text editor black box by an inexperienced person. For example, the user has entered incorrect characters at screens 4 and 5 and has discovered the mistake at screen 6. The ← key is depressed three times, screens 7–9, each invoking the transition,

> Move the cursor one position to the left.

The mistake is corrected by depressing the proper characters at screens 10 and 11, but now the cursor is positioned at n, which is a correct character. So the → key is depressed next, to invoke the transition,

> Move the cursor one position to the right

in order to resume typing. Similar mistakes and fix-ups are made at two other points in the screen sequence.

Although the black box transition sequences of Figures 2.4-1 and 2.4-2 are very different, they both create the same final response for the user, namely,

> regional sales_

In the first case, the sequence of transitions was completed quickly and efficiently, in the minimum possible number of keystrokes. In fact, every keystroke became part of the final response, with no wasted effort whatsoever. In the second case, more time and effort were required, both from the user and the black box, in terms of the many extra keystrokes entered and processed to create the final response. Stimuli that are recognized as errors by humans are just ordinary transitions to the black box, which never knows when it has accepted an erroneous stimulus or the stimulus to correct it.

So operator skill is an important factor in the functioning of a text editor black box. One operator may be able to enter a 20-line letter in five minutes, whereas another may take an hour to accomplish the same job. The slower operator may make dozens of mistakes, each of which must be corrected through reference to the display screen. Even though the slower, less-skilled operator makes many more keyboard entries than does the faster operator, the result is the same. The final response is the

Screen	Stimulus	Response
1	r	r_
2	e	re_
3	g	reg_
4	o	rego_
5	i	regoi_
6	n	regoin_
7	←	regoin
8	←	regoin
9	←	regoin
10	i	regiin
11	o	region
12	→	region_
13	a	regiona_
14	l	regional_
15	s	regionals_
16	←	regionals
17	<sb>	regional _
18	s	regional s_
19	a	regional sa_
20	l	regional sal_
21	e	regional sale_
22	d	regional saled_
23	←	regional saled
24	s	regional sales_

<sb> means "spacebar"

Figure 2.4-2. An Alternate Stimulus Sequence.

completed text of a letter displayed on the screen. Thus, the ability of a black box to provide immediate feedback to inexperienced users in a sequence of transitions is a crucial component of the user training and skill acquisition process, and is a significant measure of the utility of a black box system.

2.4.2 A Business Information System Exhibits Black Box Behavior

The same principles of black box behavior that we have described for hand calculators and personal computers apply to more complex systems as well. For example, the information system of an electronic parts business, with both people and machines as components, exhibits black box behavior in accepting stimuli from and returning responses to a variety of users, all to accomplish the many information processing tasks required in the conduct of business operations. While the overall information system itself behaves as a black box, each of its components, both persons and machines, functions as an individual black box within it, all cooperating in their work to achieve business objectives.

In illustration, consider an instance of black box behavior of people and machine components in the electronic parts business information system when a customer wants to order some memory chips for her personal computer. Figure 2.4-3 depicts a possible telephone conversation between a salesperson and a customer. The salesperson behaves as a black box, accepting verbal stimuli from the customer and providing verbal responses in return. The customer likewise exhibits black box behavior, in accepting stimuli from and providing responses to the salesperson.

Of course, both of these people have experienced long stimulus histories extending from birth, only the latest fragments of which are shown in the conversation. But these stimulus histories have led ultimately to this conversation, with one person initiating a telephone query on memory chips for a personal computer, and the other answering a telephone query for an electronic parts company.

The purpose of the black box dialogue of Figure 2.4-3 is to establish the part number of the desired memory chips. This requires question asking by the salesperson, with answers provided by the customer. A hardcopy catalog containing electronic parts information is also used by the salesperson to identify the part number from the description provided by the customer.

Figure 2.4-4 shows a new black box dialogue which occurs after the customer/salesperson dialogue. In this case, the dialogue is between the salesperson and the Computer Information System of the electronic parts business. The initial stimulus for this person/computer dialogue is the final response, that is, the output, of the prior customer/salesperson dialogue—namely, the part number of the memory chips. The salesperson

Customer Stimulus	Salesperson Response
Phone Rings	"Hello, this is ABC Electronic Parts."
"Hello, I need some information on memory chips for my computer."	"What computer do you have?"
"An IBM Personal Computer. I want to add 512K."	"Is it a regular PC or an XT model?"
"It's a regular PC with two disk drives."	"My catalog shows that you will need eight chips with 64K each. The part number is N1076-45388."
"OK. I'll take them if you have them in stock."	"I'll check our inventory. Please hold."

Figure 2.4-3. Customer/Salesperson Black Box Dialogue.

Keyboard Stimulus	Display Screen Response
Query Inventory Part = N1076-45388 Quantity = 8	Part: N1076-45388 Quantity on Hand: 296 Bin Location: A-42 Unit Price: $20.00 Total Price: $160.00
Reserve Part Quantity = 8	Part: N1076-45388 Quantity = 8 Reserved
Print Invoice	Invoice Number: 86-9471

Figure 2.4-4. Salesperson/Information System Black Box Dialogue.

enters keystrokes at a terminal for a database query on inventory status of the requested chips, to which the machine responds with a display of the quantity on hand (296), the warehouse bin where the chips can be found (A-42), the unit price ($20.00), and the total price ($160.00). The salesperson then reserves the chips for the customer, so that the machine will show only 288 on hand if an identical query is entered later on, and requests printing of an invoice.

The salesperson's inventory query is only the latest stimulus in a long and complex stimulus history of the Computer Information System (CIS). At some point in its stimulus history, the CIS must have been loaded with software to create and maintain a database of electronic parts information, and to answer queries such as this one on inventory status. A large part of the stimulus history resulted from entering the initial database contents, possibly as part of a conversion from manual to automated inventory control. In fact, the CIS may have "worked" in a completely different business at some time, and experienced a different stimulus history, which was erased by resetting the system to a fresh initial condition when it was purchased by the electronic parts company! In any case, the stimulus history of the CIS enabled it to answer the salesperson's query as part of an effective black box dialogue.

Finally, Figure 2.4-5 depicts the completion of the original customer/salesperson black box dialogue of Figure 2.4-3 to inform the customer of

Salesperson Stimulus	Customer Response
"We have it in stock. The total cost is $160.00 plus tax."	"Good, I'll pick it up right away."
"O.K. Your order number is 86-9471."	"O.K. I've got it."
"Thanks for the call."	"You're welcome."
"Good-bye."	"Good-bye."

Figure 2.4-5. Salesperson/Customer Black Box Dialogue.

the order number, and to confirm arrangements to pick up the memory chips.

In this illustration, two black boxes of the electronic parts business information system, one a person and the other a computer, cooperated to achieve the objectives of the two persons. The customer's objective was to purchase memory chips for her computer, and the salesperson's objective was to make a sale. The computer black box supported the salesperson in achieving this objective. But the salesperson also relied on training and personal knowledge, and on the business judgment of the inventory manager, who decided that memory chips for IBM Personal Computers were a good item to keep in stock!

2.4.3 People Exhibit Black Box Behavior

As we have seen, the definition of a black box as a representation of a system applies to people as well as to machines. People respond to stimuli, act as mechanisms to handle data, and generate responses. In this sense, of course, the black box of a person is not the whole human being. Rather, the black box is an expression of the person's capability to accept and respond to stimuli. The history of a human black box is a reflection of the accumulated experiences of the person.

The black box behavior of a human, then, consists of the reflexes and thoughts that respond to stimuli from the outside world and which produce motions and sounds as appropriate. In thinking of the black box behavior of people, it is necessary to be selective and precise about that portion of the person that is involved. A person is composed of many systems, such as physiological, emotional, intellectual, and others. These factors bear upon the capabilities and behavior of a person in accepting and responding to stimuli.

As a black box, a person accumulates a stimulus history that evolves continuously throughout life. Black box behavior is altered continuously by such factors as the language that is learned, the education that is absorbed, and other experiences.

The most striking difference between the black box behavior of people and of devices such as hand calculators or computers lies in the fact that the human being does not have a clear or reset key. Thus, while it is possible to begin an entirely new history in a hand calculator by pressing a single key, a person retains a history that continues to be altered by experiences without giving up any previous, cumulative effects. Understanding these characteristics of the black box behavior of people is important because people, in turn, are part of the black boxes of business systems.

As these three examples illustrate, any information system whatsoever, small or large, simple or complex, exhibits black box behavior in its operation. That is, the common behavioral property of all information systems, no matter what their function or complexity, is acceptance of stimuli from and return of responses to their users. This property of black box behavior applies whether or not the users know that their information system is behaving as a black box and whether a black box description of its behavior has ever been written down.

Much of the value of a black box description of an information system lies in the very fact that it omits details of internal processing. Instead, a black box description focuses on external behavior, that is, "what" the information system does, without discussing "how" it is done. This separation of what and how is called a **separation of concerns.** It represents a crucial strategy in information systems development, and constitutes a major theme of this book. And it is because the black box concept excludes descriptions of processing internals that it can be used to describe the behavior of systems that have human as well as machine components. As illustrated above, humans exhibit black box behavior just as do machines, in accepting stimuli from other humans or machines and returning responses to them. This behavior can be summarized in a black box description, without the necessity for difficult explanations of internal processing, that is, how humans actually process information in their minds.

> **Summary:** Black boxes are a completely general means for defining and analyzing behavior in information systems. Through their focus on stimuli and responses, black box descriptions correspond to how users actually interact with systems, with no need for discussion of processing internals.

2.5 BLACK BOX STRUCTURES

> **Preview:** Black boxes can be combined into larger black boxes by organizing them into one of four box structures. The behavior of these black box structures can be deduced from the behaviors of their component black boxes.

In black box behavior at the stimulus (S), response (R) level of description, as depicted in Figure 2.5-1, we have restricted our discussion to stimulus, response pairs which are initiated and used by people. But it is possible for a response of one device to be used as a stimulus by another

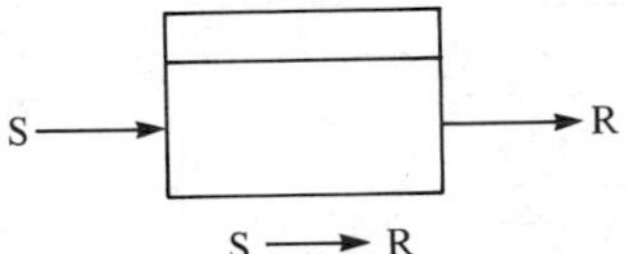

Figure 2.5-1. Stimulus/Response Description of a Black Box.

device. In this case the response may be in the form of an electrical current, or other physical interaction not necessarily visible to people. Therefore, we will consider stimuli and responses suitable for either people or machines as the occasion requires. This possibility motivates an investigation of how black boxes can be combined into larger black boxes, and how the behavior of these larger black boxes can be analyzed and understood. A **black box structure** is a description of how several black boxes are connected to achieve the behavior of a larger black box.

2.5.1 Black Box Primitive Structures

A new black box can be constructed from black boxes by one of four primitive composition steps which define:

1. Successive transitions of two black boxes (named the **sequence structure**).
2. Selected transitions of one of two black boxes (named the **alternation structure**).
3. Repeated transitions (zero or more) of a black box (named the **iteration structure**).
4. Concurrent transitions of two or more black boxes (named the **concurrent structure**).

In each of these structures, the transitions will have the same form as the initial black boxes, namely,

$S \rightarrow R$

The Black Box Sequence Structure

The sequence structure is depicted in Figure 2.5-2 below with boxes B1 and B2

B1: S1 → R1

B2: S2 → R2

and allocation

R1 is renamed S2

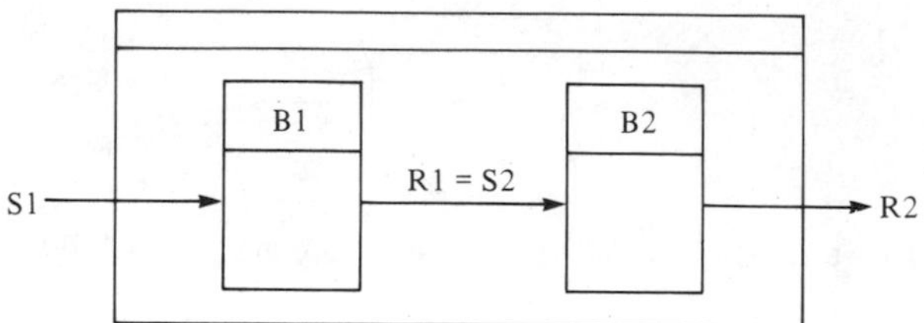

Figure 2.5-2. The Black Box Sequence Structure.

That is, the response R1 of B1 is used as the stimulus S2 for B2. In this case, a stimulus S1, submitted to B1, which returns a response R1, which is submitted as stimulus S2 to B2, which returns a response R2. In summary, if stimulus S1 is submitted to the sequence structure, the response R2 is produced.

This black box sequence structure behaves like a black box. That is, any stimulus history will produce a unique response. To see that, recall that any stimulus history for B1 will produce a unique response history (of R1's), and the unique response history becomes a unique stimulus history (of S2's) for B2 which will produce a unique response, as asserted.

The Black Box Alternation Structure

The alternation structure is depicted in Figure 2.5-3 with a special kind of black box, called C, and two black boxes, namely, B1 and B2. The

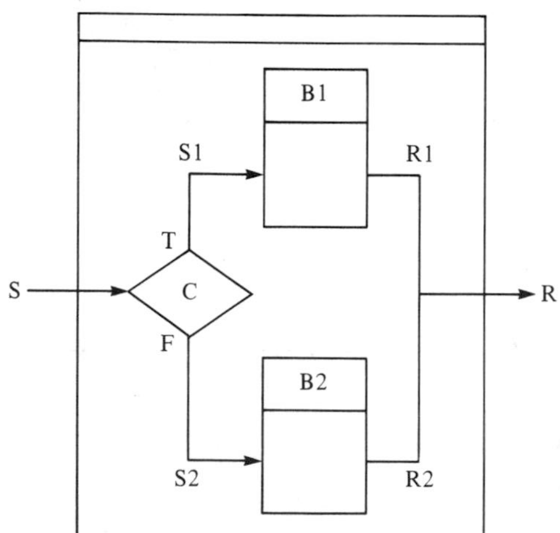

Figure 2.5-3. The Black Box Alternation Structure.

black box C (C for Condition) is denoted by a diamond and returns responses T or F (True or False). The function of C is to switch the stimulus S to exactly one of B1 or B2, that is, to rename S as either S1 or S2. The response R1 or R2 is automatically renamed R. In this case a stimulus S, submitted to C is switched (as S1 or S2) to B1 or B2, eliciting response R1 or R2 which is renamed R.

This black box alternation structure behaves like a black box. That is, any stimulus history will produce a unique response. To see that, recall that any stimulus history for C will create a unique history of switches (to B1 or B2) and two unique subhistories (of S1's and S2's). Each subhistory will produce a unique response (whichever is called for by the current stimulus), as asserted.

The Black Box Iteration Structure

The iteration structure is depicted in Figure 2.5-4 with a special kind of black box, called C, and a single black box, B. As in the alternation structure, the black box C (C for condition) is denoted by a diamond and returns responses T or F (True or False). The function of C is to switch the stimulus S to black box B or to R, that is, to rename S as either S1 or R. The response R1 of black box B is automatically renamed S for evaluation by condition C. Thus, stimulus S is switched either to R directly, or to B, which produces an internal stimulus which will in turn be switched either to R or B, continuing in this manner until the internal stimulus is switched to R.

This black box iteration structure also behaves like a black box, with any stimulus history producing a unique response. To see that, observe that any stimulus history for C will create a unique history of switches (to

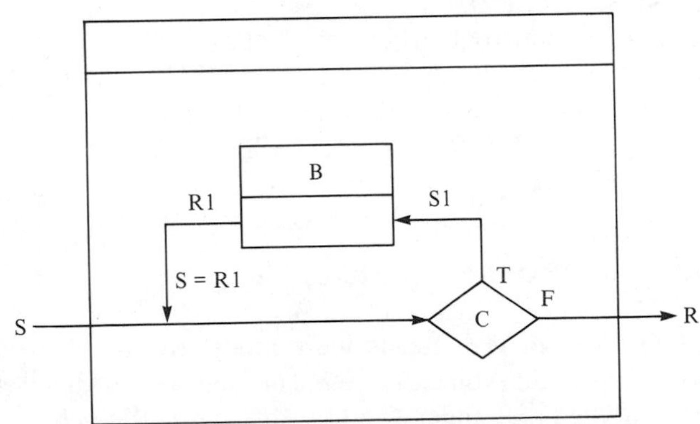

Figure 2.5-4. The Black Box Iteration Structure.

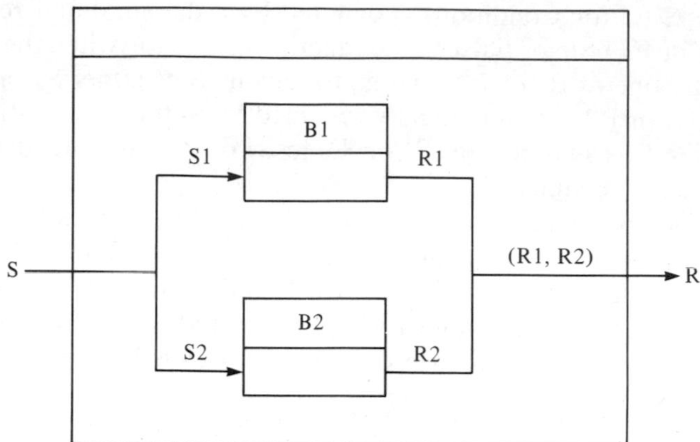

Figure 2.5-5. The Black Box Concurrent Structure.

B or R) and two unique subhistories (of S1's and R's). In the case of the S1 subhistory, each unique response will be determined by one or more internal iterations (however many are called for by the current stimulus). Thus, each subhistory produces a unique response, as asserted.

The Black Box Concurrent Structure

The concurrent structure is shown in Figure 2.5-5. The structure has two black boxes labeled B1 and B2 that execute simultaneously. The stimulus S is sent to both boxes, labeled S1 for B1 and S2 for B2. The B1 transition produces the response R1 and the B2 transition produces the response R2. The response R for the concurrent structure, then, is the complex response (R1, R2).

The black box concurrent structure behaves like a black box since any stimulus history will also be the stimulus histories to the black boxes B1 and B2. The responses R1 and R2 will be uniquely determined by the stimulus history. Therefore R = (R1, R2) will also be uniquely determined by the stimulus history.

2.5.2 Analysis of Black Box Structures

Stimulus and response patterns were analyzed above to understand the behavior of individual black boxes. The behavior of black box structures can be analyzed and understood by combining the behaviors of their component black boxes.

In illustration, recall the Add2 black box whose response R is the sum of its last two stimuli. The transitions of Add2 can be numbered 1, 2, . . . , i, where i is any integer, with corresponding stimuli S(1), S(2), . . . , S(i), and responses R(1), R(2), . . . , R(i). The formula for an Add2 transition is thus denoted by the equation

Add2 formula: R(i) = S(i) + S(i − 1)

Also, consider the black box called Max2, which produces as a response the maximum of its last two stimuli. The formula for a Max2 transition is denoted by the equation

Max2 formula: R(i) = max(S(i), S(i − 1))

As already seen, the behavior of the Add2 black box for our example stimulus history is, for initial condition 0:

Add2 history:

S	R
3	3
6	9
1	7
9	10
6	15

The behavior of the Max2 black box for the example stimulus history is, for initial condition 0:

Max2 history:

S	R
3	3
6	6
1	6
9	9
6	9

Analysis of Sequence Structures

A new black box sequence structure, called Add2;Max2, can be formed by combining Add2 and Max2 as shown in Figure 2.5-6. A semicolon is used to separate black boxes in a sequential structure. The behavior of Add2;Max2 can be worked out for our example stimulus history as follows:

Add2;Max2 history:

S1	R1 = S2	R2
3	3	3
6	9	9
1	7	9
9	10	10
6	15	15

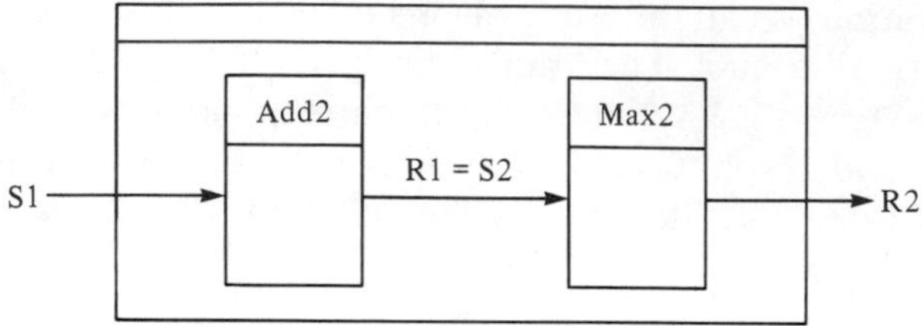

Figure 2.5-6. Add2;Max2 Black Box Sequence Structure.

That is, Add2;Max2 behaves as a new black box which produces a response R2 for every stimulus S1. The exact form of the transition equation of the new black box can be derived, step by step, starting with R2 and working back to S1 as follows:

$$
\begin{aligned}
\text{Add2;Max2 transition: } R2(i) &= \max(S2(i), S2(i-1)) \\
&= \max(R1(i), R1(i-1)) \\
&= \max(S1(i) + S1(i-1), S1(i-1) + S1(i-2)) \\
&= S1(i-1) + \max(S1(i), S1(i-2))
\end{aligned}
$$

The first line of the derivation is obtained by the definition of Max2 (since R2 is produced by Max2), the second line by the sequence structure of Add2;Max2, the third line by the definition of Add2 (since R1 is produced by Add2) and the final line by factoring the term S1(i − 1) out of the max operation. In this case, R2(i) depends on the three previous stimuli S1(i), S1(i − 1), and S1(i − 2), with the initial condition that "all previous stimuli" are zero.

The formula for Add2;Max2 can be used to obtain the values of R2 directly, without obtaining intermediate values for R1 and S2. Thus, S1 and R2 can be renamed simply S and R, respectively, and the formula rewritten as:

$$\text{Add2;Max2 transition: } R(i) = S(i-1) + \max(S(i), S(i-2)).$$

The responses of the Add2;Max2 black box sequence structure can now be computed directly from the stimulus history, and confirm the previous computation:

Add2;Max2 history:

i	S	S(i − 1) + max(S(i), S(i − 2))	R
1	3	0 + max(3, 0)	3
2	6	3 + max(6, 0)	9
3	1	6 + max(1, 3)	9
4	9	1 + max(9, 6)	10
5	6	9 + max(6, 1)	15

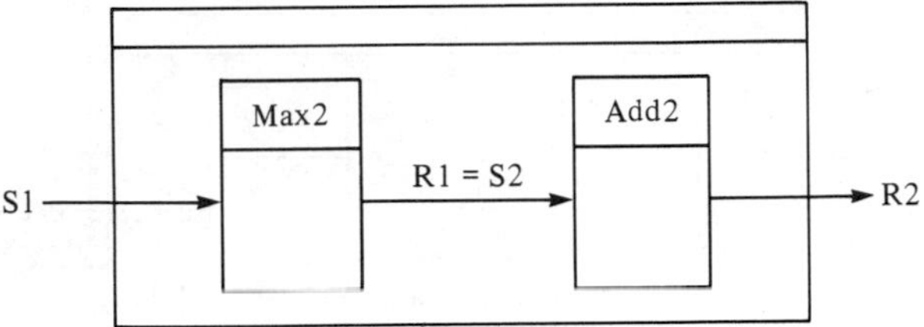

Figure 2.5-7. Max2;Add2 Black Box Sequence Structure.

Successive transition numbers are listed in the column labeled i, and the computation of responses from the Add2;Max2 formula is shown. In this case, the response values for Add2;Max2 are easy to work out mentally in an abbreviated table,

S	R
3	3
6	9
1	9
9	10
6	15

with the same result as before, in the knowledge that the computations can always be recorded for detailed analysis in more complex situations.

A different black box sequence structure can be created by reversing Add2 and Max2, as depicted above in Figure 2.5-7 with example history,

Max2;Add2 history:	S1	R1 = S2	R2
	3	3	3
	6	6	9
	1	6	12
	9	9	15
	6	9	18

a quite different result from the Add2;Max2 sequence structure. The transition formula for Max2;Add2 is

$$
\begin{aligned}
\text{Max2;Add2 transition: } R2(i) &= S2(i) + S2(i-1) \\
&= R1(i) + R1(i-1) \\
&= \max(S1(i), S1(i-1)) + \max(S1(i-1), S1(i-2))
\end{aligned}
$$

which cannot be simplified any further.

A single black box such as Add2 can be reused in a sequence structure, as depicted in Figure 2.5-8. In this case, the transition formula

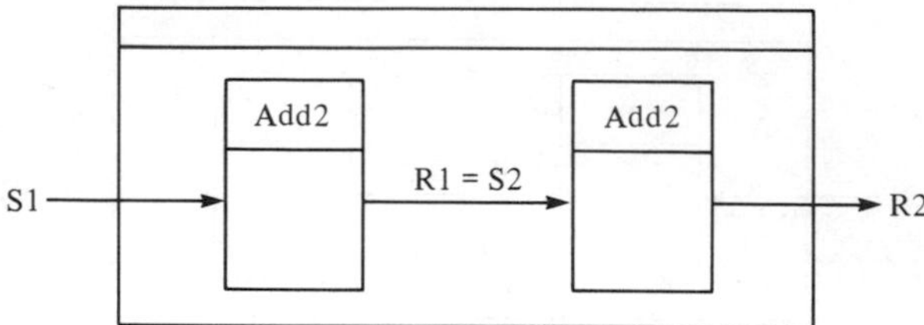

Figure 2.5-8. Add2;Add2 Black Box Sequence Structure.

is

$$
\begin{aligned}
\text{Add2;Add2 transition: } R2(i) &= S2(i) + S2(i-1) \\
&= R1(i) + R1(i-1) \\
&= (S1(i) + S1(i-1)) + (S1(i-1) + S1(i-2)) \\
&= S1(i) + 2 * S1(i-1) + S1(i-2)
\end{aligned}
$$

Reuse of the Max2 black box is depicted in its black box structure of Figure 2.5-9. The transition formula for Max2;Max2 is:

$$
\begin{aligned}
\text{Max2;Max2 transition: } R2(i) &= \max(S2(i), S2(i-1)) \\
&= \max(R1(i), R1(i-1)) \\
&= \max(\max(S1(i), S1(i-1)), \\
&\qquad \max(S1(i-1), S1(i-2))) \\
&= \max(S1(i), S1(i-1), S1(i-2))
\end{aligned}
$$

Thus, both the Add2;Add2 and Max2;Max2 sequence structures have relatively simple behavior compared to the behavior of Add2;Max2 and Max2;Add2.

Analysis of Alternation Structures

Figure 2.5-10 illustrates a black box alternation structure called Odd:Add2|Max2, where the condition Odd transfers control to the True (T) branch (Add2) if S is an odd number, and to the False (F) branch (Max2) if S is an even number. In general the notation used for alternation

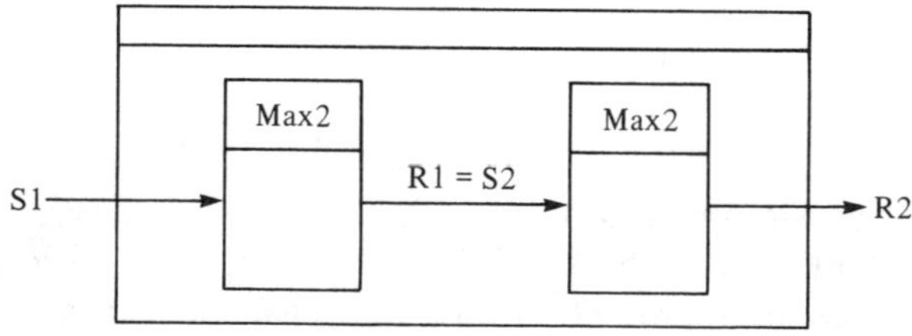

Figure 2.5-9. Max2;Max2 Black Box Sequence Structure.

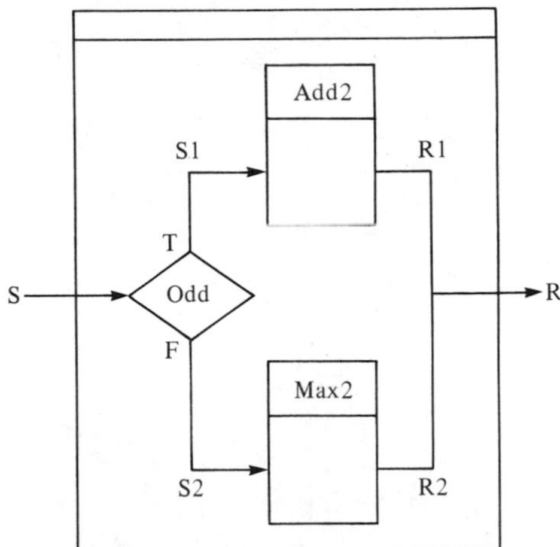

Figure 2.5-10. Odd:Add2|Max2 Black Box Alternation Structure.

structures is C:B1|B2, where C is the condition, B1 and B2 are black boxes, and the straight line denotes that one of the black boxes is executed. The example history for Odd:Add2|Max2 is as follows, where blank entries in the table represent control branches not taken. That is, S1, R1 or S2, R2 are only filled in when the condition Odd has switched control to Add2 or Max2, respectively:

Odd:Add2|Max2 history:

S	S1	R1	S2	R2	R
3	3	3			3
6			6	6	6
1	1	4			4
9	9	10			10
6			6	6	6

The transition formula for Odd:Add2|Max2 is a little harder to describe than for Add2;Max2 or Max2;Add2. With some thought the form of Odd:Add2|Max2 can be seen as

Odd:Add2|Max2 transition: If S(i) is odd then

R(i) = S(i) + S(j)

where S(j) is the last odd stimulus preceding i

If S(i) is even then

R(i) = max(S(i), S(k))

where S(k) is the last even stimulus preceding i

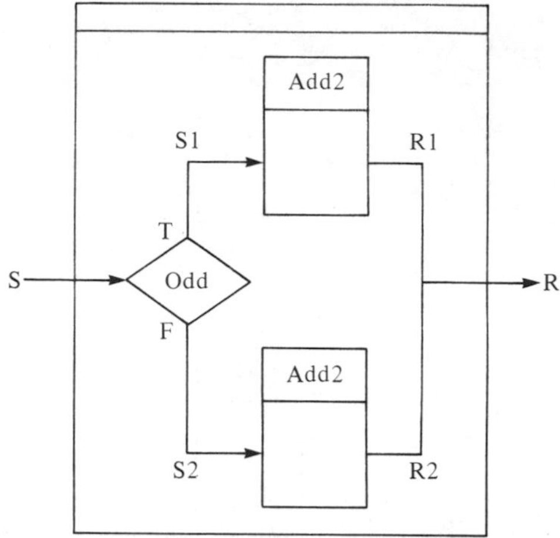

Figure 2.5-11. Odd:Add2|Add2 Black Box Alternation Structure.

The Odd:Add2|Add2 black box alternation structure depicted in Figure 2.5-11 is instructive. At first glance, Odd:Add2|Add2 may seem to do no more than Add2 alone. After all, whether S is odd or even, control is switched to Add2. However, look at the history of Odd:Add2|Add2

Odd:Add2|Add2 history:

S	S1	R1	S2	R2	R
3	3	3			3
6			6	6	6
1	1	4			4
9	9	10			10
6			6	12	12

which is quite different than for Add2 alone. The difference is that the upper Add2 is only sent odd numbers as stimuli and the lower Add2 is only sent even numbers as stimuli. As a result, the upper Add2 only adds odd numbers (after the first) and the lower Add2 only adds even numbers. Therefore, the general form of Odd:Add2|Add2 can be seen as

Odd:Add2|Add2 transition: If S(i) is odd then

R(i) is the sum of the last two odd numbers in the stimulus history

If S(i) is even then

R(i) is the sum of the last two even numbers in the stimulus history

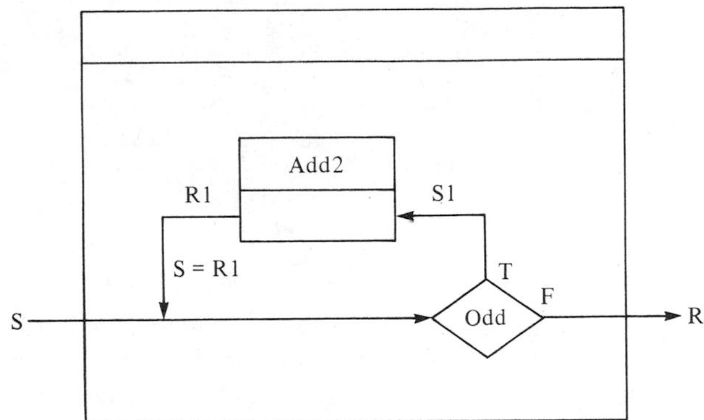

Figure 2.5-12. *Odd:(Add2) Iteration Structure.

Analysis of Iteration Structures

Figure 2.5-12 illustrates a black box iteration structure called *Odd:(Add2), where the condition Odd transfers control to the True (T) branch (Add2) if S is an odd number, and the false (F) branch (R) if S is an even number. In general, the notation used for iteration structures is *C:B, where C denotes the condition, B is a black box, and the * (star) symbol indicates that the black box (B) will be executed zero or more times while the condition is true before the response is produced. The example history for *Odd:Add2 is as follows, where blank entries in the table represent control branches not taken:

*Odd:Add2 history:

I	S	S1	S2	S3	...	R
		Iterations				
0	3	3	6			6
3	6					6
3	1	4				4
1	9	10				10
9	6					6

The initial condition I of the Add2 black box for each stimulus is shown in a column on the left of the table. This value is always the last stimulus presented to Add2 by condition Odd, and is thus guaranteed to be odd, except possibly for the first initial condition, which in this case has been defined as zero. Given this example, the general form of *Odd:Add2 is easy to see, as shown in the following table in three cases:

	Case		
	1	2	3
S(i)	Odd	Odd	Even
I	Odd	Even	—
R(i)	S(i)+I	2 * S(i)+I	S(i)
NewI	S(i)	S(i)+I	I

This formula guarantees that the *Odd:Add2 iteration structure will produce a response for any stimulus history. However, iteration structures can be defined that will not produce responses for any stimulus history. For example, the *Odd:Max2 iteration structure will no longer produce responses once Max2 has selected an odd number as the current maximum. In this case, the iteration structure cannot complete a transition.

Analysis of Concurrent Structures

Figure 2.5-13 shows the black box concurrent structure in which Add2 and Max2 are performed concurrently on the same stimulus history. The notation for this structure is Add2||Max2; in general, B1||B2 where B1 and B2 are performed concurrently. The complex response for the structure is the grouping of the individual black box responses, R = (R1, R2). The example history for the concurrent structure is easily seen:

Add2||Max2 history:

S	S1	S2	R1	R2	R
3	3	3	3	3	(3, 3)
6	6	6	9	6	(9, 6)
1	1	1	7	6	(7, 6)
9	9	9	10	9	(10, 9)
6	6	6	15	9	(15, 9)

The transition formula for a concurrent black box is simply the transition formulas of its black boxes. In this case the transition formula is

$R(i) = (R1(i), R2(i))$ where

$R1(i) = S(i) + S(i - 1)$, and
$R2(i) = \max(S(i), S(i - 1))$

These examples of sequence, alternation, iteration, and concurrency show that black box structures can produce complex behavior with simple component black boxes.

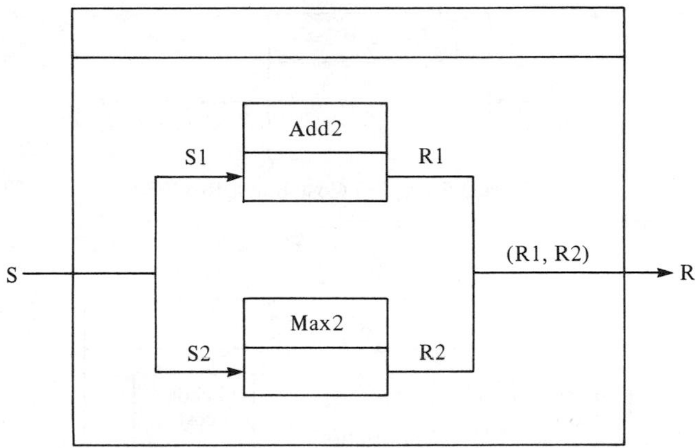

Figure 2.5-13. Add2||Max2 Concurrent Structure.

> **Fundamental Principle. Black Box Systems:** If complex behavior is required in a system, it may be possible to achieve it with sequence, alternation, iteration and concurrent structures of simpler black boxes.

2.5.3 Black Box Structures in Business Operations

As we have seen, black boxes can be combined into structures that exhibit new black box behavior. But black boxes can also be decomposed into black box structures that exhibit **equivalent behavior,** by a process called **black box expansion.**

When people or organizations deal at "arms length," that is, without a common organization objective or control, they are behaving as a black box structure. Even within an organization it may be desirable to put different units at arms length for purposes of decentralization, simplification, security, etc., with the result of creating black box structures.

In illustration, consider the job cost function for a carpet company, which can be represented as the black box depicted in Figure 2.5-14. The Job cost black box accepts a carpeting job description as input and produces the total cost of the job as output. If the cost of materials and the cost of labor are independent, the Job cost black box can be expanded into a black box sequence structure with equivalent behavior, as shown in Figure 2.5-15.

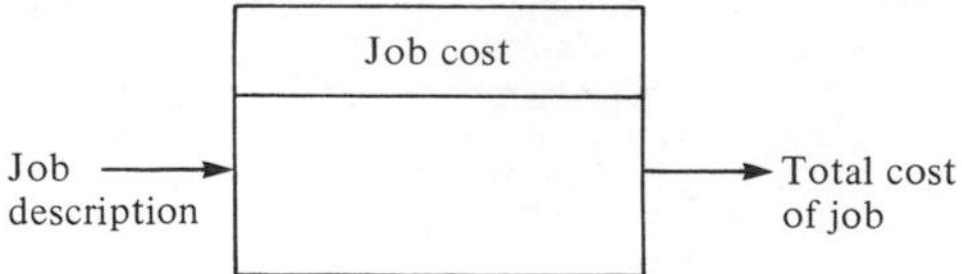

Figure 2.5-14. Job Cost Black Box.

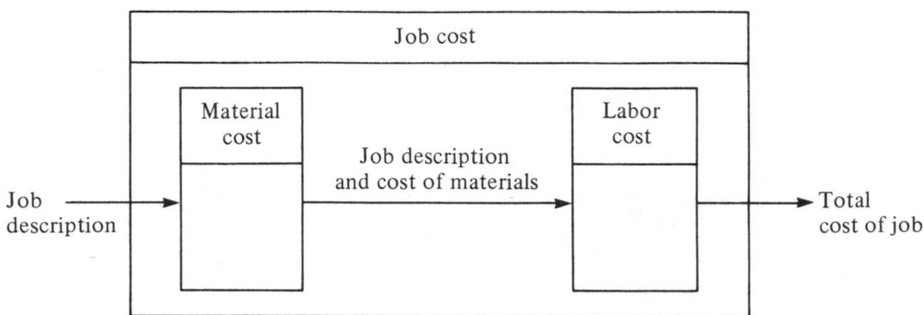

Figure 2.5-15. Sequence Expansion of the Job Cost Black Box.

The input for the Material cost black box is the job description; the output is the job description and the cost of material. The input for the Labor cost black box is the output of the Material cost black box; the output is the total cost of the job. Note that the Material cost black box must pass the job description through as input to the Labor cost black box to make the black box sequence structure function properly.

In turn, the Material cost black box may have additional structure, based on which of two suppliers may be used for the carpeting, for example, based on the size of the job. In this case, the Material cost black box can be expanded into an alternation structure with equivalent behavior, as shown in Figure 2.5-16. In turn, each of the Big supplier, Small supplier, or Labor cost black boxes might be expanded further.

Note that the communication of information in a black box structure is strictly limited to stimuli received from and responses passed to adjacent black boxes. A black box in a black box structure has no knowledge of the transitions of any other black boxes beyond the stimuli it receives, and conveys no knowledge of its own transitions beyond the responses it produces. For example, the two black boxes of Figure 2.5-16 operate at arms length on the basis of a response-to-stimulus connection, with no other sharing of information possible. Thus, the Labor cost black box cannot know which of the two suppliers was selected for a given job.

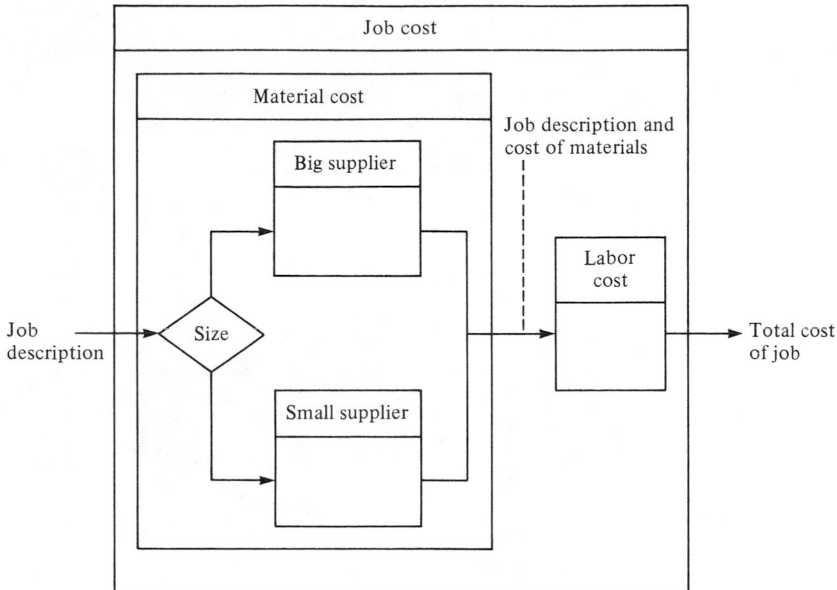

Figure 2.5-16. Alternation Expansion of the Material Cost Black Box.

> **Fundamental Principle:** In constructing a system as a black box structure, details of behavior can be localized to individual black boxes for greater security, separation of concerns, and clarity of design.

An alternate design is possible for the Job cost box structure which illustrates the use of concurrency in box expansions. First, Figure 2.5-17 depicts a new sequence expansion of the Job cost black box of Figure

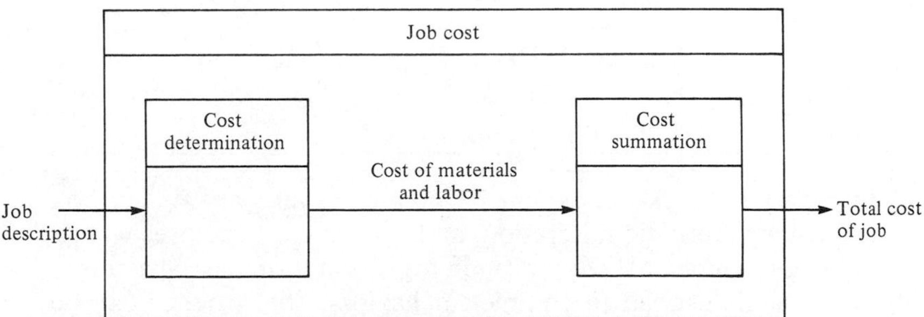

Figure 2.5-17. A New Expansion of the Job Cost Black Box.

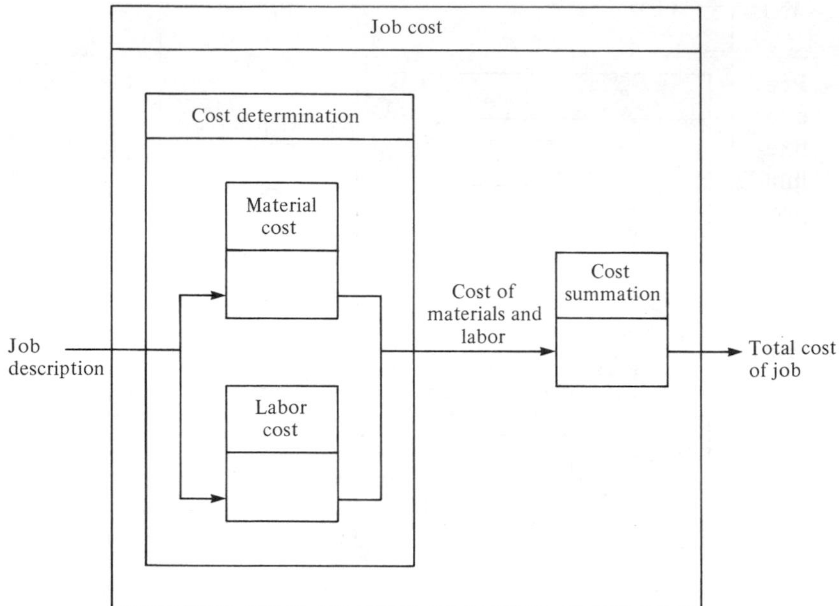

Figure 2.5-18. Concurrent Expansion of the Cost Determination Black Box.

2.5-14 in which the first part is to determine the cost of both materials and labor, the second to sum them up.

The Cost determination black box can now be expanded into a concurrent structure, as shown in Figure 2.5-18. This expansion is possible because the transitions of the Material cost and Labor cost black boxes are independent, given the common job description stimulus. This concurrent box structure reveals an opportunity to simultaneously assign the costing operations to, say, two groups within a contracting department, thereby decreasing the time required to cost a job and improving response to customers.

Fundamental Principle: The analysis of black box structures may reveal opportunities for concurrency that result in more effective business operations.

Summary: Black boxes can be combined into new black box sequence, alternation, iteration, and concurrent structures whose behavior can be analyzed through transition formulas. Structures of black boxes lead to complex behaviors. Black boxes can be expanded into black box structures that exhibit equivalent behavior.

2.6 INTRODUCTION TO BOX DESCRIPTION LANGUAGE

Preview: A Box Description Language (BDL) is introduced to express black boxes and black box structures. BDL consists of a fixed outer syntax of keywords and typography and a flexible inner syntax of business English and math appropriate to the problem at hand.

2.6.1 The Idea of Box Description Language

Because of the size and complexity of information systems and the number of people required to develop and use them, precise communication of system behavior and structure is crucial to success. Box structures provide a theoretical foundation for information systems development and use, and Box Description Language, or BDL, provides a corresponding language for recording, communicating, and analyzing box structures among developers, users, and managers of information systems.

Box Description Language is an open-ended specialization of natural language. It contains textual forms for black box, state machine, and clear box analysis and design. Each form is defined in terms of a fixed outer syntax that deals with overall structure and organization and a flexible inner syntax that deals with specific objects and operations within outer syntax structures. Outer syntax is defined by keywords in a tabular typographic format. Inner syntax is expressed in natural language or in specialized notation appropriate to the problem at hand.

2.6.2 Black Boxes in BDL

BDL accommodates two design steps for black boxes, namely, black box definition and black box invocation.

Black Box Definition. A black box definition in BDL prescribes behavior in terms of transitions from stimulus history to response, with the usual understanding that the current stimulus becomes the latest member of the stimulus history for the current transition. The BDL syntax for a black box definition is

```
define BB <BB name>
  stimulus
    <stimulus name>:<type>
  response
    <response name>:<type>
  transition
    <BB transition>
```

with outer syntax keywords **define BB, stimulus, response,** and **transition.** The fixed indentation structure serves to display the definition parts for better readability. The angle brackets (<,>) enclose generic names for parts of syntax that must satisfy further syntax rules. In this case, <BB name> is the name of the black box, <stimulus name> and <type> describe the stimulus, <response name> and <type> describe the response, and <BB transition> describes the transition of the black box. A <type> definition prescribes permissible values of a data item as, for example, shown in Table 2.6-1.

In BDL, the stimulus history of a black box has the form

<stimulus name>.0, <stimulus name>.1, <stimulus name>.2, ...

where <stimulus name>.0 refers to the current stimulus, <stimulus name>.1 refers to the previous stimulus, <stimulus name>.2 refers to the next previous stimulus, etc. This sequence is finite for finite black boxes, but otherwise can grow without limit. Where no misunderstanding can arise, references to the current stimulus can be abbreviated to <stimulus name>. Thus, for stimuli named K, K.1 and K.4 refer to the first and fourth predecessor stimuli, respectively. The advantage of this notation is that the absolute index, from the beginning of the stimulus history (e.g., K(i), K(i − 1), etc.) is not needed. The indexes used in the BDL language are relative to the current stimulus.

In illustration, the Add2 black box can be defined in BDL as

```
define BB Add2
   stimulus
      S:number
   response
      R:number
   transition
      R := S.0 + S.1
```

where the stimulus is named S of type number and the response named R, likewise of type number. In this case, the transition is defined by a **data assignment.** A data assignment in BDL has the general form

<variable> := <expression>

where the inner syntax <expression> is computed and assigned to the item represented by the inner syntax <variable>. Thus, in the Add2 example, R is assigned the value of the expression on the right, namely the sum of the current stimulus and the latest member of the stimulus

Table 2.6-1

Data Type Examples

Type	Values
Hour	1,2,...,24
Number	Digit strings
Word	Letter strings
Day	Sun.,Mon.,...,Sat.
Weekday	Mon.,Tues.,...,Fri.
Region	Northeast,Northwest, Southeast,Southwest

history. The names S and R were selected for their mnemonic value, but such interpretations can be misleading. For example,

```
define BB Sub2
  stimulus
    R:number
  response
    S:number
  transition
    S := R.0 + R.1
```

likewise defines the behavior of Add2! In short, names are simply placeholders in BDL. It is the transitions themselves that define black box behavior, and not the names used to define them. Nevertheless, names should be chosen with care, to help suggest correct interpretations to the reader.

The data assignment of the Add2 transition could likewise be expressed in natural language

transition
    Set the response to the sum of the current and previous stimuli

with no loss of equivalence. The expressive forms chosen for defining black box transitions depend on the subject matter and intended audience, just as does all human communication. Another useful representation for the Add2 transition would be the transition formula stated as an equation

transition
    $R(i) = S(i) + S(i - 1)$

As noted previously, the equal sign (=) denotes equality, not assignment.

Whatever forms are used, the objective is completeness and precision. Any transition definition that excludes possible behavior or includes impossible behavior is simply incorrect, and can lead to confusion among users and developers alike. Transition definition should be a major focus of intellectual effort in information systems development. But, however transitions are described, the same test for completeness can be applied.

> **Transition Completeness Rule:** A black box transition must define all possible responses from all possible stimulus histories.

The black box for Max2 can be defined as

define BB Max2
  stimulus
    S:number
  response
    R:number
  transition
    R := max(S.0, S.1)

where the reader is expected to know that max is a short name for an operation that produces the maximum of two arguments, namely, S.0 and S.1. An equivalent transition could be expressed as

transition
  S.0 $\geq$ S.1 $\rightarrow$ R := S.0 | S.0 < S.1 $\rightarrow$ R := S.1

read "If S.0 is greater than or equal ($\geq$) to S.1, then ($\rightarrow$) set R to the value of S.0, otherwise (|) if S.0 is less than (<) S.1, then ($\rightarrow$) set R to the value of S.1." Such an expression is known as a **conditional assignment,** with general form

condition $\rightarrow$ assignment | condition $\rightarrow$ assignment | ...

with the understanding that evaluation proceeds from left to right, and the first condition satisfied results in execution of the corresponding assignment. Conditional assignments can be easily expressed in equivalent natural language form for more general audiences, with no loss of precision.

Black boxes can be specified in BDL at various levels of abstraction. For example, an **abstract specification** of a hand calculator black box could be written as follows:

define BB Hand calculator
  stimulus
    I:proper string

response
O:number
transition
O := value of arithmetic expression in I

where proper string is defined as any valid arithmetic expression (without parenthesis) delimited by C and =. That is, the function of a hand calculator is to evaluate arithmetic expressions. A **concrete specification** for this hand calculator will describe, stimulus by stimulus, how numbers are displayed, as built up a digit at a time. But any such concrete specification only represents a method of creating an input, using good human engineering principles to assist the user in creating the input.

Note also the extra explanation of input I as a proper string. It is possible that the term arithmetic expression needs more explanation as well, for example, in describing the forms numbers can take (decimal points or not, etc.) and the arithmetic functions permitted. A complete abstract specification will settle all such questions, so that the exact meaning of the term proper string is fully defined, rather than leaving it to implementation. On the other hand, the concrete specification may be left to implementation. If the abstract specification is satisfied, the function of the hand calculator will be achieved, while the concrete specification can be designed to make the hand calculator as easy to use as possible. The variations in the concrete specifications of hand calculators represent different attempts to be user friendly in concrete specification.

In further illustration, consider the carpet company example given as a clear box structure in the preceeding section. Its black box definition is

define BB Job cost
stimulus
Job description : carpet type and dimensions
response
Total cost of the job : dollars
transition
Determine total cost of the job, including cost of labor
and materials, from job description

Note that Job cost is a finite black box of order 1 in this definition, since previous job descriptions will not affect the total effect of this job. But note, also, that a concrete specification which might be used to keystroke the stimulus job description into a terminal, with as many corrections as required, would not be a finite black box, since any number of key strokes might be required (with corrections) for entering the job description. That is, the finiteness of the black box may be relative, depending on the granularity of the data entry defined.

Black Box Invocation. Black boxes can be invoked in BDL by **black box statements** of the form

use BB <BB name> (<stimulus name>; <response name>)

where the keyword **use BB** means "carry out a transition of the black box with name <BB name>, given stimulus <stimulus name> and producing response <response name>." For example, the black box statements

use BB Add2 (i;j)

and

use BB First (i;j)

invoke black box transitions using data objects i, j for stimulus, response, respectively. Thus, for stimulus history 3 6 1 9 6 and i = 2,

use BB Add2 (2;j)

sets j to 8, and

use BB First (2;j)

sets j to 3 (sets j to the first stimulus).

In the case of the carpet Job cost black box, its invocation takes the form

use BB Job cost (job description; total cost of the job)

or, more indirectly

use BB Job cost (memo; cost)

where memo is some job description and cost will be regarded as the total cost of the job. That is, the stimulus and response can be given any names whatsoever, but must conform to the type of objects in the **define BB** statement. In contrast, the name Job cost is the one and only name which refers to the Job cost black box.

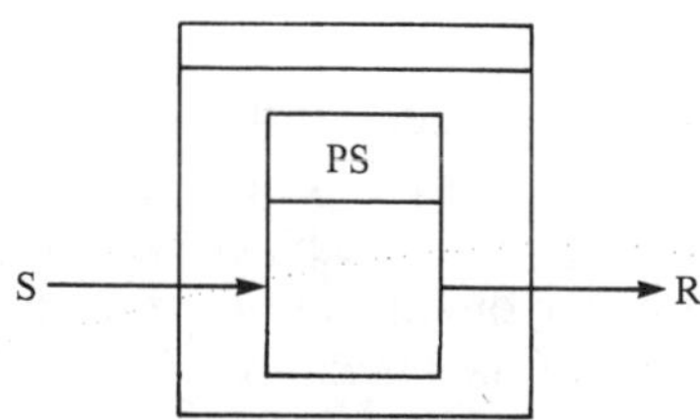

Figure 2.6-1. Procedure Statement.

2.6.3 Black Box Structures in BDL

Black box structures invoke black boxes in sequence, alternation, and iteration control structures. The BDL syntax for a black box structure is

define BB <BB name>
  stimulus
    <stimulus name> : <type>
  response
    <response name> : <type>
  proc
    <procedure statement>
  corp

where keywords **proc** and **corp** (proc, spelled backward) delimit the syntax part <procedure statement> (PS for short), and where <procedure statement> is itself subject to additional syntax rules. A procedure statement is the fundamental unit of black box structure, a single entry/single exit structure as depicted in Figure 2.6-1. A procedure statement in a black box may be a black box statement as defined above or a control statement. The black box control statements of sequence, alternation, iteration, and concurrency represent single entry/single exit control structures.

The BDL Sequence Control Structure

The BDL syntax for the sequence control structure is depicted in Figure 2.6-2 as the **do** statement. The statement is delimited by keywords **do** and **od (do,** spelled backward), and defines successive invocation of procedure statements, separated by semicolons.

Ordinarily, each keyword and procedure statement is written on a separate line, with the procedure statements indented for readability. However, the linear **do** statement

do PS1; PS2 **od**

is correct and has the same meaning. The linear form saves space, but the tabular form is more useful for documentation and description in realistic

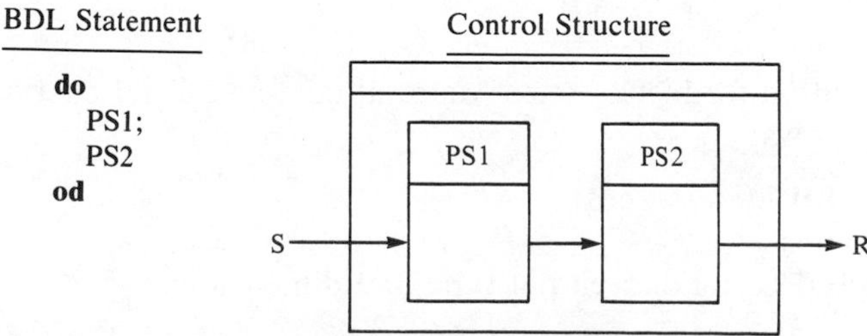

Figure 2.6-2. The BDL do Statement.

examples. The individual procedure statements PS1, PS2, ... are called part1, part2, ... of the **do** statement.

The linear statement

do PS1 **od**

is also correct, consisting of a **do** statement with only one procedure statement. It will have the effect of PS1, itself, in any procedure. The linear statement

do od

is also correct, containing no procedure statements with no effect. It will be useful to have a name for this, namely, the null statement, literally no procedure statement. In this case, **do od** contains a single null statement. In further illustration,

do PS1; PS2; PS3; **od**

contains four statements, the fourth one being a null statement, and

do; PS1 **od**

contains two statements, the first one being a null statement. As a result of this definition of null statements, semicolons may be inserted anywhere in a **do** statement without effect.

In illustration, the first decomposition of the black box structure for Job cost can be described in the form:

```
define BB Job cost
  stimulus
    Job description : carpet type and dimension
  response
    Total cost of job : dollars
  proc
    do
      use BB Material cost (job description ; job description
                                          and cost of material);
      use BB Total cost (job description and cost of material;
                                              total cost of job)
    od
  corp
```

where definition for black boxes Material cost and Total cost have been given as follows:

```
define BB Material cost
  stimulus
    Job description : carpet type and dimensions
```

response
Job description and cost of material : carpet type and dimensions and dollars
transition
Determine cost of materials from job description and pass on the description

define BB Total cost
stimulus
Job description and cost of material : carpet type and dimensions and dollars
response
Total cost of job : dollars
transition
Determine total cost of the job including labor from job description and cost of material

Note that response from Material cost must exactly match (in name) the stimulus to Total cost.

The BDL Alternation Control Structure

Procedure statements in an alternation control structure are usually written in indented text form, delimited by keywords **if, then, else, fi** (if, spelled backward), shown in Figure 2.6-3 as the **if** statement.

However, the linear statement

if C **then** PS1 **else** PS2 **fi**

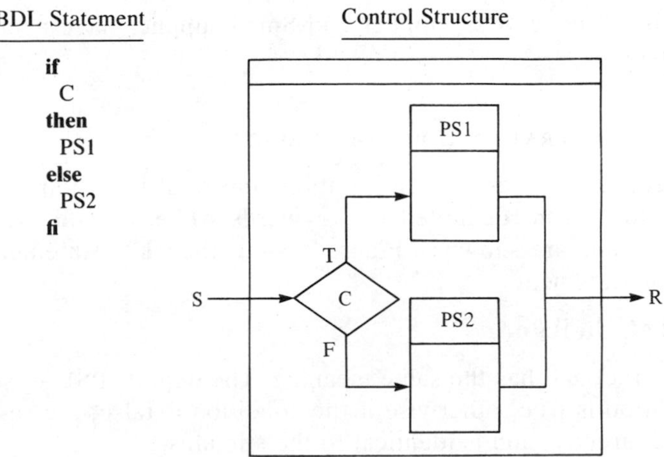

Figure 2.6-3. The BDL if Statement.

is correct and has the same meaning. The **thenpart** (PS1) is executed if the condition is true, otherwise the **elsepart** (PS2) is executed.

If the **else** branch (PS2) is a null statement, then the **else** keyword and the null statement can be omitted. That is,

if C **then** PS1 **fi**

The condition C is called the **if condition,** and the individual procedure statements PS1, PS2 are called the **thenpart, elsepart** of the **if** statement.

In the carpet company illustration, the black box Material cost was itself given as a black box alternation structure in the form:

```
define BB Material cost
  stimulus
    Job description : carpet type and dimensions
  response
    Job description and cost of material: carpet type and
                                          dimensions and dollars
  proc
    if
      size is big
    then
      use BB Big supplier (job description; job description
                                          and cost of material)
    else
      use BB Small supplier (job description; job description
                                          and cost of material)
    fi
  corp
```

and the black boxes Big supplier and Small supplier have been defined accordingly.

The BDL Iteration Control Structure

Procedure statements in an iteration control structure are written in indented text form, delimited by keywords **while, do, od** (**do,** spelled backward), and are shown in Figure 2.6-4 as the **while** statement. The linear statement

while C **do** PS **od**

is also correct and has the same meaning. The **dopart** (PS) is executed if the condition is true, otherwise if the condition is false, the response is produced directly, and is identical to the stimulus.

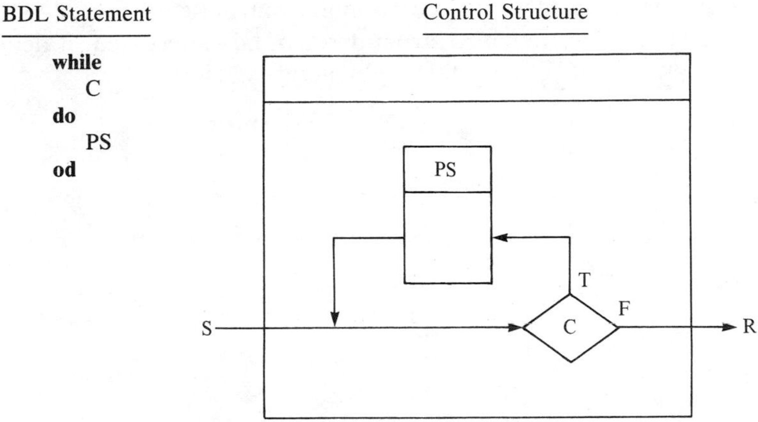

Figure 2.6-4. The BDL while Statement.

The BDL Concurrent Control Structure

The procedure statements in a concurrent control structure are delimited by the keywords **con** and **noc** as shown in Figure 2.6-5. Any number of concurrent procedure statements (PS) may be present, separated by commas. The completion of the concurrent structure is signaled by **noc,** at which point all concurrent statements are completed and all responses are formed.

Compound BDL Structures. Because the BDL control statements are themselves procedure statements, they can be nested and sequenced

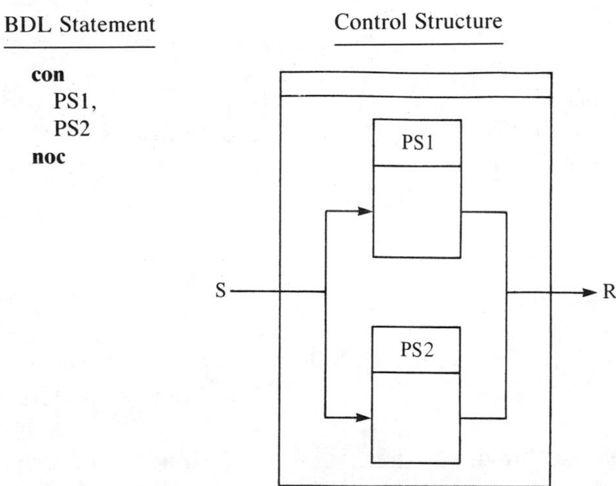

Figure 2.6-5. The BDL con Statement.

in any pattern whatsoever in compound control structures. The **do-od** delimiters of nested sequence structures can be suppressed, if delimited by the keywords of their containing structures. Thus,

```
proc                                      proc
  do                                        PS1;
    PS1;          is equivalent to          PS2
    PS2                                   corp
  od
corp
```

and

```
if                                        if
  C                                         C
then                                      then
  do                                        PS1;
    PS1;                                    PS2
    PS2           is equivalent to        else
  od                                        PS3;
else                                        PS4
  do                                      fi
    PS3;
    PS4
  od
fi
```

In illustration, the entire black box structure for Job cost is defined as follows:

```
define BB Job cost
  stimulus
    Job description : carpet type and dimensions
  response
    Total cost of job : dollars
  proc
    if
      size is big
    then
      use BB Big supplier (job description; job description
                                          and cost of material)
    else
      use BB Small supplier (job description; job description
                                            and cost of material)
    fi;
```

```
      use BB Total cost (job description and cost of material:
                                               total cost of job)
    corp
```

Sequence and alternation control structures must ultimately be expressed in terms of black box statements. For example, the following black box structure

```
define BB Sample
  stimulus
    S: number
  response
    T: number
  proc
    if
      S odd
    then
      use BB Add2(S;R)
    else
      use BB Max2(S;R)
    fi;
    if
      R even
    then
      use BB Max2(R;T)
    else
      use BB Add2(R;T)
    fi
  corp
```

is a sequence structure whose part1 and part2 are both alternation structures which invoke black boxes Add2 and Max2. What is the behavior of the black box structure? To find out, first determine the behavior of each of the sequence parts, then combine to derive the overall behavior.

> **Summary:** Black boxes can be expressed in a clear and concise format in BDL. The black box stimulus and response are related by the description of a transition in the form of English and mathematics appropriate to the problem. Black box structures are defined in BDL by the syntax rules of sequence, alternation, iteration, and concurrent structures.

EXERCISES

1. Which of the following devices exhibit black box behavior?

(a) A clock with a button to display current time
(b) A counter with a button to add one to current count
(c) A word processor
(d) An automatic chess player
(e) A human chess player
(f) A combination lock
(g) A key lock
(h) A telephone
(i) A smoke alarm

2. Discover and discuss properties of black box behavior such that

(a) response := 3rd previous stimulus
(b) response := previous response
(c) response := 3rd previous response
(d) response := number of stimuli since the maximum stimulus
(e) response := previous response plus current stimuli

3. Explain the following black box behaviors of devices with user interfaces like that of Figure 2.1-2. What part of its stimulus history must each device remember in order to exhibit consistent behavior?

Device 1

Stimulus	9	2	1	7	4	2	5	8
Response	9	11	12	10	12	13	11	15

Device 2

Stimulus	9	2	1	7	4	2	5	8
Response	9	29	12	17	47	24	25	58

Device 3

Stimulus	9	2	1	7	4	2	5	8
Response	9	2	1	1	1	1	1	1

Device 4

Stimulus	9	2	1	7	4	2	5	8
Response	4	5	1	4	5	3	3	6

Device 5

Stimulus	1	7	2	0	3	9	4	0	2	2	4	0
Response	0	0	0	1	0	0	0	3	0	0	0	2

4. Which of the following are finite black boxes, and what is the order of each finite black box?

Min:	Response is minimum of stimuli accepted
Double:	Response is double the stimulus
Add3:	Response is sum of previous 3 stimuli
AddEvens:	Response is sum of previous 2 even stimuli

5. Consider a black box MaxPD (Previous Day) that returns the maximum hourly load of the previous calendar day (not the last 24 hours), what is the order of MaxPD?

6. What is the difference between a transition and a transaction in a word processing system?

7. Is there a difference between a transition and a transaction in Add2 (assuming the user cannot add)?

8. Assuming the user has a short memory and can remember only the current stimulus, what are transactions in the black boxes of Section 2.1.4?

   (a) Echo
   (b) Previous
   (c) Constant
   (d) First
   (e) Max

9. Some hand calculators display the last operator (+,−,...,=) keyed in. Do they exhibit the same or different black box behavior as hand calculators that do not display operators?

10. Can you develop a procedure for use of a hand calculator whose clear key is missing?

11. In using a hand calculator for addition, entries are made for the first number, the operator (+), and the second number. When the = key is depressed, the hand calculator determines the result and displays it. At that point, depressing the + or = key may reveal some unanticipated black box behavior of the hand calculator. What responses do you get for the following stimulus histories?

    (a) C 3 + 9 = = =
    (b) C 3 + 9 = 6 =
    (c) C 3 + 9 + + +

    What do the responses tell you about the black box behavior of the hand calculator? Invent some stimulus histories of your own to invoke unusual black box behavior.

12. What is the black box behavior of your hand calculator in accepting and calculating numbers that overflow its display?

13. Identify three black boxes you have interacted with in the past week.

14. Discuss the role of education in a black box description of a person.

Given additional black boxes Min2 (response R is the minimum of the last two stimuli) and Prod2 (response R is the product of the last two stimuli), what is the behavior of the following black box compositions for stimulus history (3 6 1 9 6)? What is the transition formula in each case?

15. Sequence Structures
- (a) Min2;Max2
- (b) Max2;Min2
- (c) Prod2;Min2
- (d) Prod2;Max2
- (e) Max2;Prod2

16. Alternation Structures
- (a) Odd:Min2|Max2
- (b) Odd:Prod2|Prod2
- (c) Odd:Min2|Prod2
- (d) Odd:Prod2|Min2
- (e) Odd:Add2|Echo

17. Iteration and Concurrent Structures
- (a) *Odd:Prod2
- (b) *Odd:Min2
- (c) Add2||Prod2
- (d) Prod2||Min2
- (e) *Odd:First

18. Compound Structures
- (a) Odd:(Min2;Max2)|(Max2;Min2)
- (b) Odd:(Add:Min2|Max2)|Prod2
- (c) (Odd:Min2|Min2);Min2
- (d) Max2;(Odd:Max2|Max2)
- (e) (Add2;Add2);Add2

19. Expand the Labor cost black box in Figure 2.5-16 based upon the following considerations. Travel time is billed at a different rate from site time. In calculating labor costs at the site, big jobs (greater than 1000 square feet) are billed at a lower rate than small jobs.

20. Decide which of these black box structures define the same black box.
- (a) (Add2;Add2);Add2
- (b) Add2;(Add2;Add2)

(c) (Add2;Max2);Add2
(d) Add2;(Max2;Add2)
(e) (Max2;Add2);Add2

21. Define the black boxes Add2, Max2, Min2, Prod2, RA12, and Max24 using BDL syntax.

22. Describe the black box structures in exercises 15 through 18 using the appropriate BDL syntax.

23. The black box Score computes the mean of the last 10 scores (stimuli) after eliminating the maximum and minimum scores of the last 10.

(a) Use black box BDL to describe Score.
(b) Expand Score as a black box structure with the sequence and alternation structures using simpler black boxes. Use black box BDL to describe Score and black box BDL for the component black boxes.
(c) Derive Score from (b) to validate that the resulting black box transition is identical to that of the Score black box in (a).

24. Consider the following BDL black box specification containing a structured English transition.

define BB Accsmall
  stimulus
    S, T, U : number
  response
    R : number
  transition
    Find the smallest of the three stimuli and set R to the running sum of the smallest numbers.

(a) Expand a black box structure for Accsmall that contains simpler black box structures. Use BDL for the design. Provide transition formuli for all black boxes used.
(b) Derive the structure from part (a) to find a transition formula for Accsmall in terms of stimulus history and response.

25. We have seen that a computer system can be described as a black box. How then can we use a computer to simulate non black-box behaviors, such as a roulette wheel or a temperature indicator?

Chapter 3 | The State Machine Behavior of Information Systems

3.1 STATE MACHINE BEHAVIOR

> **Preview:** A state machine is an alternate form of description of a system that substitutes internal storage for the stimulus history of its black box. Any system that exhibits black box behavior can be described as a state machine.

3.1.1 Describing Black Boxes as State Machines

In Chapter 2, we learned that a black box can return different responses to the same stimulus at different points in time, and that the response of a black box depends not only on its current stimulus, but on its stimulus history as well. So our model of a black box in Chapter 2 can be restated in the transition

$$(\text{stimulus, stimulus history}) \rightarrow (\text{response, new stimulus history})$$

where "new stimulus history" is simply "stimulus" appended to "stimulus history", and is understood to replace "stimulus history" for the next transition.

Now a black box stimulus history can grow indefinitely large, and any real mechanism—a hand calculator, a personal computer, or a business information system, for example—will eventually be saturated and unable to retain too large a history. And of course, black box descriptions of calculator, computer, or information system behavior would quickly become unwieldy for stimulus histories of any size. So as a practical matter, it would be useful to have another description of black box behavior that does not depend on explicit recording of stimulus histories. Fortunately, such a description is available. It is called a **state machine.**

A state machine is defined by a **state,** which incorporates the stimulus history, and a **machine,** another (usually simpler) black box that carries out the transitions, of the state machine. So a black box that converts any stimulus history into a response can be simulated by a state machine that converts the stimulus and the state into a response and a new state. Of course, only the response will be visible to the user, not the new state. This new state will then become the current state for converting the next stimulus into a response and another new state. Such a state machine description is diagrammed as Figure 3.1-1.

Note that the machine in the figure takes in both the current stimulus and state and then produces a response and a new state which replaces the current state. The next stimulus triggers the machine to take in that stimulus and the new state it produced on the last transition, to produce the next response and the next new state. Thus, our previous model of a black box,

(stimulus, stimulus history) → (response, new stimulus history)

can be replaced by a new model,

(stimulus, state) → (response, new state)

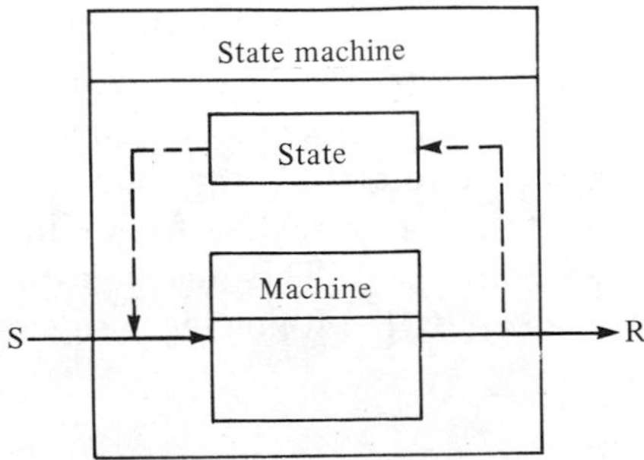

Figure 3.1-1. State Machine.

where "new state" is understood to replace "state" for the next transition. That is, the machine of a state machine is itself a black box with a complex, two-part stimulus and a complex, two-part response. The two-part stimulus of the machine black box is the stimulus and state of the state machine; the two-part response is the response and new state of the state machine.

Does every black box have a state machine description, or is there some conceivable black box for which a state machine description is insufficient? The answer is that a state machine description is possible for any black box whatsoever, because in the state machine definition

(stimulus, state) → (response, new state)

the state simply represents the stimulus history, so we can write, instead,

(stimulus, stimulus history) → (response, new stimulus history)

to recover the original definition.

> **Fundamental Principle:** Every black box can be described by a state machine.

3.1.2 State Machine Transitions

If the state of a state machine simply replicated the entire stimulus history of a black box, then a state machine description would be as cumbersome as one based directly on stimulus history. So again, as a practical matter, the state of a state machine description of a black box should be defined as a summarization, or abstraction, of the stimulus history.

In general, many different abstractions of a stimulus history are possible, and the abstraction must be defined with care, so as to permit the calculation of the black box responses. For example, consider again the Add2 black box. A natural state choice is to define the state as the previous stimulus. Given this state definition, a rule for the machine of the state machine to follow for each transition from (stimulus, state) to (response, new state) can be defined. This **machine transition** rule for the Add2 state machine is given by the following assignments:

response := stimulus + state
state := stimulus

With these definitions for the state and the machine, the successive transitions of the Add2 state machine, with initial state 0,

Stimulus	Response
3	3
6	9
1	7
9	10
6	15

can be diagrammed as in Figure 3.1-2. Note that after each transition the state value is replaced with just the right portion of the stimulus history to permit a correct computation on the following transition. The state will simply retain that value until the next transition is invoked by the user, whether in the next minute or the next month. As can be seen, this simple state definition permits behavior equivalent to an Add2 black box with access to its entire stimulus history.

The diagrams of Figure 3.1-2 are themselves cumbersome, and the transition sequences they depict are more conveniently expressed in the equivalent tabular form

(stimulus, state) → (response, new state)

as follows:

(3,0) → (3,3)
(6,3) → (9,6)
(1,6) → (7,1)
(9,1) → (10,9)
(6,9) → (15,6)

or, even more briefly, for any values X, Y,

(X, Y) → (X + Y, X)

Although this state machine for Add2 is very natural, it is not the only one possible. As a variation, consider a new state definition in which the state is double the previous stimulus. Then, for this state definition, a new machine is required, namely,

```
response := stimulus + state/2
state := 2 * stimulus
```

That is, if the state definition is changed, the machine must be changed for both the response and the new state.

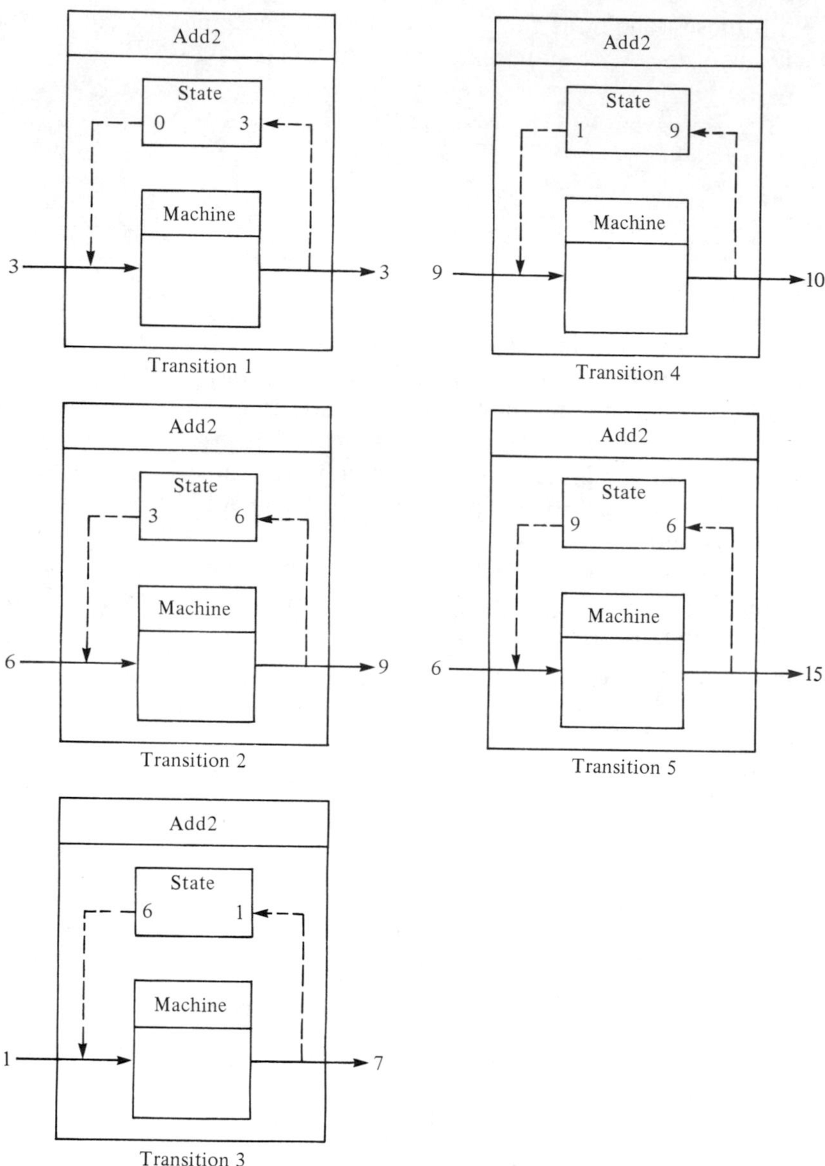

Figure 3.1-2. Transitions of the Add2 State Machine for Stimulus History 3 6 1 9 6.

3.1.3 Finite State Machines

We have already defined a finite black box as a black box in which every response can be determined from a finite stimulus history (whose minimum length is the order of the black box). A state machine with a finite number of states is called a **finite state machine.** As a practical matter, we deal with finite state machines when we implement an information system. The essential finiteness of computers and their limitations on data size force the system to use a finite state. For example, we can define a state variable, X, as an integer type. Conceptually, the set of possible values of this state is infinite. Practically, however, the magnitude of the state of X ranges from 0 to the largest integer representation in the computer.

Note that finite black boxes and finite state machines are not synonymous. A finite black box can be described by a finite state machine, but the converse is not so; a finite state machine description of a black box does not guarantee it to be a finite black box.

To see the first relation that any finite black box can be described by a finite state machine, recall the definition of a finite black box and its order k. Consider the set of all stimulus histories of order k. With a finite number of possibilities for each stimulus, this set of stimulus histories of order k will be finite, and can serve as the set of states for the state machine. Therefore, there exists a finite state machine which describes the finite black box.

To see that the converse does not hold, consider the black box First, which merely responds with its first stimulus, no matter what and how many stimuli follow. First is not a finite black box, because there is no maximum size stimulus history required to determine a response. Yet First can be described by a finite state machine, whose state merely contains the first stimulus.

It turns out to be very fortunate that the converse does not hold, because many interesting and important nonfinite black boxes can be described as finite state machines. In particular, the hand calculator, a nonfinite black box, can be described as a finite state machine, as we shall see.

3.1.4 The Master File Update State Machine

A common class of data processing system is built around a single **master file** of records. For example, retail hardware stores, as customers of wholesale distributors, would each have a record in a distributor's

accounts receivable file. Such a record could contain the amount owed, credit limit, address, and other information about the customer. Each day, the customer charges can be collected in another file, called a **transaction file.** Then, overnight, the data in the transaction file can be used to update the master file and produce a **report file.**

The master file can be used each month to create the customer's monthly bills. Then, as payments come in, they can be collected in another transaction file, and this can be used each night to update customer records, too. Thus far, we have discussed two kinds of inputs, charges and payments, and one kind of output, bills. But there are many more, for example, new customers may be added, or old customers deleted from the master file. The sales manager and credit manager may want various kinds of tabulations and reports from the master file.

Rather than attempting to name all possible inputs and outputs, we use the generic name transaction file for any input and report file for any output of a master file update. Each transaction file can begin with the kind of update required for the records that follow, and the computer program that updates the master file can be guided accordingly.

The foregoing description of a master file update data processing system defines a state machine as depicted in Figure 3.1-3. A master file update state machine has as its state a master file, such as receivables, inventory, or personnel records, and takes in an update, or transaction, file containing any additions, deletions, and changes. The state machine applies the updates to the master file and produces a file update report as output.

The master file update state machine is a completely general model of a black box. That is, any black box whatsoever can be described in the form of a master file update state machine! At first thought, this may seem

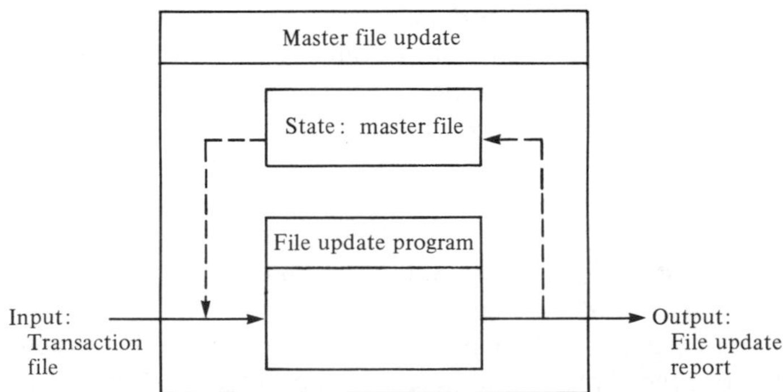

Figure 3.1-3. The Master File Update State Machine.

a surprising statement—that such a simple and standard data processing operation as master file update encompasses the entire range of data processing behavior—even the behavior of the information system of an entire business. The reason this simple idea works is that it applies to a wide range of events and information states and the update cycle can be a day, a minute, or a second in online systems. For example, when the counterperson of an electronics parts business receives an input from a customer, the state of information of the business has been changed into a new state—if ever so slightly, it is a new state. When the counterperson returns an output to a customer, one transaction of input and state to output and new state has been completed in this business information state machine.

With its wide applicability, the state machine is a general model of computer science and engineering. For example, every hardware device will behave as a state machine. For a computer, the state is the whole state of memory, including mass online storage, high speed memory, internal registers, even the instruction counter. Each instruction execution changes that state ever so slightly. Thus, an add instruction in a computer with a million bytes of storage will change only four bytes out of that million, but it is a new state, nevertheless.

3.1.5 A Business Enterprise Exhibits State Machine Behavior

The state machine model can be applied to business operations. In fact, any business has a state machine description. This is true whether or not the business is well managed or even whether or not the management of the business is conscious of the existence of state machine behavior. As an entity, a business accepts and responds to stimuli. These stimuli come from the outside world, from customers, vendors, the government, banks, and other sources. The responses produced by these stimuli are directed, in many instances, back toward the sources of the stimuli.

As a state machine, a business has many traits and characteristics similar to those of people. To illustrate, as with a person, the state of a business state machine is composed of its initial state, as altered by a continuing, ongoing history of stimuli. As with people, the learning process leads to continuing changes in state. This can be illustrated with what happens to a business when it undertakes to develop a new information system. The people associated with the information system project are subjected to a whole new series of stimuli. The stimuli, in turn, affect the states of the individual persons. Since the business itself is a composite of the experiences of the people, the changes to the states of the people also affect the state of the business. Sources for these state-changing experiences include the sales and educational actitivites of computer and software vendors and interactions with future system users in the business organization.

As a further point of similarity, the state of a business changes with the shifting of the cumulative experiences of its people. For example, as people are hired and fired, the state of the business—as well as the way in which a business will react to stimuli—is changed.

To illustrate, consider the situation of a counterperson within an electronics parts business. A different set of responses will come from this particular business state machine for customers who are served by different counterpersons. There can be vast differences in level and quality of services based on experience or attitude of different counterpeople. For example, if an experienced, conscientious counterperson leaves the business, the pattern of responses to stimuli in the form of customer orders will change dramatically. An outside observer of this particular business state machine may not know about the internal personnel status of the business. However, an outside observer, such as a dissatisfied customer, can infer the reasons for the changes in the same way that a systems analyst infers the behavior of a system from observations of its stimuli and responses.

Summary: A state machine represents the stimulus history of a system in a state that is acted upon by a machine. Any black box can be described as a state machine. The state machine view of any business system provides the insights gained by separating data (state) and processing (machine).

3.2 STRATEGIC USES OF STATE MACHINES

Preview: State machines can be used to model customer service strategies in many business processes. A state machine whose transactions and state data are sufficient to deal with all conditions of its business use exhibits the crucial property of transaction closure. State migration is an important strategy in expanding state machines into clear boxes which introduce new black box/state machine/clear box structures at the next level, and is introduced using state machines alone.

3.2.1 A State Machine Model of Customer Service

In many businesses, customer service strategies must be carefully planned in order to provide the best possible facilities. For example, in planning a new bank branch, the number of teller windows represents a potential level of customer service. When all tellers are busy, other customers must wait. The longer customers must wait, the more likely they are to take their business elsewhere. Customer waiting can be decreased by planning more teller windows, but that costs money too. The question of planning how many servers to provide for a service facility is very common. How many toll booths for an entrance to an expressway? How many reservation clerks answering an airlines telephone number? How many pump islands in a gasoline service station?

In order to see how customer service in a multiserver facility can be analyzed as a state machine, suppose that each bank customer requires one minute to be served. Then we can define a customer service state machine with a number of servers N, stimulus C (customers arriving in a given minute), response S (customers served during this minute), state W (customers waiting at the beginning of this minute) and machine M

```
M: S := min(N,W)
   W := W + C − min(N,W)
```

as shown in Figure 3.2-1. The black box of this Customer service state machine gives an overall view of how the service facility works. It is a different black box for each value of N. Given a stimulus history (of expected customer traffic), these different black boxes (and state machines) can be compared for their specific performance, and the best value of N chosen.

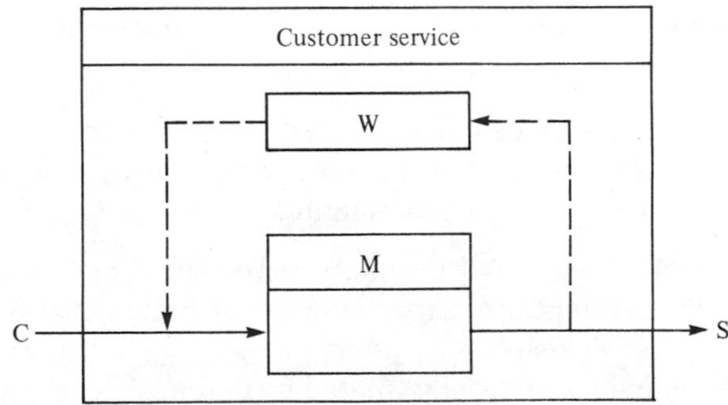

Figure 3.2-1. Customer Service State Machine.

Table 3.2-1

Customer Service at Three Levels of Service

N								
2			3			4		
C	W	S	C	W	S	C	W	S
	3			3			3	
3	4	2	3	3	3	3	3	3
5	7	2	5	5	3	5	5	3
4	9	2	4	6	3	4	5	4
1	8	2	1	4	3	1	2	4
6	12	2	6	7	3	6	6	2
2	12	2	2	6	3	2	4	4
0	10	2	0	3	3	0	0	4
3	11	2	3	3	3	3	3	0
4	13	2	4	4	3	4	4	3
2	13	2	2	3	3	2	2	4

In illustration, Table 3.2-1 shows customer service for three levels of service. The same history of ten customer arrivals over a ten-minute interval is shown for N values of 2, 3, 4. The average arrival rate over this ten-minute period is three customers per minute, varying from zero to six per minute. The initial state (customers waiting) is three in each case. The entire difference between the three cases is the maximum service rate defined by N.

Case N=2: The service units are used fully, but the customers waiting, W, is growing because the arrival rate, 3, exceeds the service rate, 2. Clearly, there is insufficient service and customer waiting times are growing, probably to an intolerable level.

Case N=3: The service units are used fully and the customer waiting is stabilized between three and seven, at an average of five, so the average waiting time in the system is 5/3 minutes.

Case N=4: The service units are used at a 75% level, but the number of customers waiting is reduced to an average of 3.4, so the average waiting time will be 3.4/4, under a minute.

Both cases N=3 and N=4 probably represent acceptable operation, with a tradeoff between idle service units and better customer service, which is a business decision.

Such state machines are easily treated by spreadsheet calculations. As shown in Chapter 1, the variables of the stimulus, state and response, can be given as headings in the spreadsheet table, initial values given for the state variables in the first row, and a column of stimuli values given. The spreadsheet calculations then fill out the rest of the table. In this case, additional calculations such as service utilization, customer waiting times, and so on can be added to the spreadsheet, as well.

3.2.2 Transaction Closure in State Machines

The analysis and description of any business process, actual or intended, is simplified by using the state machine model. The first step is to identify the transactions of the business process as a black box. The second step is to identify the state data of a state machine that are needed to calculate the outputs of the transactions from the inputs. Usually, several iterations of these two steps will be required. Any state data identified must have been acquired by previous transactions. Such previous transactions may be of types not thought of before, and must be added to the behavior of the black box. In turn, such transactions may require new state data not thought of before, and more new transactions will be required.

Eventually, a set of transactions and state data will be discovered that are self-sustaining. This condition is called **transaction closure,** because a set of transactions has been found that are closed under all conditions of the business process.

> **Fundamental Principle—transaction closure:** The condition of transaction closure is satisfied if the transactions are sufficient to generate all state data, and the state data are sufficient to generate all the transactions.

For example, consider a simple book location system for a library. The primary book location transaction is an input which is the identity of a book and an output which is the location of the book in the library. But on reflection, the state data must include a table of book identifiers and locations that must have been the result of a previous transaction, say a book entry transaction. This transaction must include a book identifier and where it will be located as input, and a confirmation of the input as output. Also, since there are now at least two types of transactions, each type must be identified explicitly. This means that the book location and book entry transaction have transaction identifiers in addition to the data used as input within each transaction type.

With a little more thought a transaction for deleting books will be apparent, but this will require no new state data. However, deleting books will lead to unusable data storage space that must be reclaimed by reorganizing the table of book identifiers and locations. Such space reclamation will require a new transaction—a system integrity transaction. If this space available is not supplied by the underlying computer system, another transaction to determine space available will be called for. In turn, the state data must be augmented to keep track of the space available, and to update it when book entry, book deletions, or space reclamation transactions are invoked.

At this point the transactions and state data may be self-sufficient and transaction closure has been reached. Beginning with a primary transaction, additional transactions have been identified to reach a self-sufficient set. The visualization of a state machine and its state data is critical for this process of transaction closure.

This pattern of systematic discovery will work for any business process for which an information system is needed. The primary transactions will be readily identified, and the additional transactions required can be discovered by constructing a state machine step by step until transaction closure is achieved.

3.2.3 State Migration between Nested State Machines

State migration is an important design strategy in expanding state machines into clear boxes which introduce new black box/state machine/clear box structures at the next level. The concept of state migration can be described using state machines alone.

The machine of a state machine, as a black box, can be described as a state machine itself with an inner state. In this case there is an inner state for the nested state machine and an outer state for the original state machine as illustrated in Figure 3.2-2.

The external stimulus S0 is combined with the outer state to make a (two part) stimulus S1 for the nested state machine. In turn, the stimulus S1 is further combined with the inner state to make a stimulus S2 for the inner machine. Conversely, the response R2 from the inner machine is decomposed into the updated inner state and response R1 of the nested state machine. In turn, response R1 is further decomposed into the updated outer state and response R0 of the original state machine.

This nesting process can be repeated to any level with corresponding inner states at each level.

The division of the state of a state machine among nested levels is entirely arbitrary. For pure state machine nestings, such nesting does not accomplish much. But when state machines are expanded into clear boxes, good divisions of the state among levels can simplify descriptions and improve designs. Fortunately, such good divisions need not be arrived at immediately in the development of a box structured system. As ideas and insights arise, parts of the state can be migrated from one level

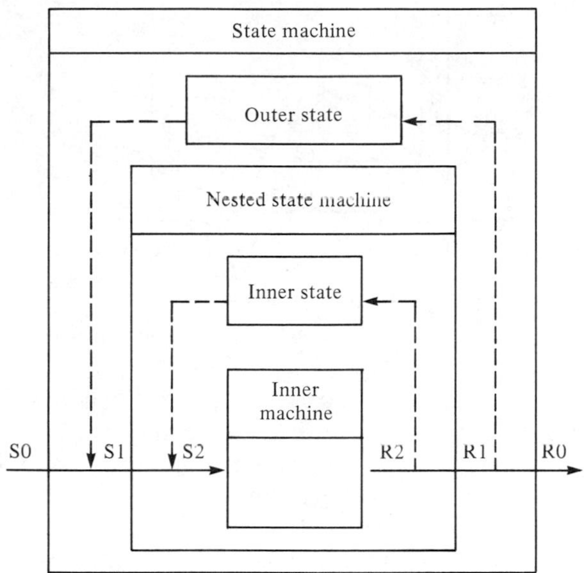

Figure 3.2-2. Nested State Machines.

to another to place data for effective storage and processing.

We can use state migration to demonstrate that two special cases of state machine structures are always possible. In the first case, the machine can be the very black box that is represented by the state machine. That is, the state is completely migrated into the machine, as in Figure 3.2-3. In this case, the state shown in Figure 3.2-3 is trivial and nothing is added to the stimulus to reach this inner black box. Even so, the form of a state machine is shown as a special case.

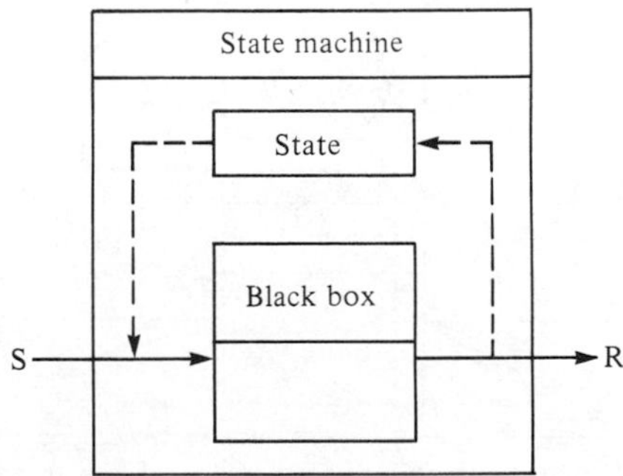

Figure 3.2-3. State Machine with Trivial State.

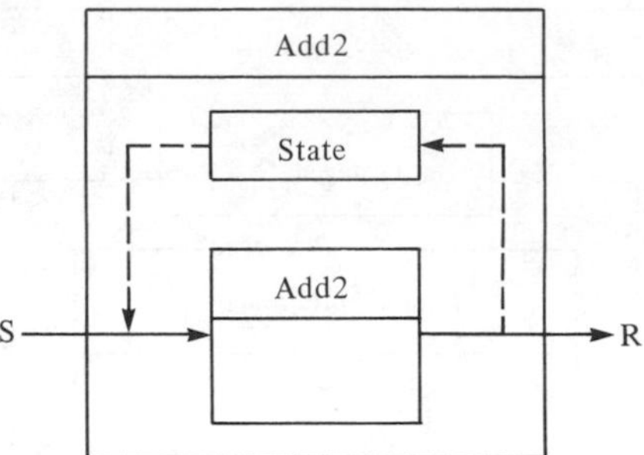

Figure 3.2-4. Add2 State Machine with Add2 Machine.

For example, an Add2 black box used as a machine in a state machine that ignores its state will produce the same behavior as the Add2 black box, as shown in Figure 3.2-4. This case is not a useful one, because the machine is no simpler than the black box being represented. But, it is a possible extreme case under the definition.

The second special case is the simplest possible, in terms of the stimulus history required. For every black box, there is a state machine whose machine needs only its stimulus and no previous history to determine its response. That is, the machine of a state machine can always be a black box of order 1.

The reason such a machine is always possible is that whatever history the machine might seem to need can in fact be migrated into the state of the state machine. Such a reconstruction of Figure 3.2-2 is depicted in Figure 3.2-5.

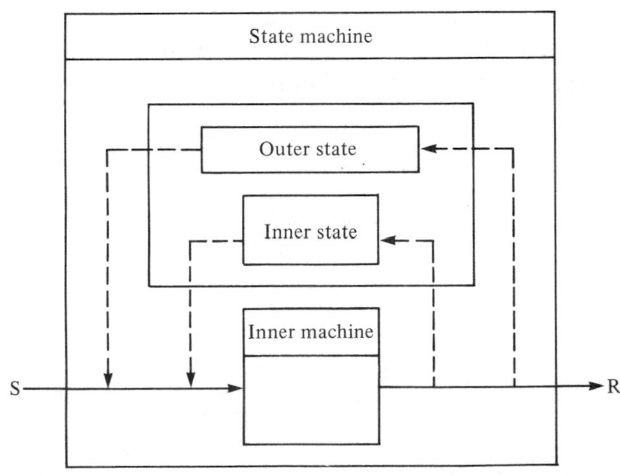

Figure 3.2-5. A State Machine with Machine State Inside the State.

We can show that this reconstruction is possible as follows. The separate machine state can be incorporated into the original state because the machine itself is unconcerned with the origin of the three items (a stimulus and two states) it receives and the destination of the three items (a response and two new states) it produces.

Therefore, the diagram of Figure 3.2-5, which depicts a state within a state construction, is equivalent in behavior to the diagram of Figure 3.2-2. But now, the state within a state is equivalent to a single state which contains the two substates as identifiable parts.

In illustration, the file update program (the machine of the state machine) need not retain any history data about its previous executions (transitions). All necessary data can be retained in the master file (the state), even data about the executions themselves. For example, the master file could contain a record of how many times the master file update program had been executed. Of course, the file update program would have to update that record along with the rest of the master file.

Summary: Behavior of state machines that model customer service can be represented and analyzed by spreadsheet calculations. Transaction closure can be achieved by identifying all business transactions required to provide the state data of a state machine. State migration is always possible between levels of nested state machines and will be useful in box structure design. As special cases, state data can always be migrated completely into the state or the machine of a nested state machine.

3.3 ANALYSIS OF BLACK BOX BEHAVIOR FROM STATE MACHINES

Preview: State machine behavior can be abstracted to equivalent state-free black box behavior, in a process called black box derivation.

3.3.1 The Black Box Behavior of State Machines

A state machine can be analyzed to determine its black box behavior by eliminating references to its state in the calculation of its responses.

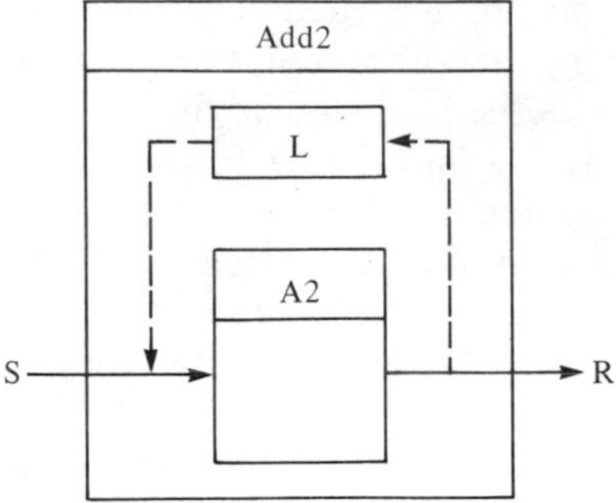

Figure 3.3-1. A State Machine for the Add2 Black Box.

Consider the state machine just discussed for the Add2 black box with machine A2 and state L that holds the last stimulus as shown in Figure 3.3-1. The machine A2 must perform on each transition from stimulus S to response R:

```
machine A2:
    R := S + L
    L := S
```

We can analyze state machine behavior as we did black box behavior, by considering the values of stimuli, responses, and states for various transitions. For example, for the Add2 state machine of Figure 3.3-1,

$$R(i) = S(i) + L(i - 1)$$
$$L(i) = S(i)$$

and, by substitution of $i - 1$ for i in the second equation, then $S(i - 1)$ for $L(i - 1)$ in the first equation, we find

$$R(i) = S(i) + S(i - 1)$$

which is the black box behavior of Add2. By eliminating references to the state L, we have derived the black box behavior, and verified that this state machine is a correct implementation of the Add2 black box.

Likewise, consider a state machine for the black box Max2 with machine M2 and state L that holds the last stimulus, as depicted in Figure 3.3-2. The machine M2 must perform on each transition as follows:

```
machine M2:
    R := max(S,L)
    L := S
```

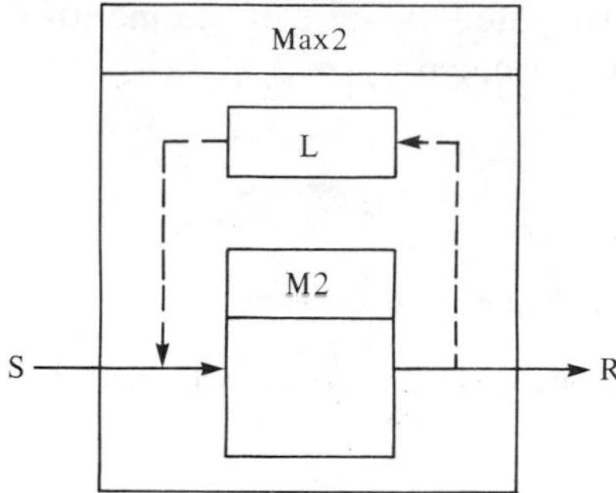

Figure 3.3-2. A State Machine for the Max2 Black Box.

For the Max2 state machine of Figure 3.3-2,

$$R(i) = \max(S(i), L(i - 1))$$
$$L(i) = S(i)$$

and, by substitution, we find

$$R(i) = \max(S(i), S(i - 1))$$

which is the black box behavior of Max2.

Note that the machines A2 and M2 completely define the black box behavior of the state machines. The states play a passive role of storing whatever the machines produce to be stored. However, as illustrated above in the variation of the Add2 state machine, the state definition determines how the machine must operate to achieve a given black box behavior.

3.3.2 Black Box Derivation of an Inventory Reorder State Machine

Business rules are often found in original form as state machines rather than as black boxes. For example, the k months of supply reorder policy, discussed in Chapter 1, proved to have the undesirable property of amplifying demand variations in its reorders when reduced to its black box behavior. The analysis that led to that discovery also led to a new kind of reorder policy that reduces demand variations to the greatest extent possible. This new reorder policy, called the exponential smoothing reorder policy (for reasons that will be clear later), is based on the choice of a single smoothing parameter s, a fraction between 0 and 1.

Each reorder R is a weighted average of the month's demand D and last month's reorder L, determined by s,

$$R := (1 - s) * D + s * L$$

If s = 0, then

$$R := (1 - 0) * D + 0 * L = D$$

so no smoothing takes place and R follows D through every variation and trend. If s = 1, then

$$R := (1 - 1) * D + 1 * L = L$$

and every R is exactly L so complete smoothing takes place, independently of variations and trends in D (of course, inventory may be piling up or disappearing). Neither of these extreme cases is useful or recommended. Instead, an inventory manager must choose s for each item to reflect the best balance between smoothing reorders and tracking demands. The inventory manager must also determine an initial inventory level because the exponential smoothing reorder policy does not determine the inventory level itself. A k months of supply calculation could be used periodically for this purpose, say, once a year.

A state machine for the exponential smoothing reorder policy can be defined by the machine:

$$R := (1 - s) * D + s * L$$
$$L := R$$

Next, references to state L can be eliminated to derive its black box behavior. That is, for month m,

$$R(m) = (1 - s) * D(m) + s * L(m - 1)$$
$$L(m) = R(m)$$

In this case, state L can be eliminated from the expression for response R directly, as

$$R(m) = (1 - s) * D(m) + s * R(m - 1)$$

but now R(m) is defined in terms of R(m − 1). But since (replacing m by m − 1)

$$R(m - 1) = (1 - s) * D(m - 1) + s * R(m - 2)$$

R(m − 1) can be eliminated in the expression for R(m) by substitution,

$$R(m) = (1 - s) * D(m) + s * ((1 - s) * D(m - 1) + s * R(m - 2))$$

so R(m) is now defined in terms of R(m − 2). Collecting the terms in D, the expression for R becomes

$$R(m) = (1 - s) * (D(m) + s * D(m - 1)) + s^2 * R(m - 2).$$

This substitution process can be continued for R(m − 2) to get

$$\begin{aligned} R(m) &= (1 - s) * (D(m) + s * D(m - 1)) + s^2 * ((1 - s) \\ &\quad * D(m - 2) + s * R(m - 3)) \\ &= (1 - s) * (D(m) + s * D(m - 1) + s^2 * D(m - 2)) + s^3 \\ &\quad * R(m - 3). \end{aligned}$$

By now, the pattern is clear, and after n substitutions

$$\begin{aligned} R(m) &= (1 - s) * (D(m) + s * D(m - 1) + \cdots + s^n * D(m - n)) \\ &\quad + s^{n+1} * R(m - n - 1). \end{aligned}$$

If $s < 1$ and n goes to infinity (becomes indefinitely large), then s^{n+1} goes to zero, so the final term of this expression for R goes to zero also. Therefore, the black box behavior of the exponential smoothing reorder policy is given by

$$R(m) = (1 - s) * (D(m) + s * D(m - 1) + s^2 * D(m - 2) + \cdots).$$

The coefficients for the demands, $(1 - s)$, $(1 - s) * s$, $(1 - s) * s^2, \cdots$, are decreasing exponentially, which explains the name. Furthermore, the sum of these coefficients is 1, because if $0 < s < 1$, then

$$(1 + s + s^2 + \cdots) = 1/(1 - s).$$

In illustration, consider the same demands discussed in Chapter 1 of 75, 100, and 125, each with equal probability. Then, in BDL stimulus history notation,

$$R = (1 - s)(D.0 + s * D.1 + s^2 * D.2 + \cdots)$$

where each D.i is 75, 100, or 125. Since R is a weighted average of the D.i's (their coefficient's sum to 1), R must be between 75 and 125, in contrast with the k months of supply example in which R could vary between 25 and 175. When s is close to 1, say .9, the form of R is

$$R = .1 * D.0 + .09 * D.1 + .081 * D.2 + \cdots$$

and no single demand, even the most recent, represents a large term in the expression for R. Therefore, R is a weighted average of many nearly equal terms and will be near the average of demands, 100, with high probability. The exponential smoothing reorder policy lives up to its name; it smooths the demand variation in the reorders as specified by its parameter s. It can be shown theoretically to be the best possible in simultaneously smoothing reorders and inventory levels.

3.3.3 Sales Forecast State Machines

The black box for RA12 (running average of 12 months sales) provides a sales forecast. However, RA12 will follow a sales trend very slowly, in fact, RA12 will lag a sales trend by 6 months because the average is taken over 12 months. In order to follow a sales trend more closely the forecast should weight recent sales more heavily than older sales. One way to attempt such a forecast is by a straight line method of weighting, such that a forecast F, in terms of a sales history S.0, S.1, S.2, ... is given as

$$F = (12 * S.0 + 11 * S.1 + \cdots + 2 * S.10 + S.11)/78$$

where the denominator 78 is the sum of digits 1 to 12.

Another way to weight recent sales more heavily is by exponential smoothing, as used in the exponential smoothing reorder policy. In fact, the exponential smoothing reorder policy can be regarded as a demand forecaster instead of a reorder policy (recall that inventory levels did not appear in its analysis). Its parameter s can be used to balance the needs between a stable forecast and tracking sales directly.

Summary: A black box is derived from a state machine by eliminating state references in the calculation of responses. Derivation of black boxes from state machines can reveal unsuspected state machine behavior, and permit more systematic analysis of user problems.

3.4 DESIGN OF STATE MACHINES FOR BLACK BOX BEHAVIOR

Preview: The state and machine transitions of a state machine can be designed from black box behavior. The state machine of a hand calculator is finite even though its black box is not. State machines must be designed to deal with both proper and improper use.

3.4.1 State Machine Design for Black Box Behavior

As already noted, some business rules are stated naturally as state machines, for example, the exponential smoothing sales forecast policy.

Other business rules are stated more naturally as black boxes, for example, a running average forecast of sales. In many cases, a black box behavior is desired but it may not be obvious how to achieve it with a state machine. That is the problem of state machine design. State machine design requires intellectual invention, but there are three principles that make such invention possible and practical.

The first two principles are based on the black box behavior required:

> **Principle 1:** The black box stimulus history defines sufficient state data.
>
> **Principle 2:** The black box response defines necessary state data.

The running average black box RA12 provides an illustration of these principles. A stimulus history of more than 12 stimuli is sufficient, in fact, more than sufficient. But the response requires the average of the last 12 stimuli, so at least that much information is necessary. This suggests a state consisting of the last 11 stimuli (the current stimulus makes the 12th).

The third principle of state machine design is based on state machine behavior.

> **Principle 3:** The state machine transition must define both the black box response and next state with sufficient data for future transitions.

In the state machine with RA12 whose state contains the last 11 stimuli, the black box response is obtained by averaging these last 11 stimuli with the current stimulus. However, unless the state is updated, these 11 stimuli are no longer the last stimuli, so there will be insufficient data for future transitions. The answer, of course, is to replace the oldest of the 11 stimuli with the current stimulus. Then, at the next transition, the state will again contain the last 11 stimuli.

The problem of how to replace the oldest of the 11 stimuli with the current stimulus is one of design. If each of the stimuli is given a distinct name, the one replaced this time will not be the one to replace the next time, because the current stimulus this time won't be the oldest the next time. A simple solution is to store past stimuli by their ages, say S1, S2, ..., S11, so that S11 is the oldest stimulus. Then, to update the state, S11 is not replaced by the current stimulus S, but by S10, then S10 is replaced by S9, S9 by S8, and so on to S2 by S1 and S1 by S. This update involves 11 replacements instead of 1, but automatically keeps the oldest stimulus in S11, the most recent in S1, etc. In this case the RA12 state machine can be defined as shown in Figure 3.4-1 with machine M.

The foregoing design of a state machine for RA12 is straightforward and maintains the necessary history of the 11 most recent stimuli in a sequence S1, S2, ..., S11 of increasing age. Each new transition requires that all 11 members of the sequence be updated, because each is now one period older than before.

There is a simple alternative to this design in which the new stimulus literally replaces the oldest stimuli at each transition. However, an additional data item is required which keeps track of which stimuli is the oldest in each new state. That is, consider a state with 11 previous stimuli, in a circular list with members T1, T2, ..., T11 (T1 follows T11 in the circular list), and a data item called Oldest which identifies which of T1, T2, ... T11 is the oldest stimulus. Oldest always has a value between 1 and 11. In a transition, the new stimulus is used with the 11 members T1, T2, ..., T11 to calculate the running average, the stimulus designated oldest by Oldest is replaced by the new stimulus, and the designation of Oldest is advanced one member in the circular list (Oldest goes through the cycle 1,2,3, ..., 11,1,2, ... in sucessive transitions).

In this case, the RA12 state machine is given in Figure 3.4-2 with M2 given as follows. That is, the RA12 state machine M2 changes only two data items (instead of 11) on each transition, but the description is less straightforward. At every transition, the last 11 stimuli are contained in the state, but a stimulus of given age is in no fixed place in the circular list.

The two state machines for RA12 already discussed calculated the running average from the last 12 stimuli (counting the new stimulus) at each transition. Still another alternative is to divide each new stimulus by

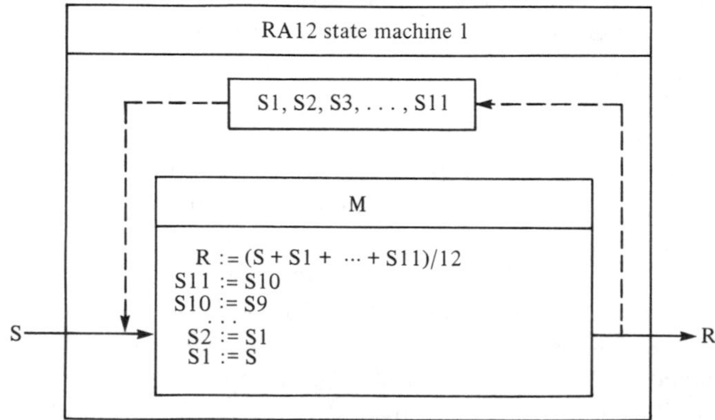

Figure 3.4-1. RA12 State Machine.

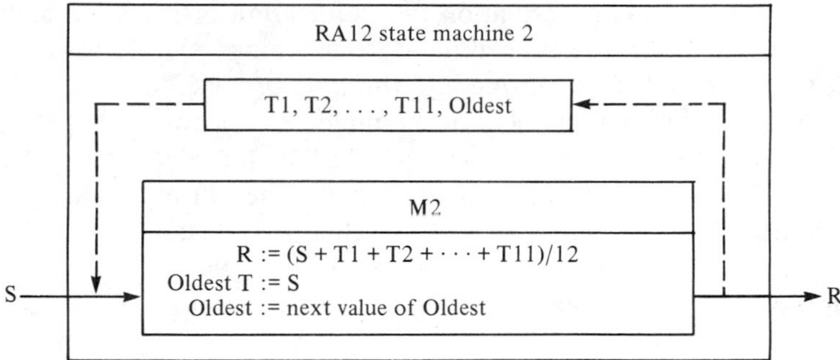

Figure 3.4-2. RA12 State Machine 2.

12 before putting it into the state. Then the running average can be calculated by simply adding the 12 values obtained from the last 12 stimuli. This leads to RA12 State Machine 3 in Figure 3.4-3.

The lesson in these three alternative state machines for RA12 is that there are many ways to simulate a black box with a state machine. The three principles provide a systematic way to think about the invention of states and transitions. In particular, principle 3 reminds us that not only must the correct response be calculated, but also a correct state for use in the next transition.

State Machine Verification

We have already seen how to derive the black box of a state machine by eliminating its state from the expression for its responses. If we have invented a state machine to simulate a black box, then we should be able to rederive that very black box from the state machine we have just

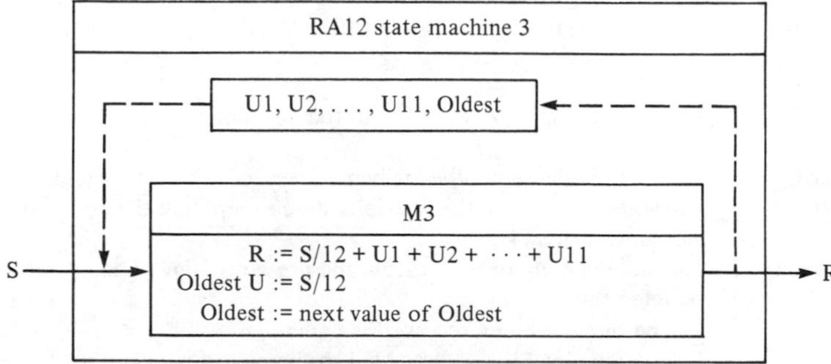

Figure 3.4-3. RA12 State Machine 3.

invented. We call this rederivation the **verification** of the state machine. That is, any of the three preceding state machines should lead back to RA12, whether or not the design and thinking process that created them from RA12 was known or not. For example, the machine of RA12 State Machine 3 can be reexpressed to eliminate the state data in U1, U2, ..., U11, and Oldest by the following argument. The variable Oldest cycles through the values 1,2, ..., 11,1,2, ... and therefore designates each of U1, U2, ..., U11 as the oldest value added to the state. Since the stimulus S is always divided by 12, the U's always contain, in no fixed places, the values S.1/12, S.2/12, ..., S.11/12, where S.1, S.2, ..., S.11 are the 11 most recent stimuli.

Therefore, the response R has the form

$$
\begin{aligned}
R &= S.0/12 + U1 + U2 + \cdots + U11 \\
&= S.0/12 + S.1/12 + S.2/12 + \cdots + S.11/12 \\
&= (S.0 + S.1 + S.2 + \cdots + S.11)/12
\end{aligned}
$$

which is the response required for RA12. Note that we have used two simple facts of arithmetic in this derivation

1. The term 1/12 can be factored out of the sum.
2. The values S.0/12, U1, U2, ..., U11 can be added in any order with the same result.

3.4.2 State Machine Design for the Hand Calculator Black Box

The black box behavior of a hand calculator can be described as a state machine. But even more, we shall find a finite state machine description even though a hand calculator cannot be described as a finite black box.

Recall the problem discussed in Chapter 2 of finding the sum of 14 and 43 with stimulus history C14+43=. In state machine terms, the hand calculator is using and then changing its state with every stimulus as well as producing its response. For example,

After C:	The state is cleared, expecting to receive a number in a series of stimuli to follow
After C1:	The state is that a number is being received whose first digit is 1
After C14:	The state is similar to the previous state except that the first two digits of the number are 14
After C14+:	The state is quite different from the previous state. The + stimulus has signified three changes: 1. The number being received is ended and is 14 2. The number 14 is to be added to the next number to be received 3. A new number is expected in a series of stimuli to follow

The state information must be retained in some form. In order to express such a state, we invent a description for it in a set of variables. In this case we call these variables registers. These registers are pure inventions, to explain the black box behavior of the hand calculator, not based on its physical construction. Specifically, we invent three registers which seem required by the preceding analysis:

Visible Register (VR). A register of digits which are identical to whatever digits are currently displayed.

Hidden Register (HR). A register of digits which can retain a number for future calculation.

Function Register (FR). A register which holds a single arithmetic function requested by the user,such as + or −.

The foregoing states can now be expressed by the values in the registers, for example,

After C14: VR = 14, HR = ?, FR = ?
After C14+: VR = 14, HR = 14, FR = +

where the question marks mean the current register contents are unknown.

Now, the next stimulus of the second 4 leads to a problem we have not yet solved, because after receiving the stimulus history C14+4 we know that VR must become 4. How does the black box machine produce this part of the state? For example, if it continued as before, any digit entered would simply be added to the end of the number already in VR. In this case, C144 would produce VR = 144, but C14+4 would produce VR = 4. The answer has to be contained in the intervening +, which says, in effect, "start a new number." One way to record this information is to invent a new register:

Begin Register (BR). A register which holds the character B (for Begin) or C (for Continue).

The Begin Register will solve another problem we have not noted until now. At the very beginning, we have the sequence

After C: VR = 0
After C1: VR = 1

Why isn't VR = 01 after C1? After all, other digits are simply appended to the end of the preceding number in VR. But the behavior of the hand calculator is such that the first digit of a number must overwrite the initial 0 in the display.

We can use BR to denote this effect of stimulus C, as well, to get

After C: VR = 0, BR = B (Begin)
After C1: VR = 1, BR = C (Continue)

Now, when a + stimulus appears, the state machine can, among other things, change BR from C to B. Then the state machine can be constructed as follows:

> If BR = B and the stimulus is a digit, disregard the contents of VR and start a new number with that digit; otherwise, if BR = C and the stimulus is a digit, append that digit to the current contents of VR.

With this new understanding we can express the successive states of the hand calculator in this problem as shown in Table 3.4-1.

Note that each state in the Table is slightly different from its predecessor, but more importantly, a **machine black box of order 1** can be devised to go from each state to the next. That is, the machine required to carry out these state transitions need not itself refer to a state.

The question marks in the Table mean that no particular values in those registers are required to explain the behavior of the hand calculator. Of course, there will be values in those registers where each question mark appears. But whatever the values are, they will simply reflect construction details of the particular hand calculator in use, and are incidental to our state machine description of black box behavior. In recognition of these two categories of values, we call the definite values in the Table **intentional data,** and the question mark values **accidental data.**

Notice one surprising aspect of this transaction with input C14+43= and output 57. The state machine works just as hard and in the same way with each stimulus, as it does in producing the response (adding 14 and 43 to get 57 is just part of a transition to the hand calculator). That is, the designation of the input as C14+43= is relative to the user of the hand

Table 3.4-1

States of a Hand Calculator Computing
C 14 + 43 = 57

After:	BR	VR	FR	HR
C	B	0	?	?
C1	C	1	?	?
C14	C	14	?	?
C14+	B	14	+	14
C14+4	C	4	+	14
C14+43	C	43	+	14
C14+43=	?	57	?	?

Table 3.4-2

Machine Transitions for a Hand Calculator

	(S ,	OS			) → (R ,	NS			)
row		BR	VR	FR	HR		BR	VR	FR	HR
1	C					0	B	0		
2	any D	B		f	y	D	C	D	f	y
3	any D	C	x	f	y	D+10x	C	D+10x	f	y
4	any F		x	f	y	yfx	B	yfx	F	yfx
5	=		x	f	y	yfx		yfx		

calculator—there is no new information for the user in this sequence of stimuli—but every one of the stimuli is new information for the hand calculator. These machine transitions can be organized and summarized as shown in Table 3.4-2. In each row, the left side of Table 3.4-2 represents a stimulus and old state, the right side the response and new state. Blanks in the Table mean that data is irrelevant. Note that yfx is an arithmetic expression, where x and y represent numbers and f represents an arithmetic function, for example 14 + 37. Why does the Table define R and VR to be the same? Why not just have R? Because VR is available and needed as part of the state for the next transition, but R is not. The machine would not operate correctly without VR.

This analysis assumes perfect arithmetic, but of course a real implementation with finite registers VR and HR would need to cope with overflow both from key entry by the user and from arithmetic operations. The problems of overflow will not change the structure of the state machine, only the rules which modify VR in rows 3 and 5 of the Table. In any case, the state of this machine is finite, even though the hand calculator it describes is a nonfinite black box.

3.4.3 State Machine Design to Deal with Improper Use

The hand calculator problem we have analyzed in Chapters 2 and 3 expresses a proper question in arithmetic with a correct expression C14+43=. However, a hand calculator can accept any sequence of stimuli whatsoever, whether it constitutes a correct expression or not. That is, no matter which key is pressed next, the display will return some response—the hand calculator won't blow up or stop operating under any possible circumstances of use. For example, given a gibberish sequence of, say,

+ − 3 = 4 2 + = 5 = 2

what responses will be produced? Those responses will, in fact, be unpredictably different for different brands of hand calculators because they will depend on accidental data that is generated in their state registers.

All brands of hand calculators will satisfy the same abstract specification, namely of converting input arithmetic expressions (suitably delimited) into output numbers which correctly evaluate such arithmetic expressions. But the abstract specification is not defined for stimulus histories that do not represent arithmetic expressions. Yet each brand of hand calculator will realize some concrete specification as a by-product of its design and manufacture. Part of this concrete specification will be intentional, as the designers seek to incorporate good human factors into the hand calculator. But stimulus histories that are gibberish will be of no interest to the abstract specification, and therefore treated as accidental.

As a simple example, consider the sequence of stimuli previously analyzed, with an additional = stimuli added to the end:

C 14 + 43 ==

What output will this sequence produce? It turns out that many hand calculators will produce one of two answers—either 57 or 71. The 57 is simply a repetition of the preceding output, while the 71 is the result of adding 14 (in HR) to 57 (in VR). Thus, the internal rule for creating the sum can be stated in several ways:

1. "Display HR + VR"
2. "Store HR + VR in VR and display VR"
3. "Store HR + VR in HR and display HR"

In the first case, the second = stimuli simply repeats the action of the first =, but in the second case, the second = actually changes the state. Note that our state machine explanation of the hand calculator would produce the second case.

The accidental differences of hand calculators represent ideal opportunities for treating hand calculators as laboratory machines in the study of black box behavior and the construction of clear box explanations for this black box behavior. It is an interesting problem in black box analysis to discover how a hand calculator really works for any sequence of stimuli, not just for proper questions of arithmetic. In order to go about that, a set of state registers is required, together with a machine description that will explain exactly what state change and response will occur for every possible stimulus combined with every possible state. This total state and machine description must explain all the observable behavior of the hand calculator.

The lesson in this is that it is important to distinguish between what a black box does and its proper use. A black box may behave in strange ways when presented with unusual stimulus histories. Thus, a definition of the behavior of a black box typically depends on a definition of its proper use. In the case of the hand calculator, proper use is defined by the rules of arithmetic, which preclude gibberish stimuli altogether.

Nevertheless, the designer of a hand calculator must be prepared to deal with both proper and improper use, to avoid failure to complete a transition. This requirement leads to the following fundamental principle.

Fundamental Principle: A state machine must be prepared to respond to any stimulus in any state to produce a response and a new state.

In further illustration, a common problem in interactive systems is that an unexpected illegal input can bring the system to an unanticipated halt. If, in fact, the fundamental principle had been applied, this situation could not arise. Thus, in design the possibility for every stimulus, proper or improper, must be accommodated from every possible state.

Summary: The state of a state machine must be sufficient for the machine to produce every possible response required by its black box behavior. A state definition of four registers is sufficient to simulate the black box behavior of a hand calculator. The concrete implementation of a hand calculator will handle any gibberish history of stimuli with definite behavior which will be irrelevant to the abstract specification of its state machine. A state machine design is insufficient if it deals solely with proper use.

3.5 STATE MACHINES IN BOX DESCRIPTION LANGUAGE

Preview: A state machine in BDL substitutes a state definition for the black box stimulus history and a machine transition for the black box transition.

BDL accommodates two design steps for state machines, namely, state machine definition and state machine invocation.

State Machine Definition. A state machine defines black box behavior at the stimulus, response level in terms of state transitions, expressed in BDL syntax as

```
define SM <SM name>
  stimulus
    <stimulus name>: <type>
  response
    <response name>: <type>
  state
    <state data>
  machine
    <SM transition>
```

with outer syntax keywords **define SM, stimulus, response, state,** and **machine,** shown here in indented text form. <SM name> is the name of the state machine, usually, the name of the corresponding black box, and <stimulus name>, <response name>, and <type> have meanings as before. In this case, **state** denotes the <state data>, which abstracts the stimulus history of the corresponding black box in terms of named state data, and **machine** denotes the <SM transition>, which defines the machine transition.

In illustration, the Add2 state machine corresponding to the Add2 black box can be described as

```
define SM Add2
  stimulus
    S:number
  response
    R:number
  state
    L:number
  machine
    R := S + L
    L := S
```

where S, R, and the single state item L are all defined as numbers. The transition is given by an assignment of the stimulus plus the old state to the response, and by an assignment of the stimulus to the state in preparation for the next transition.

Is the Add2 state machine a faithful description of the Add2 black box? As already seen, the answer can be found by eliminating the state, and expressing the response in terms of stimulus history. The response of the ith state machine transition is

$$R(i) = S(i) + L(i - 1)$$

and the value of L(i − 1) from the previous transition is

L(i − 1) = S(i − 1)

Thus, the results of the transition can be rewritten as

R(i) = S(i) + S(i − 1)
L(i) = S(i)

where the assignment to R(i) is the transition formula of the Add2 black box expressed in terms of stimulus history, and the assignment to L(i) ensures the availability of the required stimulus history for the next transition.

The state machine for Max2 is

```
define SM Max2
  stimulus
    S:number
  response
    R:number
  state
    L:number
  machine
    if L > S then R := L else R := S fi;
    L := S
```

where the state is a single number L and the transition is defined as a conditional assignment of the maximum of the old state and the stimulus to the response, and by an assignment of the stimulus to the state for the next transition.

Similarly, the Odd:Add2|Max2 state machine is

```
define SM Odd:Add2|Max2
  stimulus
    S:number
  response
    R:number
  state
    L1, L2:number
  machine
    if S odd
      then R  := L1 + S;
           L1 := S
      else R  := max(L2, S);
           L2 := S
    fi
```

The hand calculator provides a special, trivial example of a state machine specification at an abstract level, since no data is to be stored between transactions from clear key to clear key. Therefore, its state machine is trivial because no state data is to be maintained:

define SM hand calculator
  stimulus
    I:arithmetic expression
  response
    O:number
  state
    (none)
  machine
    O := value of arithmetic expression in I

However, the state machine of the hand calculator is nontrivial at the concrete level, since data is stored between transitions from keystroke to response, as informally described in Table 3.4-2:

define SM hand calculator
  stimulus
    S:key
  response
    R:display
  state
    BR:(B,C)
    VR:number
    FR:(+, −, *, /)
    HR:number
  machine
    As described in Table 3.4-2

State Machine Invocation. State machines can be invoked by BDL procedure statements of the form

use SM <SM name> (<stimulus name>; <response name>)

where keyword **use SM** means carry out a transition of the state machine with name <SM name>, given stimulus <stimulus name> and producing response <response name>.

For example, the procedure statements

use SM Add2(i; j)

and

use SM First(i; j)

invoke state machine transitions using data objects i, j for stimulus, response, respectively. Thus, for stimulus history 3 6 1 9 6 and i = 2,

use SM Add2(2; j)

sets j to 2 plus the old state of Add2 and sets the new state of Add2 to 2, and

use SM First(2, j)

sets j to 3 and leaves the old state (3) unchanged.

> **Summary:** State machine BDL requires both state and machine definitions. State data are given names and type definitions. The machine transition is defined in terms of new state and response as a function of stimulus and old state.

EXERCISES

1. Define state machine descriptions for the following black box structures:

(a) Add2;Max2
(b) Max2;Add2
(c) Max2;Max2
(d) Odd:Add2|Max2
(e) Odd:Add2|Add2
(f) Odd: * (Add2:Odd)
(g) Add2||Max2

2. Determine the black box behavior of state machines with stimulus S, response R, state L, and rules:

(a) R := S + L
L := 2 * S
(b) R := L
L := S
(c) R := S + L
L := S

3. Consider a state machine with stimulus S, response F, state L, and rule:

F := max(S, L)
L := F

Determine its black box behavior and discuss whether this state machine is a good sales forecaster, if not, can you think of any other use for it?

4. The (s,S) reorder policy is defined by two numbers s and S which define lower levels and upper levels in the inventory level of an item. That is, when inventory falls below s, reorder enough to bring it up to S. Determine the state machine for the (s,S) policy and discuss the nature of its black box behavior.

5. Use Table 3.4-2 to work out the state machine transitions in the form of Table 3.4-1 for the following stimulus histories:

(a) C 324 + 19 + 1 =
(b) C 27 − 94 + 82 =
(c) C 18 * 4 − 36 =

6. Consider a hand calculator which has a clear entry key (CE) that permits the user to start over in keying in a number. Modify the hand calculator state machine to include the clear entry key.

7. Consider a hand calculator which has no decimal point key (DP) and permits arithmetic to no more than 8 significant digits. Work out a consistent way to deal with overflow on key entry and arithmetic operations. Modify the hand calculator state machine to include your solution.

8. Given black box Max24 (the response is the maximum of the last 24 stimuli), develop two state machines in analogy to the first two state machines in the chapter for RA12, namely, one state machine whose state maintains stimuli by age in fixed places and one which minimizes the state changes during a transition.

9. A 6 months' sum of digits forecast forecasts sales next month as a weighted average of the past 6 months of sales, which weights the oldest sales (sales 6 months ago) 1, next oldest 2, ..., most recent 6. Note that a common denominator must be found to make the weight add up to 1. Develop a state machine to describe a 6 months' sum of digits forecast black box.

10. Can you devise a state machine for black box RA12 in which only previous responses are retained in the state and not previous stimuli?

11. A 5-day stock trend indicator declares the trend +, 0, or − depending on whether the last price is above, same, or below the median of the previous 5 prices. That is, it is a black box with daily stock prices as stimuli, and the indicators +, 0, − as responses. Develop a state machine for the 5-day stock trend indicator black box.

12. Discover, if possible, a foolproof way to do reliable calculations without using the clear key. To state the problem more clearly, imagine you are presented with a hand calculator with unknown history of use, whose clear key is missing! For example, suppose the display held 14, and you entered −14 to get the display to 0. Can you consider the calculator cleared? What would you do if the sequence −14 produced the display −14?

13. At the end of the week a manufacturer decides how many units of a product to make during the next week. The decision is based upon the product's current quantity on hand and the average number of orders for that product over the past two weeks. Outstanding orders for the product are indicated by a negative value for quantity on hand.

The following state machine BDL specifies the above decision rule:

```
define SM Product
    stimulus
        S: number {orders during week}
    response
        R: number {products to make next week}
    state
        made: number {products made during week}
        prev : number {previous week's orders}
        qoh  : number {current quantity on hand}
    machine
        do
            qoh := qoh + made − S;
            R := (S + prev)/2 − qoh;
            prev := S;
            made := R
        od
```

Derive the black box transition formula for this state machine by eliminating all state variables. Comment on the usefulness of the decision rule.

14. Describe a system that is a

(a) finite black box–finite state machine,
(b) finite black box–nonfinite state machine,
(c) nonfinite black box–finite state machine,
(d) nonfinite black box–nonfinite state machine.

15. Design a state machine for a simple word processing system. Develop a transition table similar to the one for a hand calculator.

16. Discuss the difference between the bank customer service state machine and a toll booth state machine where movement between lanes is not possible.

17. Use a state machine to model the behavior of a traffic intersection with a stop light. Assume initially that there is one lane of traffic going in each direction. How would the state machine change with two lanes or three lanes in each direction, turn lanes?

18. Describe a database management system as a state machine. What are the inputs and outputs for this system? What transactions would the machine perform?

Chapter 4 The Clear Box Behavior of Information Systems

4.1 CLEAR BOX BEHAVIOR

> **Preview:** The machine of a state machine can be expanded into a structure of machines by one of four primitive steps, resulting in a clear box with the same external behavior as the state machine. In turn, a clear box machine can be expanded into an equivalent structure of simpler machines.

In Chapter 3 we introduced an equivalent description of black box behavior at the state machine level, as depicted in Figure 4.1-1, where the machine of the black box transforms a stimulus (S) and old state (OS) into a response (R) and a new state (NS).

A clear box is an expansion of a state machine in which the machine is replaced by a structure of component machines. In turn, any component machine of a clear box can be replaced by another machine structure. Such a clear box will have the behavior of some state machine, with transitions of the form

(S,OS) → (R,NS)

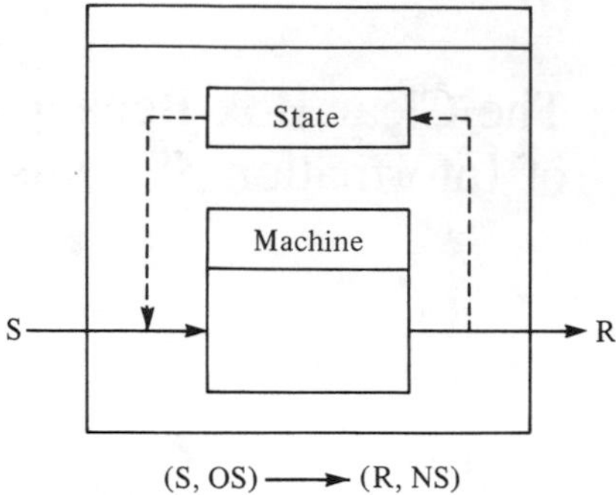

Figure 4.1-1. State Machine Description of a Black Box.

4.1.1 Clear Box Syntax

As discussed in Chapter 2, a **black box structure** is a compound structure of black boxes expressed in sequence, alternation, iteration, and concurrency primitive structures. By definition, the black boxes of a black box structure share no state. We now introduce a **clear box** in BDL that admits the possibility of shared states among its component machines.

A clear box defines a procedure for state machine behavior in terms of operations and tests on state and working data. Clear boxes in BDL are defined by the following syntax structure:

```
define CB <CB name>
  stimulus
    <stimulus name>:<type>
  response
    <response name>:<type>
  state
    <state data>
  machine
    data
      <procedure data>
    proc
      <procedure>
    corp
```

where <CB name> is the name of the clear box, usually the name of the corresponding state machine and <stimulus name>, <response name>, and <type> have the same meanings as before. The keyword **state** denotes <state data>, defined by a declaration of data objects that correspond to the state structure of the state machine, and that participate in the operations and tests of the clear box.

The clear box procedure, denoted by keyword **machine,** is defined in two parts delimited by keywords **proc** and **corp.** The first part, denoted by keyword **data,** defines <procedure data>, the local, working data (if any) used by the machine procedure. The second part is the <procedure> itself, which defines a structure of nested and sequenced operations and tests that carry out the transition of the corresponding state machine.

Note in comparison that the black box structure of Chapter 2 specified no state data nor procedure data, only a transition procedure which referenced component black boxes.

The procedure statements (PS) of a clear box procedure include BDL statements previously discussed:

Assignment Statement: <variable> := <expression>
Black Box Statement: use BB <BB name> (<stimulus name>; <response name>)
State Machine Statement: use SM <SM name> (<stimulus name>; <response name>)

and the sequence, alternation, iteration, and concurrent control structures, likewise previously discussed. In addition, BDL clear box procedure statements include the **case** control structure (a generalization of the alternation control structure).

4.1.2 Clear Box Structures

The sequence clear box is determined by a sequence of two (or more) machines M1, M2, as shown in Figure 4.1-2. The resulting structure is a

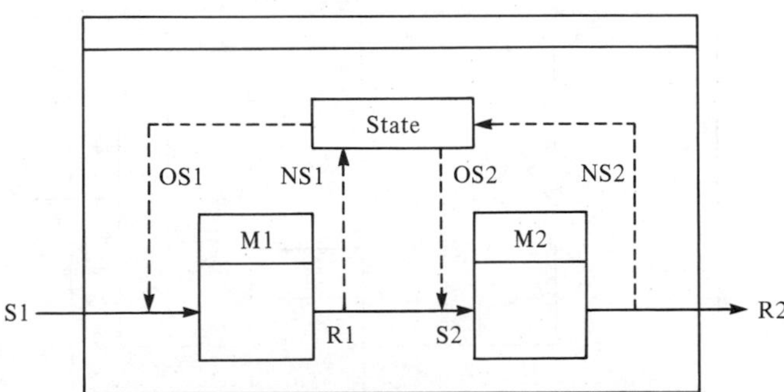

Figure 4.1-2. The Sequence Clear Box Structure.

clear box, and will have the same external behavior as the state machine whose machine is decomposed into the sequence of M1 and M2. In effect, M1 passes its response to, and creates a new state for M2 to use as a stimulus and an old state to produce the response of the clear box and its new state. More precisely,

Sequence Clear Box Execution. A transition of machine M1 from (S1, OS1) to (R1, NS1) is invoked, R1 is renamed S2 and NS1 is renamed OS2, then a transition of machine M2 from (S2, OS2) to (R2, NS2) is invoked, and NS2 is renamed OS1 for the next clear box transition.

The BDL syntax for the sequence (or **do**) procedure statement is

do
PS1;
PS2
od

where any number of procedure statements (including none) may be present, separated by semicolons and the **do, od** keywords may be omitted where no misunderstanding can arise.

The alternation clear box is determined by a conditional test, denoted

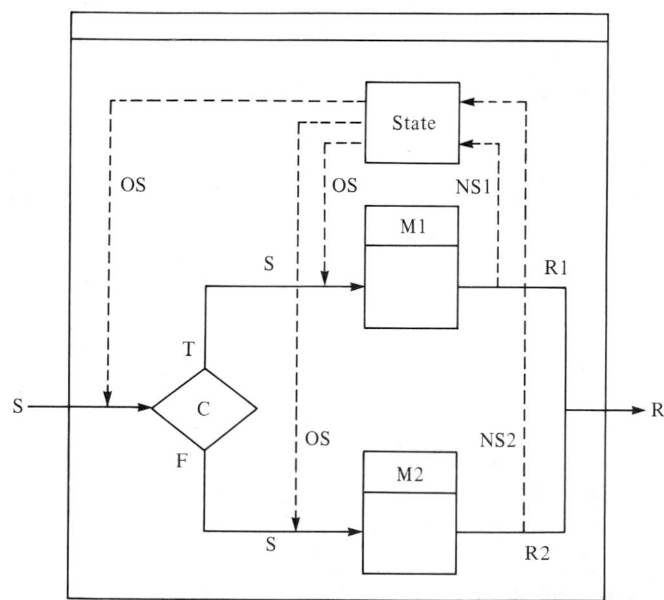

Figure 4.1-3. The Alternation Clear Box Structure.

C, which serves to switch the stimulus to exactly one of the machines M1 or M2. The test is denoted by a diamond, and may depend on the state as well as the stimulus, as shown in Figure 4.1-3.

The resulting structure is a clear box, and will have the same external behavior as the state machine whose machine is decomposed into cases M1 and M2 by C. In effect, C decides on the basis of the stimulus and the old state which machine, M1 or M2, to use for each transition.

More precisely,

Alternation Clear Box Execution. Condition C is determined by reference to stimulus S and state OS. If C evaluates to T (true), a transition of machine M1 from (S, OS) to (R1, NS1) is invoked, R1 is renamed R, and NS1 is renamed OS for the next transition. If C evaluates to F (false), a transition of machine M2 from (S, OS) to (R2, NS2) is invoked, R2 is renamed R, and NS2 is renamed OS for the next transition.

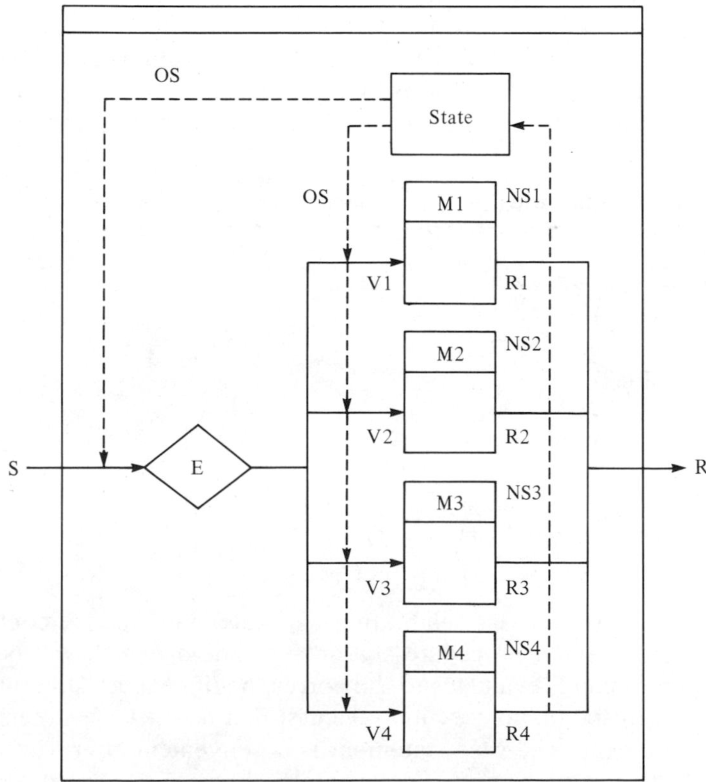

Figure 4.1-4. The Case Clear Box Structure.

The BDL syntax for the alternation procedure statement (or **if** statement) is

```
if
  C
then
  PS1
else
  PS2
fi
```

The **case** clear box structure, depicted in Figure 4.1-4, is a generalization of the alternation structure. It provides a convenient means for carrying out one of a fixed number of alternative machine transitions (fixed at four in the example shown), corresponding to values (V1 to V4) of expression E. If no case value corresponds to the value of E, response R is set to S. More precisely,

Case Clear Box Execution. Expression E is determined by reference to stimulus S and state OS. If the expression evaluates to any value Vi associated with a machine Mi, a transition of that machine from (S, OS) to (Ri, NSi) is invoked, Ri is renamed R, and NSi is renamed OS for the next transition. Otherwise, the response R is set to S.

The BDL syntax for the case procedure statement is

```
case
  E
part (value1)
  PS1
part (value2)
  PS2
part (value3)
  PS3
part (value4)
  PS4
esac
```

Keywords **case** and **esac** delimit the case statement, and the contained **part** keywords delimit procedure statements, one of which will be executed if expression E evaluates to the corresponding value. Otherwise, if no values correspond, no execution occurs; that is, the case statement is the null statement. The case statement is a convenient abbreviation for nested alternation statements. For example, the case statement

```
case
  E
part (1)
  PS1
part (7)
  PS2
part (4)
  PS3
esac
```

corresponds to the following nested alternations

```
if
  E = 1
then
  PS1
else
  if
    E = 7
  then
    PS2
  else
    if
      E = 4
    then
      PS3
    fi
  fi
fi
```

Here E values 1, 7, or 4 result in execution of PS1, PS2, or PS3, respectively. Any other E value results in no procedure statement execution.

The clear box iteration structure is depicted in Figure 4.1-5, with case C and machine M1. The effect of the iteration is to invoke transitions of machine M1 repeatedly while case C is satisfied. More precisely,

Iteration Clear Box Execution. Condition C is determined by reference to stimulus S and state OS; if the condition evaluates to F (false), the response R is set to S and iteration is terminated; if the condition evaluates to T (true), the stimulus S is renamed S1, a transition of machine M1 from (S1, OS) to (R1, NS) is invoked, NS is renamed OS for the next transition, R1 is renamed S, and the iteration clear box is invoked again.

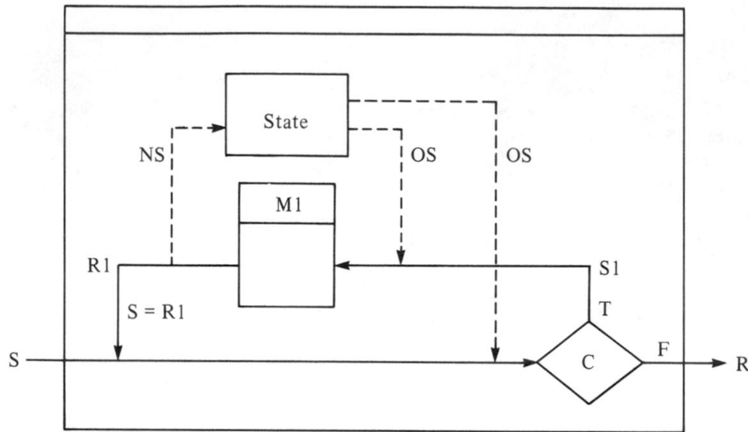

Figure 4.1-5. The Iteration Clear Box.

The BDL syntax for the iteration procedure statement (or **while** statement) is

while
  C
do
  PS1
od

where the keyword **while** is used with the previously defined keywords **do, od.** The condition C must eventually be set to false in PS1, otherwise, an endless loop will result. Note that if the condition is initially false, then PS1 will not be executed at all.

The while statement

while C **do** PS1 **od**

can be viewed as a convenient abbreviation for a nested pattern of alternation structures

if C **then** PS1; **if** C **then** PS1; **if...fi fi fi**

In this nested pattern of alternation structures, PS1 will be executed repeatedly while C is true, just as stated for the iteration control structure. The condition C is called the **while condition,** and the procedure statement PS1 is called the **do statement** (of the **while** statement).

As surprising as it may seem, the three BDL control structures above, namely, sequence, alternation, and iteration, are sufficient to express the

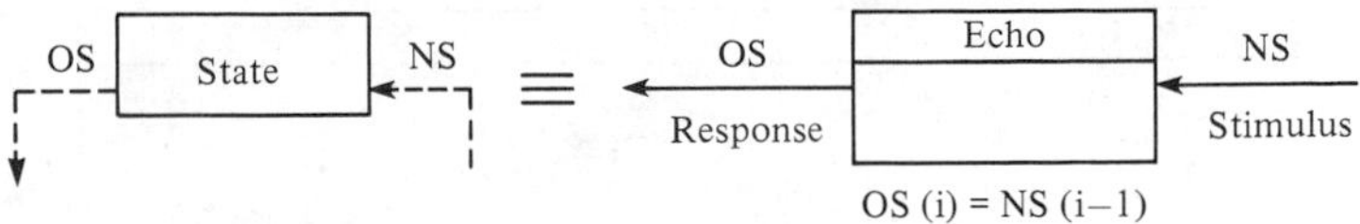

Figure 4.1-6. The State as an Echo Black Box.

design of any clear box procedure whatsoever. This fact was not always known, and other complex and arbitrary control structures have been in use since the early days of programming. But the opportunity now is to design procedures with simpler and more understandable control structures than was heretofore possible.

The concurrent control structure interacts with the state in a more complex manner than do the previous control structures. For the sequence, alternation, and iteration structures, the state acts as a black box that executes the Echo transition. That is, the state accepts the new state, NS, as a stimulus in one transition and produces it, now called the old state, OS, as a response in the next transition, where OS = NS. Figure 4.1-6 shows this equivalent behavior.

The concurrent control structure, however, does not interact with the state as an Echo black box. Recall that the concurrent black box structure accepts a stimulus and directs it to each concurrent black box (Figure 2.5-5). The response of the concurrent control structure is a complex multi-part response from the concurrent black boxes.

In the clear box concurrent structure, each machine, Mi, (black box) not only produces a response, Ri, but also a new state, NSi. A structure of n concurrent machines produces a multi-part response (R1, R2,...,Rn) and a group of new states (NS1, NS2,..., NSn). The state, as a black box, must resolve the multiple new states into a single state. We call this black box Resolve and illustrate its use in Figure 4.1-7. By the definition of a state machine, the Resolve black box will behave like the Echo black box when the stimulus (NS1, NS2,..., NSn) is simple, that is, when n = 1.

With this understanding of the state, the concurrent clear box is shown in Figure 4.1-8, with two component machines. More precisely,

Concurrent Clear Box Execution. The stimulus S and the old state

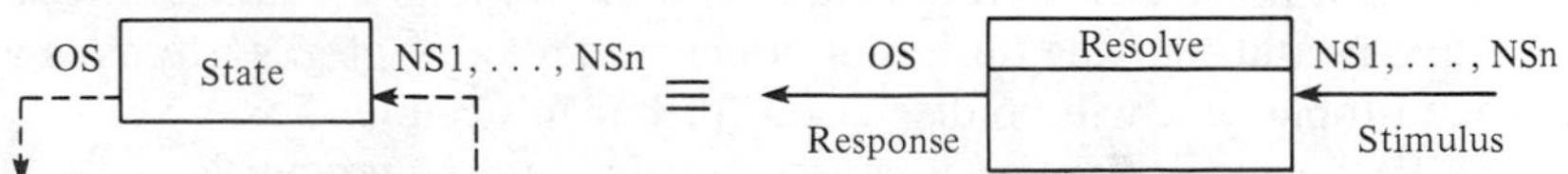

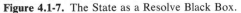

Figure 4.1-7. The State as a Resolve Black Box.

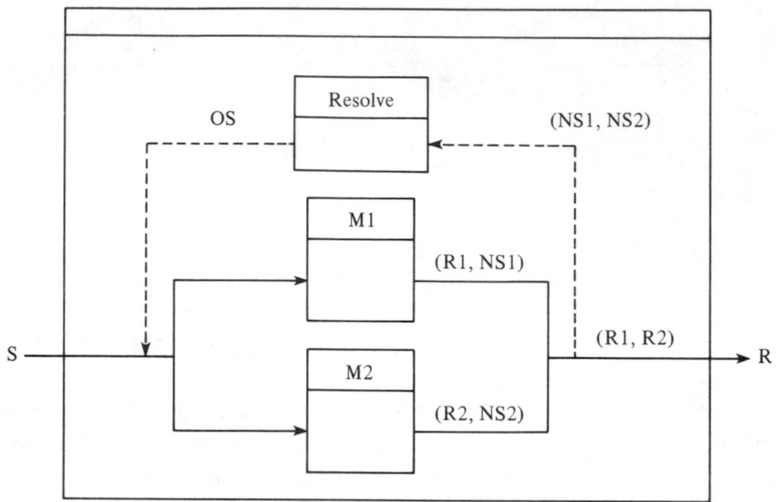

Figure 4.1-8. The Concurrent Clear Box Structure.

OS are referenced by all concurrent machines. The complex response is a grouping of the responses from the individual machines (R1, R2). The new state is determined by the black box Resolve with the complex stimulus (NS1, NS2).

The concurrent clear box synchronizes the responses of its component machines. Note that the state black box Resolve must be specified and designed as well, to meet a variety of intentions in the use of the concurrent black boxes in its clear box. For example, **serializability** is a well-known design tactic. Serializability requires that the behavior of the concurrent structure be equivalent to one of the possible orderings of its component machines.

As defined, the execution of a concurrent clear box is in part determined by the design of Resolve unless both (all) execution sequences define the same behavior for all stimuli. However, there are situations in which ambiguous behavior represents necessary reality. For example, suppose M1 and M2 represent two reservation attempts for the last seat on a flight. Then the order in which they execute determines who gets the last seat. But these cases are typically settled in the fine details of design and implementation. The design of concurrency control in systems is an advanced topic that will be discussed further in Chapter 5.

The BDL syntax for the concurrent procedure statement is

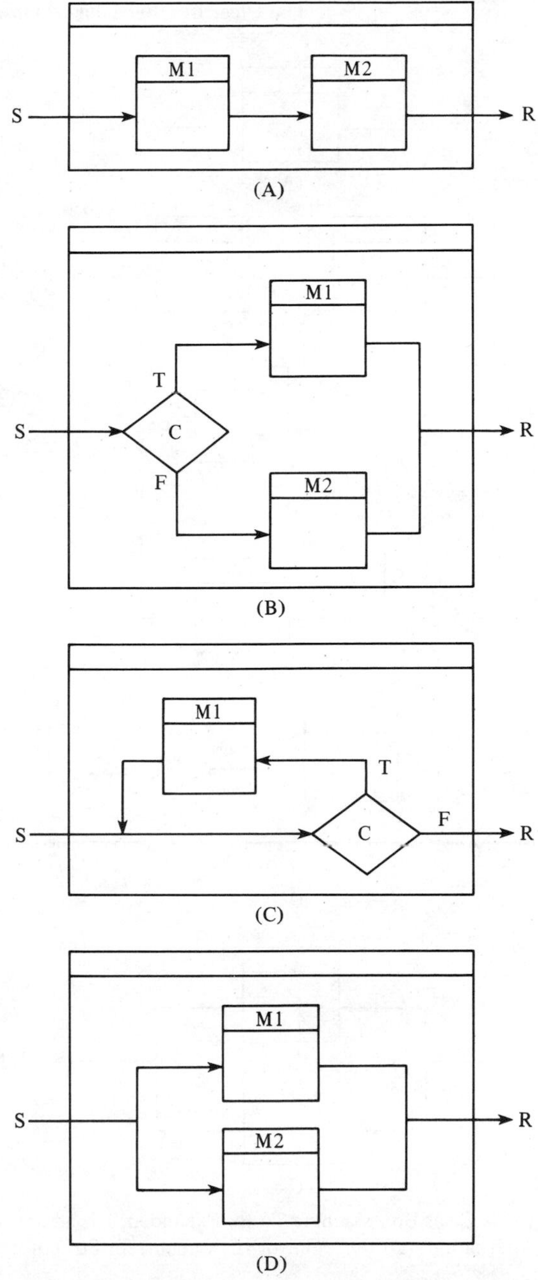

Figure 4.1-9. Clear Box Abbreviations, with State References Omitted. (A) Sequence Abbreviation, (B) Alternation Abbreviation, (C) Iteration Abbreviation, and (D) Concurrency Abbreviation.

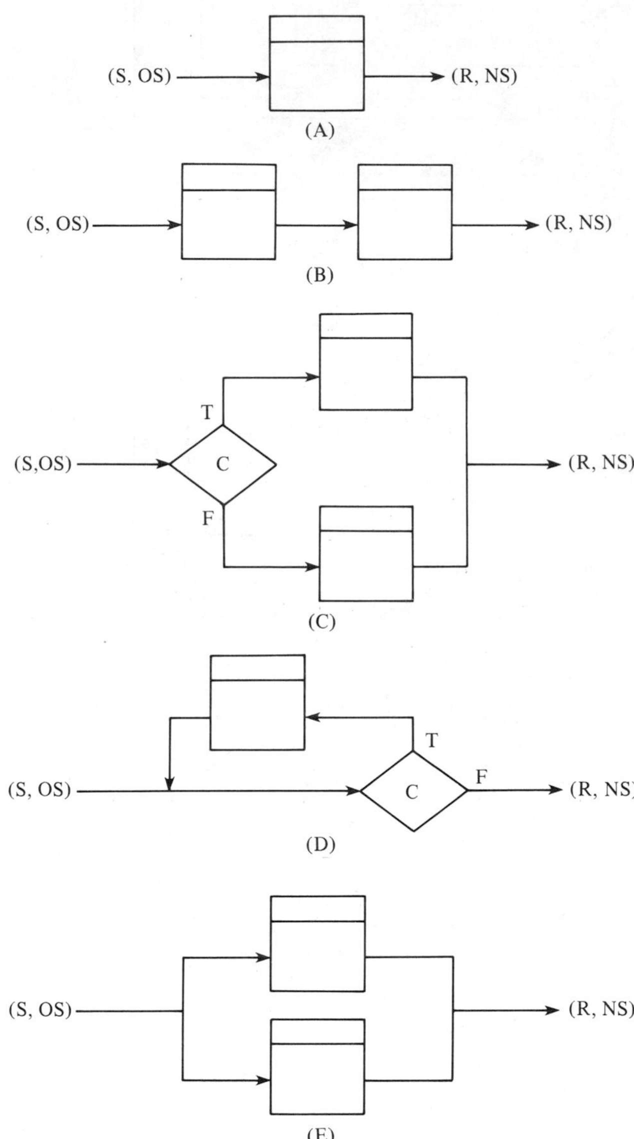

Figure 4.1-10. (A) A Clear Box Machine To Be Expanded. Expanding a Machine into a (B) Sequence, (C) Alternation, (D) Iteration, or (E) Concurrent Structure.

```
with
  PS0
con
  PS1,
  PS2
noc
```

where PS0 plays the role of Resolve.

Once the state machine properties of a clear box are defined and understood, the clear boxes can be abbreviated to omit the common state references, in the realization that the full structure can always be recalled for specific analyses and arguments. The clear box abbreviations for sequence, alternation, iteration, and concurrency are shown in Figure 4.1-9.

4.1.3 Clear Box Expansion

Since each machine in a clear box plays exactly the role of a machine in a state machine (converting its stimulus and an old state into its response and a new state), the sequence, alternation (or case), iteration, or concurrent decomposition can be applied to each such machine. Such a step expands the clear box by one or more additional internal machines, permitting any original state machine to be redefined by clear boxes with smaller and simpler individual machines. That is, any machine of a state machine or clear box, such as depicted in Figure 4.1-10 Part A, can be expanded, at any stage, into a sequence, alternation, iteration, or concurrent structure, as depicted in Parts B, C, D, and E, respectively.

> **Summary:** The machine of a state machine can be expanded into sequence, alternation (or case), iteration, and concurrent clear box structures. These structures can be expressed in BDL, and in turn expanded into more detailed BDL procedures.

4.2 STRATEGIC USES OF CLEAR BOXES

> **Preview:** Complex business operations may be directly explained or specified as clear box structures. Such clear box structures can be used as a basis for organizing related descriptions in user manuals and instruction guides. Procedures with arbitrary structure can be transformed into clear boxes with equivalent behavior and more systematic structure for better understandability.

4.2.1 Clear Box Business Procedures

As we have seen, business operations behave as black boxes and can be described as state machines. However, business operations are often most conveniently defined directly in clear boxes. Clear box business procedures codify explicit rules in terms of tests and actions required to carry out business operations.

Clear boxes can evolve naturally out of day-to-day operations in any business, in the effective organization and distribution of work. Such clear boxes may never be written down, being transmitted informally through on-the-job training, but they are clear boxes nevertheless. For example, in many businesses, a verbal description of required telephone answering procedures for new employees may be sufficient. However, telephone answering in operations such as mail order or airline reservations will be subjected to a good deal of analysis and experimentation, to arrive at optimum clear box procedures for structuring efficient conversations, maximizing information flow, and minimizing connect time. Such procedures will be explicitly defined and taught to new employees, together with procedures for using online information systems to answer customer questions on costs, availability, etc., during the conversation. Such clear box procedures may be carried out by hundreds of employees, hundreds of times every working day, and even minor improvements in their design can have major effects on business efficiency and competitive advantage.

Other clear box procedures may be prescribed by law as conditions and requirements on business operations, for example, in tax and labor laws that specify explicit practices that businesses must follow. Clear box procedures in these areas can be extraordinarily complex, requiring extensive study and analysis in both design and execution. While such procedures indeed represent state machines, their transitions, typically complicated by large numbers of special cases and exceptions to more general rules, are most easily described directly in clear box form.

The tax accounting required of a business enterprise can be defined by a set of clear box procedures that specify a complex transition of the business information system (of people and machines), to produce tax computations for the previous year. These clear boxes are specified by instructions compiled by federal, state, and local governments, in guides, manuals, and forms, all written in natural language. The state to which these clear boxes refer is the state of the business itself, and their procedures involve operations on the old state to compute tax liabilities and corresponding updates to produce a new state that reflects these liabilities.

Over the course of a tax year, the information system of a business must accumulate the state data required by the tax clear box. Many decisions will be made in the conduct of the business which will affect the tax computation. For example, methods of cash management, inventory valuation, and asset depreciation, as well as decisions on capital investment and investment credits, will all be influenced by tax laws and reflected in the state of the information system.

Because of the close coupling between tax laws and business operations, the information system of a business is usually designed to explicitly capture and retain tax-related information in its state. In fact, the tax clear box may be incorporated directly into the information system. In this case, periodic changes in the tax laws will be a major source of modification to an information system. If the state of a business information system does not contain the right historical data, the transition required by the tax clear box cannot be carried out, and the tax computation becomes difficult indeed.

4.2.2 The Clear Box of Schedule C

In illustration, consider a small business that must account for tax liability on Schedule C (Profit or (Loss) From Business or Profession) of Form 1040, depicted in Figure 4.2-1. Such a business would depend on an information system that maintains business records of inventory, sales, costs, etc., for day-to-day management and control of operations, as well as for annual tax computations. A clear box of such an information system is depicted in Figure 4.2-2. Component black boxes are shown for updating business records maintained in the state, and for computing Schedule C based upon state data that has accumulated over a year of business operations. In what follows, we focus on the clear box expansion of the Schedule C computation.

The clear box procedure for completing Schedule C is specified in *Tax Guide For Small Business,* published by the Internal Revenue Service. Much of the information content of *Tax Guide For Small Business* is devoted to specifying the Schedule C clear box. While the Schedule C information is complete and comprehensive, it is distributed throughout the guide into categories of decisions and computations, that are likewise distributed in Schedule C. As a result, the information in the guide is not directly expressed in the form of a clear box. However, the structure and organization of Schedule C itself provides a basis for reorganizing the Schedule C information in the guide into a clear box. In fact, such a reorganization is derived on the fly every time a Schedule C is filled out, in

SCHEDULE C (Form 1040)
Department of the Treasury
Internal Revenue Service (0)

Profit or (Loss) From Business or Profession
(Sole Proprietorship)
Partnerships, Joint Ventures, etc., Must File Form 1065.
▶ **Attach to Form 1040 or Form 1041.** ▶ **See Instructions for Schedule C (Form 1040).**

OMB No. 1545-0074

Name of proprietor | **Social security number of proprietor**

A Main business activity (see Instructions) ▶ ; product ▶

B Business name and address ▶ | C Employer identification number

D Method(s) used to value closing inventory:
(1) ☐ Cost (2) ☐ Lower of cost or market (3) ☐ Other (attach explanation)

E Accounting method: (1) ☐ Cash (2) ☐ Accrual (3) ☐ Other (specify) ▶ | **Yes** | **No**

F Was there any major change in determining quantities, costs, or valuations between opening and closing inventory?
If "Yes," attach explanation.

G Did you deduct expenses for an office in your home?

PART I.—Income

1 a Gross receipts or sales	1a	
b Less: Returns and allowances	1b	
c Subtract line 1b from line 1a and enter the balance here	1c	
2 Cost of goods sold and/or operations (Part III, line 8)	2	
3 Subtract line 2 from line 1c and enter the **gross profit** here	3	
4 a Windfall Profit Tax Credit or Refund received in 1983 (see Instructions)	4a	
b Other income	4b	
5 Add lines 3, 4a, and 4b. This is the **gross income** ▶	5	

PART II.—Deductions

6 Advertising		23 Repairs	
7 Bad debts from sales or services (Cash method taxpayers, see Instructions)		24 Supplies (not included in Part III)	
8 Bank service charges		25 Taxes (Do not include Windfall Profit Tax here. See line 29.)	
9 Car and truck expenses		26 Travel and entertainment	
10 Commissions		27 Utilities and telephone	
11 Depletion		28 a Wages	
12 Depreciation and Section 179 deduction from Form 4562 (not included in Part III)		b Jobs credit	
		c Subtract line 28b from 28a	
		29 Windfall Profit Tax withheld in 1983	
13 Dues and publications		30 Other expenses (specify):	
14 Employee benefit programs		a	
15 Freight (not included in Part III)		b	
16 Insurance		c	
17 Interest on business indebtedness		d	
18 Laundry and cleaning		e	
19 Legal and professional services		f	
20 Office expense		g	
21 Pension and profit-sharing plans		h	
22 Rent on business property		i	

31 Add amounts in columns for lines 6 through 30i. These are the **total deductions** ▶	31	
32 Net profit or (loss). Subtract line 31 from line 5 and enter the result. If a profit, enter on Form 1040, line 12, and on Schedule SE, Part I, line 2 (or Form 1041, line 6). If a loss, go on to line 33	32	

33 If you have a loss, you must answer this question: "Do you have amounts for which you are not at risk in this business (see Instructions)?" ☐ Yes ☐ No
If "Yes," you must attach Form 6198. If "No," enter the loss on Form 1040, line 12, and on Schedule SE, Part I, line 2 (or Form 1041, line 6).

PART III.—Cost of Goods Sold and/or Operations (See Schedule C Instructions for Part III)

1 Inventory at beginning of year (if different from last year's closing inventory, attach explanation)	1	
2 Purchases less cost of items withdrawn for personal use	2	
3 Cost of labor (do not include salary paid to yourself)	3	
4 Materials and supplies	4	
5 Other costs	5	
6 Add lines 1 through 5	6	
7 Less: Inventory at end of year	7	
8 **Cost of goods sold and/or operations.** Subtract line 7 from line 6. Enter here and in Part I, line 2, above.	8	

For Paperwork Reduction Act Notice, see Form 1040 Instructions. **Schedule C (Form 1040)**

☆U.S. G.P.O. 1983-390-080 E.I. 43-0787287

Figure 4.2-1. Schedule C, Form 1040 (Tax Guide for Small Business, Internal Revenue Service).

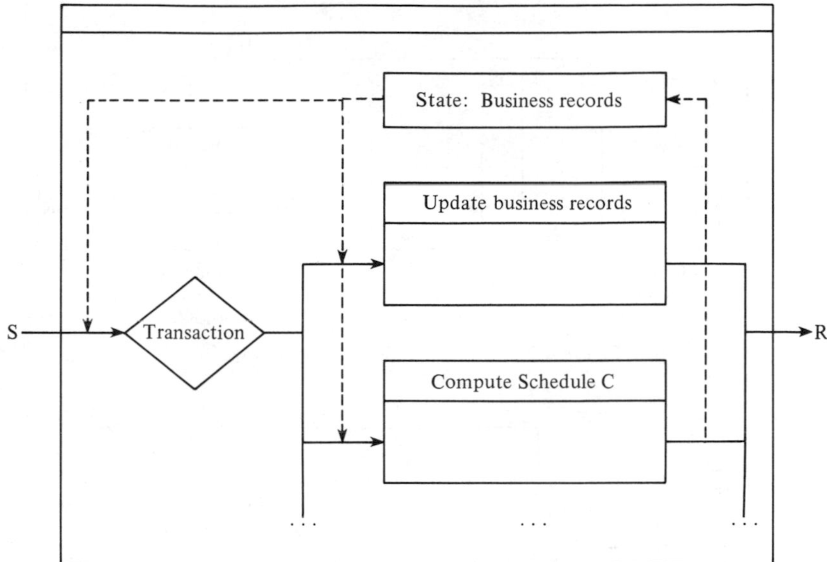

Figure 4.2-2. A Clear Box of a Schedule C System.

the step-by-step application of rules and regulations from different parts of the guide required at each point in the computation.

Because Schedule C is intended to record successive intermediate stages of a cumulative computation, it is, of necessity, a logically structured form. In fact, a procedure for computing Schedule C can be defined solely in terms of BDL control structures. The Schedule is organized into parts, I, II, and III, dealing with income, deductions, and costs, respectively. However, the parts cannot be completed in that order since Part I depends on the outcome of Part III. In essence, Part III computes cost of goods sold, which is then subtracted in Part I from gross sales to produce gross income. The deductions computed in Part II are then subtracted from gross income to arrive at net profit (or loss). Thus, the Schedule C clear box has the structure depicted in Figure 4.2-3, namely, a sequence of three operations on shared state data.

The state items accessed and stored by the firstpart operation of the sequence, shown in Figure 4.2-4, can be identified by examining the Part III computations of Schedule C. The old state, OS1, contains opening and closing inventory valuations and various cost items; the new state, NS1, contains the computed cost of goods sold.

Figures 4.2-5 and 4.2-6 depict state items accessed and stored for the secondpart and thirdpart operations, respectively. The secondpart old

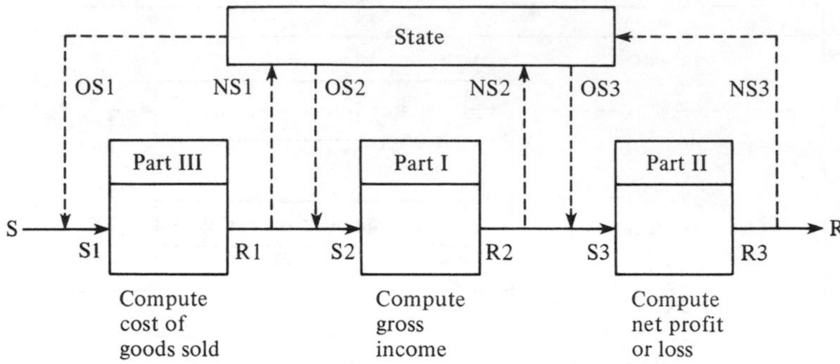

Figure 4.2-3. The Schedule C Clear Box.

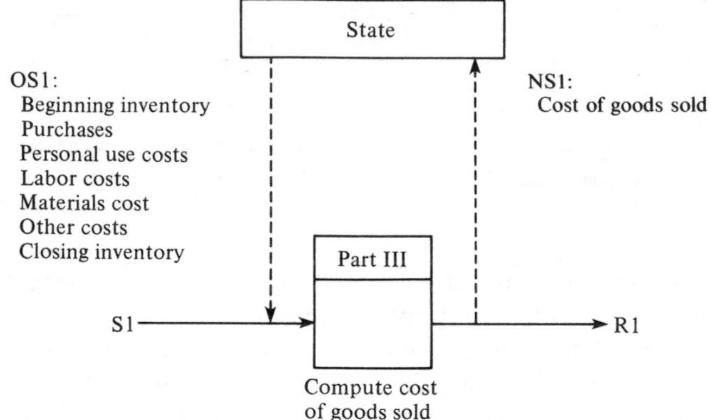

Figure 4.2-4. Firstpart State Items.

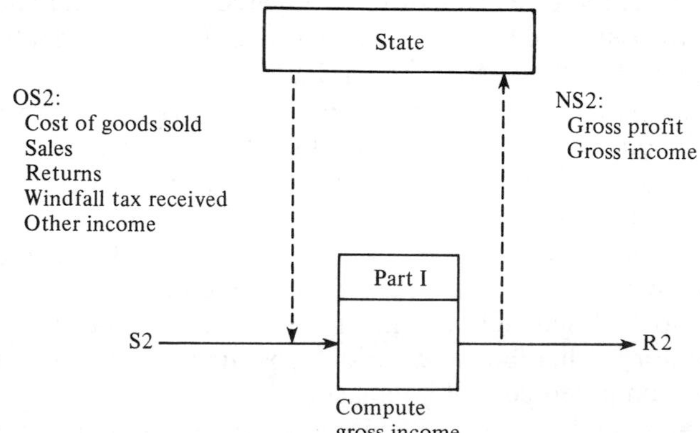

Figure 4.2-5. Secondpart State Items.

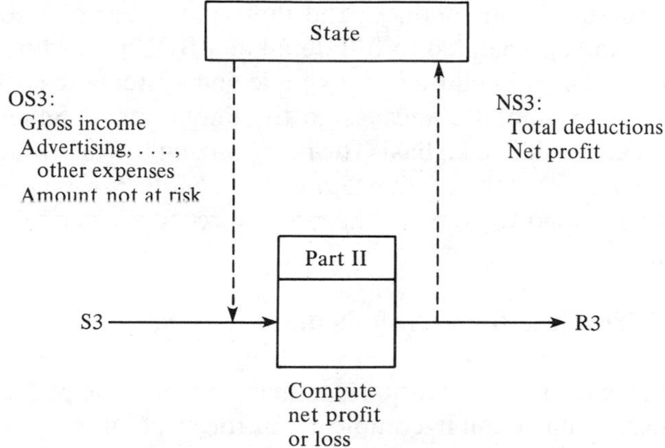

Figure 4.2-6. Thirdpart State Items.

state, OS2, contains the cost of goods sold computed by the firstpart, together with various income items; the new state, NS2, contains gross profit and gross income. Gross profit is not used by the thirdpart operation, but is a useful item to retain for unforeseen needs in other clear boxes in the information system. Finally, the thirdpart old state, OS3, contains gross income and various deduction items; the new state, NS3, contains total deductions and net profit, both likewise useful to retain for unforeseen needs. These state items are enumerated in the BDL clear box definition of Figure 4.2-7.

define CB Schedule C
  stimulus
    Compute Schedule C
  response
    Schedule C
  state

beginning inventory	returns
purchases	windfall tax received
personal use costs	other income
labor costs	gross profit
materials costs	gross income
other costs	advertising, ..., other expenses*
closing inventory	total deductions
cost of goods sold	net profit
sales	amount not at risk

  machine
    (see Figure 4.2-8)

* See Part II of Schedule C, items 6 to 30, Figure 4.2-1, for full enumeration of these items.

Figure 4.2-7. Schedule C Clear Box.

With required state items defined, the three part clear box sequence of Figure 4.2-3 can be expanded to full detail in a BDL procedure as shown in Figure 4.2-8. The procedure has a simple and systematic structure that uses state data to compute and assign the line items of Schedule C. In fact, the procedure defines a basis for reorganizing the distributed Schedule C information content of *Tax Guide For Small Business* into logical units that correspond to steps in the procedure, for easier reference and application.

4.2.3 Deriving Clear Boxes from Natural Procedures

In analyzing business operations, human and machine procedures may be encountered that exhibit complex control structures, with arbitrary and confusing connections among operations and tests. Because such structures are not expressed in nested and sequenced BDL control structures, they are difficult to understand and deal with as a foundation for new information system development. We call such structures **natural procedures.**

In illustration, consider the natural procedure of Figure 4.2-9, depicted in flowchart form, which defines operations and tests in processing job applicants for a business enterprise. Such a clear box could emerge from interviews of personnel employees as an explanation of existing operations, prior to designing an information system for applicant processing.

Even in this miniature example, it is difficult to identify and analyze possible paths of applicant processing. Fortunately, a systematic process exists to transform a natural procedure into a clear box expressed solely in BDL control structures. Such a transformation can help answer critical questions of completeness and correctness of the natural procedure. And be forewarned that the transformation will reveal some surprising behavior in this case. The transformation process is defined in six steps (which are elaborated below):

1. Convert to proper form.
2. Structure abstraction.
3. Sequence and alternation construction.
4. Clear box construction.
5. Clear box simplification.
6. Clear box expansion.

Step 1: *Convert to proper form.* A natural procedure is in **proper form** if it has a single entry and single exit. If the natural procedure is not in proper form, a case structure can be used to collect multiple entries and/or exits into a proper form.

```
proc
  do [compute cost of goods sold (part III)]
    line III.1 := beginning inventory
    line III.2 := purchases − personal use costs
    line III.3 := labor costs
    line III.4 := materials costs
    line III.5 := other costs
    line III.6 := line III.1 +··· + line III.5
    line III.7 := closing inventory
    cost of goods sold := line III.6 − line III.7
    line III.8 := cost of goods sold
  od
  do [compute gross income (part I)]
    line I.1a := sales
    line I.1b := returns
    line I.1c := line I.1a − line I.1b
    line I.2 := cost of goods sold
    gross profit := line I.1c − line I.2
    line I.3 := gross profit
    line I.4a := windfall tax received
    line I.4b := other income
    gross income := line I.3 + line I.4a + line I.4b
    line I.5 := gross income
  od
  do [compute net profit or loss (part II)]
    line II.6,...,line II.30i := advertising, ..., other
                                  expenses
    total deductions := line II.6 +···+ line II.30i
    line II.31 := total deductions
    net profit := gross income − total deductions
    line II.32 := net profit
    if
      net profit > 0
    then
      enter net profit on Form 1040, line 12, and on Schedule SE,
         line I.2 or on Form 1041, line 6
    else
      if
         amount not at risk > 0
      then
         check "yes"
         attach Form 6198
      else
         check "no"
         enter net profit on Form 1040, line 12, and on
            Schedule SE, line I.2 or Form 1041, line 6
      fi
    fi
  od
corp
```

Figure 4.2-8. Schedule C Clear Box Procedure.

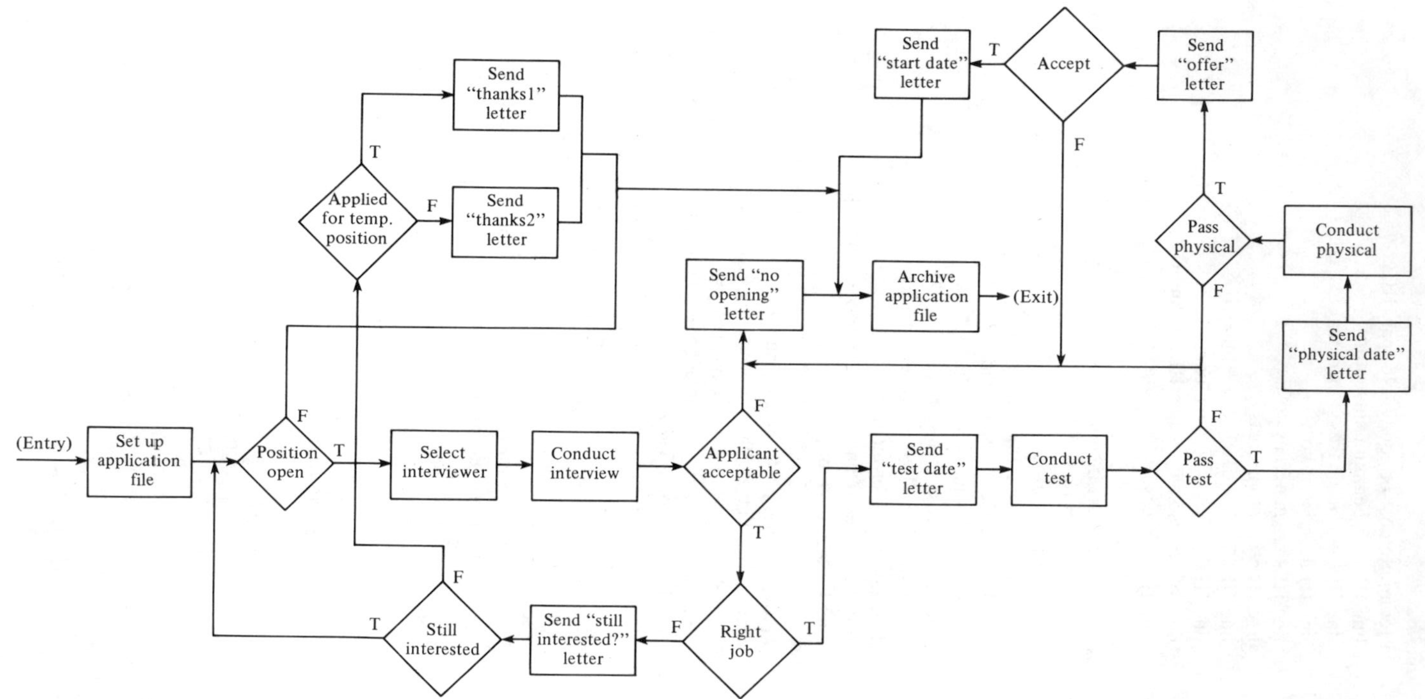

Figure 4.2-9. A Natural Procedure for Job Applicant Processing.

Step 2: *Structure abstraction.* The second step in simplifying natural procedures is to replace any BDL control structures (sequence, alternation, iteration, concurrency) in the flowchart with **abstract procedure statements.** Such control structures are readily understandable as is; they will be removed and saved in this step and reinserted in step 6, once the arbitrary, less understandable parts of the procedure have been dealt with in steps 3 through 5.

In flowchart terms, abstract procedure statements are simply named boxes that represent the single-entry/single-exit BDL control structures that have been removed. For example, with reference to Figure 4.2-9, the sequence control structure

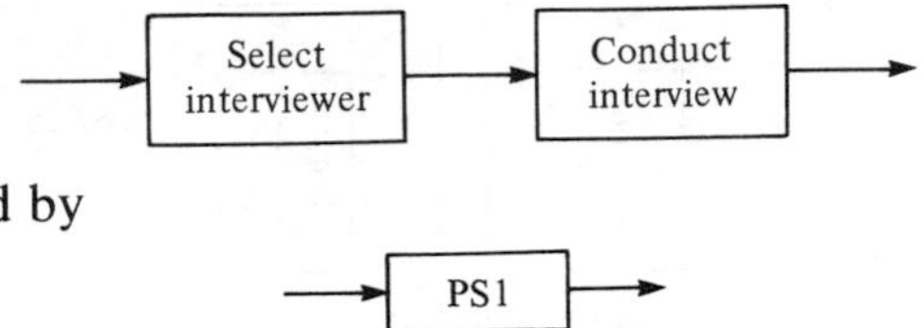

can be replaced by

PS1

Figure 4.2-10 depicts the statement abstractions possible in the job applicant procedure. Each abstraction is named (PS1, ..., PS4) and delimited by a dashed line. In general, a statement abstraction, once made, may permit another abstraction not possible before. For example, a structure such as

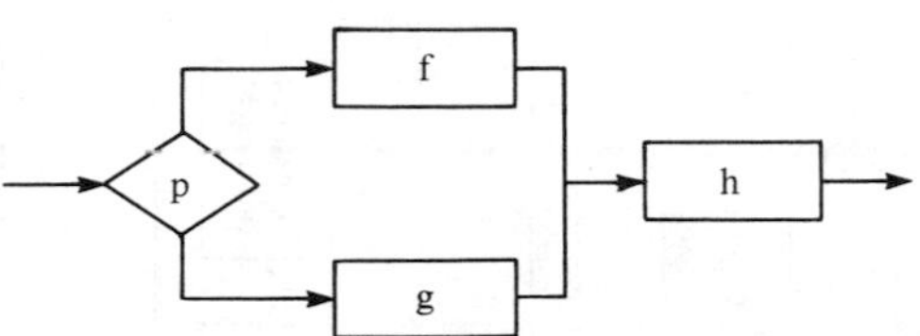

cannot be directly abstracted, but by abstracting the alternation a sequence emerges

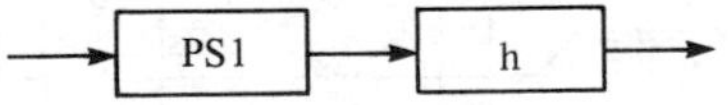

which can in turn be abstracted to a single procedure statement:

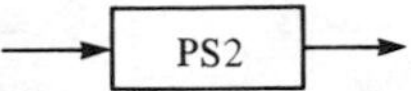

Step 3: *Sequence and alternation construction.* The third step is to construct a set of new sequence and alternation structures, one for every operation and test, respectively, in the reduced natural procedure. First, the operations and tests of the procedure must be numbered, in any arbitrary order, and the exit line numbered 0. Next, a new variable, known as the label variable, is introduced. Referring to Figure 4.2-10, the

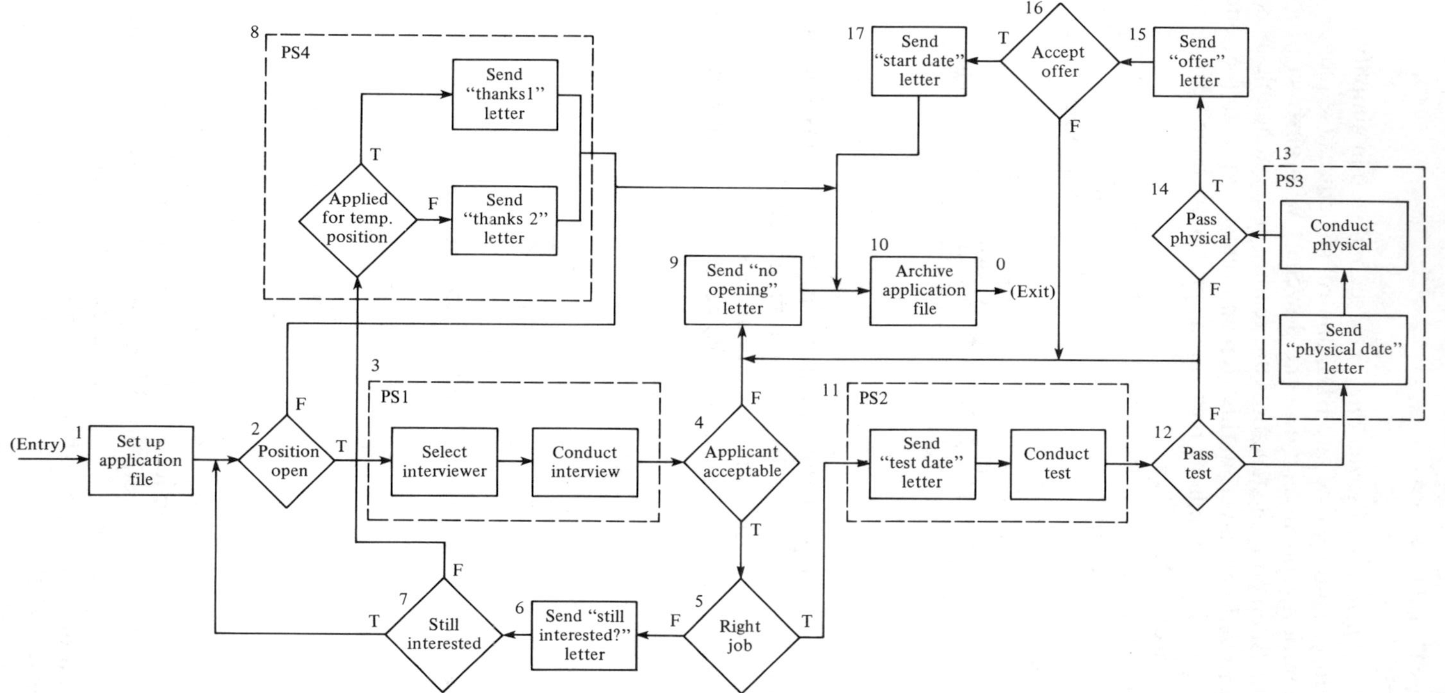

Figure 4.2-10. Statement Abstractions in the Job Applicant Procedure.

new sequence and alternation structures can be constructed as shown in Figure 4.2-11. The firstpart of every new sequence is an operation from the procedure, the secondpart is an assignment to the label variable, L, of the number of the next operation or test to be visited in the procedure. Similarly, the condition of every new alternation is a test from the procedure, with assignments to the label variable, L, of the number of the next operation or test to be visited in the procedure on the true and false branches.

New Sequence Structures

```
(1) do
      set up application file;
      L := 2
    od

(3) do
      PS1;
      L := 4
    od

(6) do
      send "still interested?"
        letter;
      L := 7
    od

(8) do
      PS4;
      L := 10
    od

(9) do
      send "no opening" letter;
      L := 10
    od

(10) do
       archive application file;
       L := 0
     od

(11) do
       PS2;
       L := 12
     od

(13) do
       PS3;
       L := 14
     od

(15) do
       send "offer" letter;
       L := 16
     od

(17) do
       send "start date" letter;
       L := 10
     od
```

New Alternation Structures

```
(2) if
      position open
    then
      L := 3
    else
      L := 10
    fi

(12) if
       pass test
     then
       L := 13
     else
       L :=  9
     fi
```

Figure 4.2-11. New Sequence and Alternation Structures (*continues*).

```
(4) if
      applicant acceptable
    then
      L := 5
    else
      L := 9
    fi

(5) if
      right job
    then
      L := 11
    else
      L := 6
    fi

(7) if
      still interested
    then
      L := 2
    else
      L := 8
    fi

(14) if
       pass physical
     then
       L := 15
     else
       L :=  9
     fi

(16) if
       accept offer
     then
       L := 17
     else
       L :=  9
     fi
```

Figure 4.2-11 (*Continued*)

Step 4. *Clear box construction.* The fourth step is to construct a new clear box with the structure of an initialized whiledo

```
proc Applicant processing
  L := number of first operation or test in procedure;
  while
     L > 0
  do
     ...
  od
corp
```

which terminates if L is 0 (corresponding to the natural procedure exit line), otherwise executes the dopart and repeats. Note that the dopart must eventually set L to 0, or an infinite loop will result. For the dopart, a case structure is constructed which tests and branches on values of L to corresponding caseparts composed of the new sequence and alternation structures of Figure 4.2-11. Such a construction is depicted in Figure 4.2-12. This clear box exhibits behavior identical to that of the natural procedure of Figure 4.2-9; that is, they are execution equivalent. (Try some sample executions to verify this assertion!) Although the clear box of Figure 4.2-12 is composed solely of nested and sequenced BDL state-

```
proc Applicant processing
  L := 1;
  while
    L > 0
  do
    case
      L
    part (1)
      set up application file;
      L := 2
    part (2)
      if
        position open
      then
        L := 3
      else
        L := 10
      fi
    part (3)
      PS1;
      L := 4
    part (4)
      if
        applicant acceptable
      then
        L := 5
      else
        L := 9
      fi
    part (5)
      if
        right job
      then
        L := 11
      else
        L := 6
      fi
    part (6)
      send "still interested?" letter;
      L := 7
    part (7)
      if
        still interested
      then
        L := 2
      else
        L := 8
      fi
    part (8)
      PS4;
      L := 10
    part (9)
      send "no opening" letter;
      L := 10
    part (10)
      archive application file;
      L := 0
    part (11)
      PS2;
      L := 12
    part (12)
      if
        pass test
      then
        L := 13
      else
        L := 9
      fi
    part (13)
      PS3;
      L := 14
    part (14)
      if
        pass physical
      then
        L := 15
      else
        L := 9
      fi
    part (15)
      send "offer" letter;
      L := 16
    part (16)
      if
        accept offer
      then
        L := 17
      else
        L := 9
      fi
    part (17)
      send "start date" letter;
      L := 10
    esac
  od
corp
```

Figure 4.2-12. New Clear Box Construction for Applicant Processing. (**proc** continues in the right-hand column.)

ments, it is likely not any more understandable than its natural procedure counterpart. However, while it is not obvious how to simplify the natural procedure, the structure of this clear box can indeed be simplified in a systematic manner, as we shall see in the next step.

Step 5. *Clear box simplification.* The clear box of Figure 4.2-12 can now be simplified by substituting the text of caseparts for occurrences of corresponding part number assignments to the label (L) variable, wherever they occur, and then eliminating the substituted caseparts. For example, casepart 3

```
...
PS1;
L := 4
...
```

is referenced in one label variable assignment, namely, L := 3 in casepart 2. Thus, the text of casepart 3 can be substituted for the assignment L := 3

```
...
part(2)
  if
    position open
  then
    PS1;
    L := 4
  else
    L := 10
  fi
...
```

and casepart 3 eliminated, since it is no longer referenced by other caseparts. Next, casepart 4, say, can be substituted for the assignment L := 4 (and eliminated) to get

```
...
part (2)
  if
    position open
  then
    PS1;
    if
      applicant acceptable
    then
      L := 5
```

```
            else
              L := 9
            fi
          else
            L := 10
          fi
        ...
```

Casepart 5 can now be substituted for the assignment L := 5 to get

```
...
part(2)
  if
    position open
  then
    PS1;
    if
      applicant acceptable
    then
      if
        right job
      then
        L := 11
      else
        L := 6
      fi
    else
      L := 9
    fi
  else
    L := 10
  fi
...
```

This process of substitution and elimination of caseparts can continue until the only caseparts remaining reference themselves, in which case they cannot be eliminated. For example, if casepart 2 above contained an assignment L := 2, it could not be substituted anywhere (and eliminated) since the L := 2 reference would no longer correspond to an existing case.

Also, a casepart substitution, if elected, must be made for all current references to the casepart. Thus, a large casepart could end up replicated in a number of places, possibly resulting in an increase in complexity. However, replication of small caseparts usually results in overall simplification. Thus, judgement is required in selecting caseparts for substitution.

Figure 4.2-13 depicts an intermediate step in simplification of the clear box. The clear box is shown as a main procedure (Applicant processing), which invokes a nested procedure (Applicant status) with a **run** statement. The case structure has been simplified to two parts. Part 1 sets up an

```
proc Applicant processing
  L := 1;
  while
    L > 0
  do
    case
      L
    part (1)
      set up application file;
      L := 2
    part (2)
      if position open
      then
        PS1;
        if applicant acceptable
        then
          if right job
          then
            run Applicant status
          else
            send "still interested?" letter;
            if still interested
            then
              L := 2
            else
              PS4;
              archive application file;
              L := 0
            fi
          fi
        else
          send "no opening" letter;
          archive application file;
          L := 0
        fi
      else
        archive application file;
        L := 0
      fi
    esac
  od
corp
```

```
proc Applicant status
  PS2;
  if pass test
  then
    PS3;
    if pass physical
    then
      send "offer" letter;
      if accept offer
      then
        send "start date" letter;
        archive application file;
        L := 0
      else
        send "no opening" letter;
        archive application file;
        L := 0
      fi
    else
      send "no opening" letter;
      archive application file;
      L := 0
    fi
  else
    send "no opening" letter;
    archive application file;
    L := 0
  fi
corp
```

Figure 4.2-13. An Intermediate Substitution Step.

```
proc Applicant processing
  set up application file;
  L := 2;
  while
     L > 0
  do
     if position open
     then
        PS1;
        if applicant acceptable
        then
           if right job
           then
              run Applicant status
           else
              send ''still interested?'' letter;
              if still interested
              then
                 L := 2
              else
                 PS4;
                 archive application file;
                 L := 0
              fi
           fi
        else
           send ''no opening'' letter;
           archive application file;
           L := 0
        fi
     else
        archive application file;
        L := 0
     fi
  od
corp
```

```
proc Applicant status
   PS2;
   if pass test
   then
      PS3;
      if pass physical
      then
         send ''offer'' letter;
         if accept offer
         then
            send ''start date'' letter;
            archive application file;
            L := 0
         else
            send ''no opening'' letter;
            archive application file;
            L := 0
         fi
      else
         send ''no opening'' letter;
         archive application file;
         L := 0
      fi
   else
      send ''no opening'' letter;
      archive application file;
      L := 0
   fi
corp
```

Figure 4.2-14. The Final Substitution Step.

application file and identifies the part 2 process to be executed next. Part 2 references itself, and so cannot be substituted for the assignment L := 2 in part 1. Casepart 1 can now be substituted for the L := 1 initialization assignment outside the iteration. Because the case statement now contains only a single part, it can be eliminated as well, to obtain the fully substituted clear box of Figure 4.2-14.

Next, observe in Figure 4.2-14 that the operation ''archive application file'' always appears with the operation ''L := 0.'' Also, a pattern of the form

```
if
   ...
then
   ...
   A
else
   ...
   A
fi
```

can be rewritten with A factored out at the end:

```
if
   ...
```

```
proc Applicant processing
   set up application file;
   L := 2;
   while
      L > 0
   do
      if position open
      then
         PS1;
         if applicant acceptable
         then
            if right job
            then
               run Applicant status
            else
               send "still interested?" letter;
               if still interested
               then
                  L := 2
               else
                  PS4;
                  archive application file;
                  L := 0
               fi
            fi
         else
            send "no opening" letter;
            archive application file;
            L := 0
         fi
      else
         archive application file;
         L := 0
      fi
   od
corp
```

```
proc Applicant status
   PS2;
   if pass test
   then
      PS3;
      if pass physical
      then
         send "offer" letter;
         if accept offer
         then
            send "start date" letter
         else
            send "no opening" letter
         fi
      else
         send "no opening" letter
      fi
   else
      send "no opening" letter
   fi;
   archive application file;
   L := 0
corp
```

Figure 4.2-15. A Further Simplification.

```
then
   ...
else
   ...
fi;
A
```

These observations permit a further simplification of Figure 4.2-14 to factor out the common operations of "archive application file" and "L := 0" in four places, as shown in Figure 4.2-15.

Now an additional simplification becomes possible. By adding a test for L = 0 at the end of the procedure, the remaining operations to "archive application file" can be factored out, as shown in Figure 4.2-16.

```
proc Applicant processing
   set up application file;
   L := 2;
   while
      L > 0
   do
      if position open
      then
         PS1;
         if applicant acceptable
         then
            if right job
            then
               run Applicant status
            else
               send "still interested?" letter;
               if still interested
               then
                  L := 2
               else
                  PS4;
                  L := 0
               fi
            fi
         else
            send "no opening" letter;
            L := 0
         fi
      else
         L := 0
      fi;
      if L = 0
      then
         archive application file
      fi
   od
corp
```

```
proc Applicant status
   PS2;
   if pass test
   then
      PS3;
      if pass physical
      then
         send "offer" letter;
         if accept offer
         then
            send "start date" letter
         else
            send "no opening" letter
         fi
      else
         send "no opening" letter
      fi
   else
      send "no opening" letter
   fi;
   L := 0
corp
```

Figure 4.2-16. Still Further Simplification.

With a little thought, it is clear that a final simplification is possible. By presetting L to 0, all of the individual assignments of 0 to L become redundant, and can be eliminated, as depicted in Figure 4.2-17.

Step 6. *Clear box expansion.* The final step in construction of the new clear box is expansion of the abstract procedure statements (PS1 to PS4) into the corresponding control structures saved in Step 1. Figure 4.2-18 depicts the fully expanded clear box.

The clear box of Figure 4.2-18 can be read and understood in a systematic manner, in sharp contrast to its natural procedure counterpart of Figure 4.2-9. The nested alternation structures explicitly define all applicant processing possibilities, and reveal some questionable actions, as

```
proc Applicant processing
   set up application file;
   L := 2;
   while
      L > 0
   do
      L := 0
      if position open
      then
         PS1;
         if applicant acceptable
         then
            if right job
            then
               run Applicant status
            else
               send "still interested?" letter;
               if still interested
               then
                  L := 2
               else
                  PS4
               fi
            fi
         else
            send "no opening" letter
         fi
      fi;
      if L = 0
      then
         archive application file
      fi
   od
corp
```

```
proc Applicant status
   PS2;
   if pass test
   then
      PS3;
      if pass physical
      then
         send "offer" letter;
         if accept offer
         then
            send "start date" letter
         else
            send "no opening" letter
         fi
      else
         send "no opening" letter
      fi
   else
      send "no opening" letter
   fi
corp
```

Figure 4.2-17. A Final Simplification.

```
proc Applicant processing
   set up application file;
   L := 2;
   while
      L > 0
   do
      L := 0;
      if position open
      then
         select interviewer;
         conduct interview;
         if applicant acceptable
         then
            if right job
            then
               run Applicant status
            else
               send "still interested?" letter;
               if still interested
               then
                  L := 2
               else
                  if applied for temp position
                  then
                     send "thanks1" letter
                  else
                     send "thanks2" letter
                  fi
               fi
            fi
         else
            send "no opening" letter
         fi
      fi;
      if L = 0
      then
         archive application file
      fi
   od
corp
```

```
proc Applicant status
   send "test date" letter;
   conduct test;
   if pass test
   then
      send "physical date" letter;
      conduct physical;
      if pass physical
      then
         send "offer" letter;
         if accept offer
         then
            send "start date" letter
         else
            send "no opening" letter
         fi
      else
         send "no opening" letter
      fi
   else
      send "no opening" letter
   fi
corp
```

Figure 4.2-18. The Fully Simplified and Expanded Applicant Processing Clear Box.

well. For example, if no position is open, the application is archived, but the applicant is not notified, and if an applicant turns down an offer, he/she is sent a "no opening" letter. Also, a "no opening" letter is sent if the physical is not passed. Such procedures obviously make little sense when seen in their true context in a well structured clear box, but can be difficult to identify and correct when embedded in the contextual confusion of a natural procedure.

Summary: Direct use of clear boxes can help simplify and clarify system behavior in complex situations. Schedule C of Form 1040 has a simple clear box structure, despite the complexity of its explanatory materials. State items and procedure steps in the Schedule C clear box are natural units of documentation and refinement in instruction guides for tax preparation. The transformation of natural procedures into clear boxes provides both a systematic process for analyzing complex business operations, and a foundation for new system design.

4.3 ANALYSIS OF STATE MACHINE BEHAVIOR FROM CLEAR BOXES

Preview: Clear boxes can be abstracted to equivalent procedure-free state machines to better study their transition behavior, in a process called state machine derivation. The Schedule C clear box can be abstracted to a state machine with equivalent behavior. An iteration clear box can also be described as a simpler alternation clear box that is used to verify correct behavior of the iteration clear box.

4.3.1 The Behavior of BDL Procedure Statements

Clear box procedures can become quite large in complex applications, and systematic methods are required in analyzing and understanding their effect on data. The key to systematic analysis is the fact that any clear box procedure, no matter how large, is composed solely of nested and sequenced BDL procedure statements. Every BDL procedure statement has a single entry line and a single exit line. While this single entry/single exit property is crucial for the nesting and sequencing of procedure statements, it has a deeper significance, as well. Because it has no other entries or exits, a BDL procedure statement simply alters data, in executing from its entry line to its exit line, with no other unforeseen effects possible. For example, the alternation structure,

```
if
  a > b
```

```
then
  m := a
else
  m := b
fi
```

sets m to max(a,b). It always does exactly this, no more, no less. What a procedure statement does to data is called its **statement function.** A procedure statement can be read and its effect on data analyzed, to arrive at the equivalent statement function. The statement function can be recorded as a **function comment,** delimited by square brackets ([ ]), immediately preceding the procedure statement text. The forms of function commentary for the procedure statements of BDL are as follows:

Sequence:

```
do [sequence function]
   PS1;
   PS2
od
```

Alternation:

```
if [alternation function]
  condition
then [thenpart function]
  PS1
else [elsepart function]
  PS2
fi
```

Case:

```
case [case function]
  condition
part (value 1) [part1 function]
  PS1
part (value 2) [part2 function]
  PS2
part (value 3) [part3 function]
  PS3
part (value 4) [part4 function]
  PS4
esac
```

Iteration:

```
while [iteration function]
  condition
do [dopart function]
  PS1
od
```

Concurrency:

```
with
   Resolve
con [concurrent function]
  PS1, [PS1 function]
  PS2  [PS2 function]
noc
```

In illustration of statement functions, with a little thought the effect on data of the sequence

```
do
   temp := price;
   price := cost;
   cost := temp
od
```

can be seen as an exchange of the values of price and cost, which also sets incidental data item temp to the value of cost. This statement function can be documented in a function comment as

```
do [exchange price, cost]
   temp := price;
   price := cost;
   cost := temp
od
```

The sequence

```
do
   price := price + cost;
   cost := price − cost;
   price := price − cost
od
```

also exchanges the values of price and cost, without use of an incidental date item. (Try some values for price and cost to see how it works!)

In further illustration, the alternation structure below sets loss carryover to the absolute value of net loss

```
if [loss carryover := absolute value (net loss)]
  net loss < 0
then
  loss carryover := −net loss
else
  loss carryover := net loss
fi
```

and the iteration structure below adds loan advances in $100.00 increments, if necessary, to an account balance until it becomes nonnegative:

```
while [if balance is negative, add $100 increments until it
        becomes non-negative]
  balance < 0
do
  balance := balance + 100
od
```

Note that in each instance the procedure statement does exactly what the function comment says, and vice versa. Thus, the function comments and their procedure statements are **function equivalent.** That is, they both exhibit the same behavior.

Function commentary can be expressed in whatever language and notation is appropriate to the problem at hand. In many instances, precise natural language may be sufficient. **Conditional assignments** are also useful, particularly in expressing statement functions of alternation structures. A conditional assignment is given by a sequence of conditions paired by arrows (→) with assignments and separated by bars (|) in which the first condition that evaluates TRUE denotes the assignment to be used; an assignment with no condition is always to be used if no preceding conditions evaluate TRUE. For example, the alternation above can be commented with a conditional rule as

```
if [net loss < 0 → loss carryover := −net loss |
                   loss carryover := net loss]
  net loss < 0
then
  loss carryover := −net loss
else
  loss carryover := net loss
fi
```

The conditional rule gives a branch-free abstraction of the alternation.

Simultaneous assignments are also useful in expressing function comments. Simultaneous assignment statements extend the idea of data assignment to several variables concurrently, denoted by a list (of equal length) on both sides of the assignment symbol,

<variable>, <variable>, ... := <expression>, <expression>, ...

where all <expression>s are evaluated, then simultaneously assigned to the respective <variable>s. For example

price, cost := list − discount, labor + material

means to compute the values of list − discount and labor + material, then simultaneously assign these values to price and cost, respectively. In the concurrent assignment

a, b := c + d, a

note that the initial value of a is assigned to b, not the value a becomes in this assignment.

Simultaneous assignments can be used to define a sequence-free abstraction of a sequence of assignments. For example, the sequence

```
do
  a := b;
  b := c;
  c := d
od
```

will have the total effect of the simultaneous assignment

a, b, c := b, c, d

and can be documented with the simultaneous assignment as a function comment, as

```
do [a, b, c := b, c, d]
  a := b;
  b := c;
  c := d
od
```

These assignments in reverse sequence, namely,

```
do
  c := d;
  b := c;
  a := b
od
```

have quite a different simultaneous assignment, namely,

```
a, b, c := d, d, d
```

Simultaneous assignments can also be placed in sequences themselves, for example in

```
do
  a, b := c + d, a;
  b, c := a, c − d
od
```

whose effect can be determined to be

```
a, b, c := c + d , c + d, c − d
```

and the sequence documented accordingly,

```
do [a, b, c := c + d, c + d, c − d]
  a, b := c + d, a;
  b, c := a, c − d
od
```

The exchange can be written as a simultaneous assignment, e.g., as

```
do [exchange x and y]
  x, y := y, x
od
```

or the simultaneous assignment used in a function comment, as

```
do [x, y := y, x]
  x := x + y;
  y := x − y;
  x := x − y
od
```

The nested and sequenced procedure statements in a clear box can be successively abstracted to statement functions, to eventually arrive at the statement function of the clear box itself. In illustration, the procedure

```
do
  if
    x ≥ 0
  then
    w := x
  else
    w := −x
  fi;
  if
    w > y
```

```
    then
      z := w
    else
      z := y
    fi
od
```

is composed of three procedure statements, specifically, two if statements comprising the firstpart and secondpart of a sequence statement. With a little thought, the statement function of the first if statement can be seen as

```
w := absolute value(x)
```

and the second if statement as

```
z := maximum(w, y)
```

to give the following branch-free abstraction:

```
do
   w := absolute value(x);
   z := maximum(w, y)
od
```

Next, the sequence can be abstracted by substituting the firstpart w value for the secondpart occurrence of w to arrive at a sequence-free statement function that defines the overall effect on data of the original procedure:

```
do
   z := maximum(absolute value(x), y)
od
```

The variable w does not appear in the final abstraction, since it is incidental to the computation of a value for z. We note that this analysis process can be reversed to show a design process of successive procedure statement expansions, in going from

```
do
   z := maximum(absolute value(x), y)
od
```

to an intermediate expansion

```
do [z := maximum(absolute value(x), y)]
   w := absolute value(x);
   z := maximum(w, y)
od
```

and then to a final expansion:

```
do [z := maximum(absolute value(x), y)]
  if [w := absolute value(x)]
    x ≥ 0
  then
    w := x
  else
    w := −x
  fi;
  if [z := maximum(w, y)]
    w > y
  then
    z := w
  else
    z := y
  fi
od
```

At each stage, successive statement functions are carried into the corresponding expansions as function comments to document the design during its expansion.

4.3.2 The State Machine Behavior of Clear Boxes

Clear boxes can be analyzed to determine their equivalent state machines, in a process called state machine derivation. In illustration, consider a sequence clear box in which machines M1, M2 do the following (See Figure 4.1-2):

```
M1:  R1  := OS1;
     NS1 := OS1 + S1

M2:  R2  := OS2;
     NS2 := OS2 − S2
```

In this case, we can derive expressions for R2 and NS2 in terms of S1 and OS1 as

```
R2  = OS2               by M2
    = NS1               by allocation
    = OS1 + S1          by M1
NS2 = OS2 − S2          by M2
    = NS1 − R1          by allocation
    = OS1 + S1 − OS1    by M1
    = S1                by simplification
```

That is, the clear box behaves like a state machine with the rule

```
R  := OS + S;
NS := S
```

In turn, the behavior of this state machine can be recognized as that of black box Add2, because, for transition i,

$$\begin{aligned} R(i) &= OS(i) + S(i) \\ &= NS(i - 1) + S(i) \\ &= S(i - 1) + S(i) \end{aligned}$$

That is, response R(i) is the sum of the last two stimuli S(i) and S(i − 1).

As a second example, consider an alternation clear box in which C tests if S is odd and M1, M2 are defined as above. The equivalent state machine is then given by the rule (see Figure 4.1-3)

```
if S is odd
then
   R  := OS;
   NS := OS + S
else (S is even)
   R  := OS;
   NS := OS − S
fi
```

That is:

```
do
   R  := OS;
   NS := OS − ((−1)^S * S)
od
```

(since $(-1)^S = -1$ if S is odd,
$= 1$ if S is even)

The black box behavior of this state machine can be described quite simply if the initial state OS = 0; in this case the black box response is the sum of previous odd stimuli minus the sum of previous even stimuli.

4.3.3 State Machine Derivation from the Schedule C Clear Box

The state machine of Schedule C can be derived from the clear box definition given in Section 4.2.2. First, the state of the state machine is identical to that of the clear box, as defined in Figure 4.2-7. Next, the state machine transition can be derived from the three part clear box sequence structure of Figure 4.2-3, as expanded in the BDL procedure of Figure 4.2-8. Thus, the first **do** statement of the procedure (compute cost of goods sold (Part III)) of Figure 4.2-8 can be abstracted to the statement function

cost of goods sold := beginning inventory + purchases
− personal use costs + labor costs
+ materials costs + other costs
− closing inventory

by substituting state values assigned to line items for occurrences of the same line items in right sides of subsequent assignments. The second **do** statement of the procedure (compute gross income (Part I)) can be similarly abstracted to a simultaneous assignment

gross profit,
gross income := sales − returns − cost of goods sold,
sales − returns − cost of goods sold
+ windfall tax received + other income

The abstraction of the third **do** statement of the procedure (compute net profit (Part II)) can be carried out in steps. First, the computation of total deductions and net profit abstracts to a simultaneous assignment

total deductions,
net profit := advertising + ... + other expenses,
gross income − (advertising + ... + other
expenses)

These three assignments define the computation up to the present point of analysis, as depicted in Figure 4.3-1.

The sequence of Figure 4.3-1 can itself be abstracted to the single simultaneous assignment of Figure 4.3-2, by the same substitution process of values for variables. This final sequence abstraction is especially illuminating, in its fully elaborated definition of computations and assignments from old state to new state carried out by the Schedule C clear box.

Next, with a little thought, it can be seen that the nested alternation structures of the procedure of Figure 4.2-8 specify three possible outcomes, as defined by the conditional rule of Figure 4.3-3.

The procedure of Figure 4.2-8 has now been abstracted to a two part sequence, namely a firstpart given by the concurrent assignment of Figure 4.3-2, and a secondpart given by the conditional rule of Figure 4.3-3. This new sequence can itself be abstracted to a single operation, the transition rule of the Schedule C state machine, as follows.

```
do
   cost of goods sold := beginning inventory + purchases
                         − personal use costs + labor costs
                         + materials costs + other costs
                         − closing inventory
   gross profit,
         gross income := sales − returns − cost of goods sold,
                         sales − returns − cost of goods sold
                         + windfall tax received + other income
   total deductions,
           net profit := advertising + ... + other expenses,
                         gross income − (advertising + ... + other
                         expenses)
od;
do
      (remainder of computation)
od
```

Figure 4.3-1. Intermediate Abstraction of the Schedule C Clear Box.

First, the conditions of the conditional rule of Figure 4.3-3 must be expressed in terms of old state items. Amount at risk is an item in the old state. However, net profit is a new state item which is not available for testing until the transition is partially completed. Figure 4.3-2 gives an expression for net profit in terms of old state items which can be used directly in the conditional rule. We name this expression E:

```
E := (sales − returns − (beginning inventory
     + purchases − personal use costs
     + labor costs + materials costs
     + other costs − closing inventory)
     + windfall tax received + other income
     − (advertising + ... + other expenses))
```

Next, observe that the simultaneous assignment of Figure 4.3-2 will always be carried out, no matter which part of the conditional rule of Figure 4.3-3 is executed. Thus, the concurrent assignment and the conditional rule can be combined into a new conditional rule by replicating the concurrent assignment in three new sequence structures, one for each part of the rule. The firstpart of each new sequence is the concurrent assignment of Figure 4.3-2, and the secondpart is the operation from the conditional rule of Figure 4.3-3. Such a construction is shown in the full Schedule C state machine abstraction of Figure 4.3-4. This procedure-free derivation contains all possibilities for the state transition in a single conditional rule. This rule could be carried out by many possible clear box

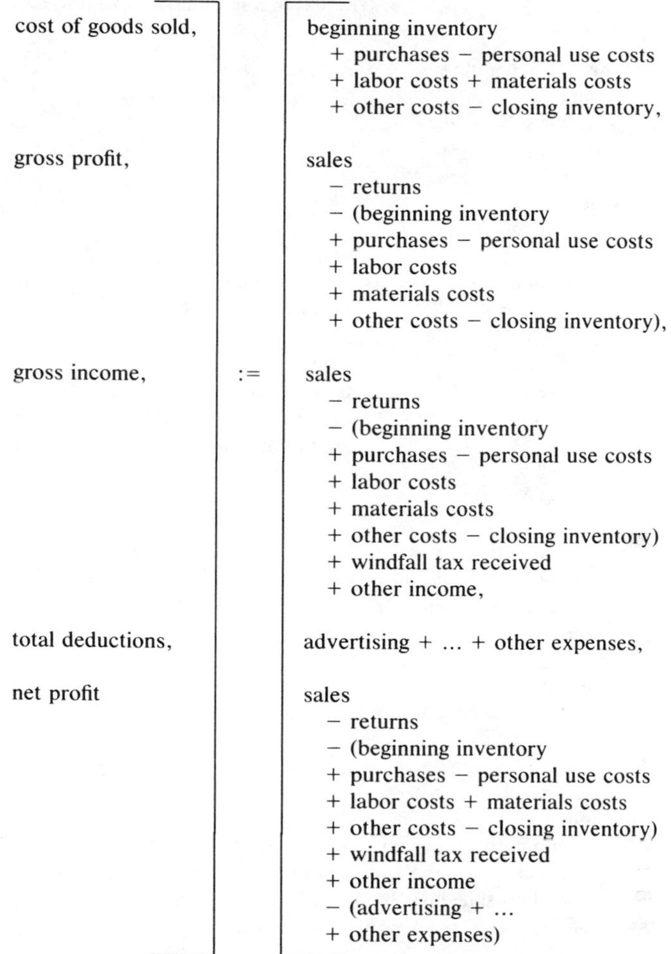

cost of goods sold,
gross profit,
gross income,
total deductions,
net profit

:=

beginning inventory
+ purchases − personal use costs
+ labor costs + materials costs
+ other costs − closing inventory,

sales
− returns
− (beginning inventory
+ purchases − personal use costs
+ labor costs
+ materials costs
+ other costs − closing inventory),

sales
− returns
− (beginning inventory
+ purchases − personal use costs
+ labor costs
+ materials costs
+ other costs − closing inventory)
+ windfall tax received
+ other income,

advertising + ... + other expenses,

sales
− returns
− (beginning inventory
+ purchases − personal use costs
+ labor costs + materials costs
+ other costs − closing inventory)
+ windfall tax received
+ other income
− (advertising + ...
+ other expenses)

Figure 4.3-2. The Simultaneous Assignment Abstraction of the Sequence of Figure 4.3-1.

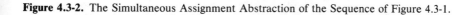

(net profit > 0 →
enter net profit on Form 1040, line 12, and on Schedule
SE, line I.2 or on Form 1041, line 6

| (amount not at risk > 0 →
check "yes", attach Form 6198

| amount not at risk = 0 →
check "no", enter net profit on Form 1040, line 12 and on
Schedule SE, line I.2 or on Form 1041, line 6))

Figure 4.3-3. Conditional Rule Abstraction of Nested Alternations of Figure 4.2-8.

```
define SM Schedule C
  stimulus
    complete Schedule C
  response
    Schedule C
  state
      beginning inventory       returns
      purchases                 windfall tax received
      personal use costs        other income
      labor costs               gross profit
      materials costs           gross income
      other costs               advertising, ..., other expenses
      closing inventory         total deductions
      cost of goods sold        net profit
      sales                     amount not at risk
  machine
  (E > 0 →
      do
        compute concurrent assignment of Figure 4.3-2;
        enter net profit on Form 1040, line 12, and on
        schedule SE, line I.2 or on Form 1041, line 6
      od
 | (amount not at risk > 0 →
      do
        compute concurrent assignment of Figure 4.3-2;
        check ''yes'' and attach Form 6198
      od
 | amount not at risk = 0 →
      do
        compute concurrent assignment of Figure 4.3-2;
        check ''no'' and enter net profit on Form 1040, line 12,
        and on schedule SE, line I.2 or on Form 1041, line 6
      od))
```

Figure 4.3-4. Schedule C State Machine.

designs, of which the clear box of Figure 4.2-8, the source of this state machine abstraction, is but one example.

4.3.4 The Behavior of Iteration Clear Boxes

Thus far in our discussion of clear box analysis, we have demonstrated methods of eliminating procedurality from clear box sequence, alternation, and case structures. Such analyses are facilitated by knowing that each machine is executed once in a structure. Analyzing the iteration structure is more difficult because the number of iterations of the machine in the structure is unknown during analysis. One method of analyzing iteration structures is to transform the iteration into an equivalent case structure, which we know how to analyze. This transformation procedure is studied in the remainder of this section.

An iteration clear box defines the behavior of a state machine, obtained by abstracting the number of its iterations out and discovering the resulting transition from stimulus and initial state to response and final state. Once this transition is discovered, it can be used to represent the effect of the iteration. Since this discovery can be difficult and subject to human error, a method of verification of a candidate transition is useful and is given below. We begin with an example to illustrate an iteration clear box that requires a variable number of iterations to complete a transition.

Consider an iteration clear box in which condition C tests if S is odd and machine M1 is defined as before, namely,

```
M1:  R1  := OS1;
     NS1 := OS1 + S1
```

Then if S = 5, and OS1 = 3, for example, the transition of this iteration clear box can be determined as follows:

iteration	S	OS1	S1	R1	NS1	R
0	5	3	5	3	8	
1	3	8	3	8	11	
2	8	11				8

namely, the transition (5,3) → (8,11). (By starting the iteration count at 0, it counts the number of times the machine M1 is invoked.)

As a second example, if S = 4, OS1 = 3, the transition is

iteration	S	OS1	S1	R1	NS1	R
0	4	3				4

namely the transition (4,3) → (4,3). As a third example, if S = 5, OS1 = 4 the transition is

iteration	S	OS1	S1	R1	NS1	R
0	5	4	5	4	9	
1	4	9				4

namely, the transition (5,4) → (4,9).

With a little thought, it can be seen that these three transitions represent all possible transitions in the following sense:

if S is even
  R, NS := S, OS [in iteration 0]

if S is odd,
  if OS is even
    R, NS := OS, OS + S [in iteration 1]
  if OS is odd
    R, NS := OS + S, 2 * OS + S [in iteration 2]

These three cases represent all possibilities for S and OS to be odd or even (if S is even, the transition occurs in iteration 0 whether OS is odd or even), and the three examples above are models of such transitions.

This example illustrates a general procedure for determining the transitions of an iteration clear box. It is to discover the conditions on S and OS for transitions to occur at iterations 0,1,2, ... and then work out what the transitions will be in each such case.

As a result, such an iteration clear box can be identified with an equivalent state machine. In this case, the state machine has the transitions as given, for S and OS odd or even.

This example shows how an iteration clear box can be determined as equivalent to a state machine. Such a state machine will have a single machine (different, of course, than the machine M1 of the clear box iteration), which can depend only on the definitions for C and M1 of the iteration clear box. Let such a machine be denoted as M(C,M1), where

M(C,M1): if S is even: R := S; NS := OS
  if S is odd:
    if OS is even: R := OS; NS := OS + S
    if OS is odd: R := OS + S; NS := 2 * OS + S

The state machine (Abbreviated) of Figure 4.3-5A and the iteration clear box of Figure 4.3-5B (Abbreviated) will have identical transitions.

A Box Structure Identity

Consider next an alternation clear box in which the foregoing iteration clear box is embedded, as shown in Figure 4.3-6A, which can be seen to carry out the first iteration, if necessary, of the iteration clear box above, then enter the iteration clear box again, inside the alternation clear box.

With a little thought, it can be seen that this alternation clear box will have the same transitions as the iteration clear box above, as the follow-

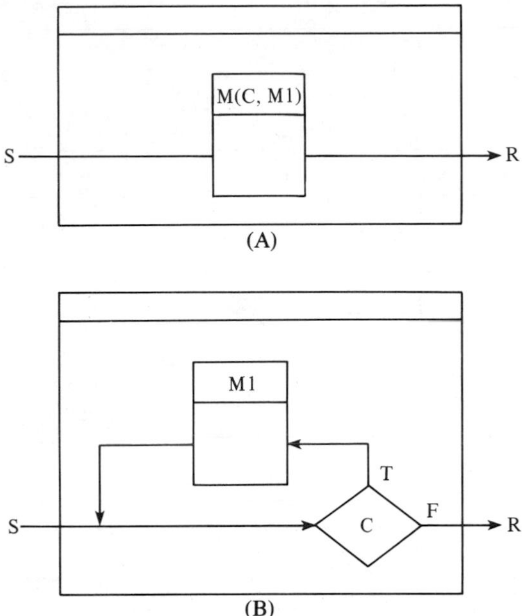

Figure 4.3-5. State Machine (A) and Iteration Clear Box (B) with Identical Behaviors.

ing analysis shows. On entry if the outcome of case C is F, the transition in either clear box is simply

(S,OS) → (S,OS)

On the other hand, if the outcome of condition C is T, the effect in either clear box is to invoke a transition of M1, then reenter the iteration clear box, from which point the iteration will be identical, with identical results.

But now, in the alternation clear box the iteration clear box can be replaced by its equivalent state machine, since its transitions will be identical, to get a new alternation clear box shown in Figure 4.3-6B, which by its construction must have transitions identical with the original iteration clear box, and therefore, with the state machine. This result is summarized as the following Theorem, with a general machine M2 in place of M(C,M1) in the example.

Iteration Theorem: For any condition definition C and machines M1, M2, if the iteration clear box of Figure 4.3-7A and state machine of Figure 4.3-7B are equivalent, then both are equivalent to the alternation clear box of Figure 4.3-7C.

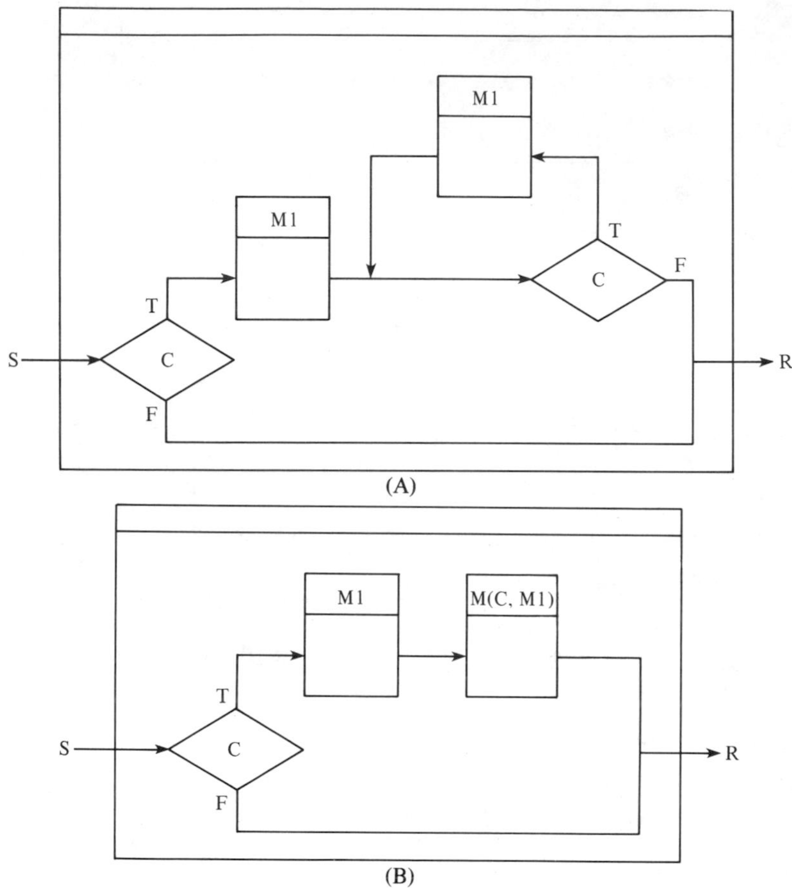

Figure 4.3-6. Alternation Clear Box with Embedded Iteration Clear Box (A) and Alternation Clear Box (B) Exhibiting Identical Behaviors.

The significance of the Iteration Theorem is that an iteration clear box is described as both a state machine and a simpler alternation clear box. Thus, the hypothesis that machine M2 describes the iterative effect of C and M1 in an iteration clear box, can be verified (or not) by reducing the alternation case clear box to a state machine and comparing its machine with M2. We summarize this in the following theorem;

Theorem (Verification Theorem): If the state machine of Figure 4.3-8A describes the behavior of the iteration clear box of Figure 4.3-8B, then the behavior of the alternation clear box of Figure 4.3-8C can be reduced to the behavior of the state machine of Figure 4.3-8A.

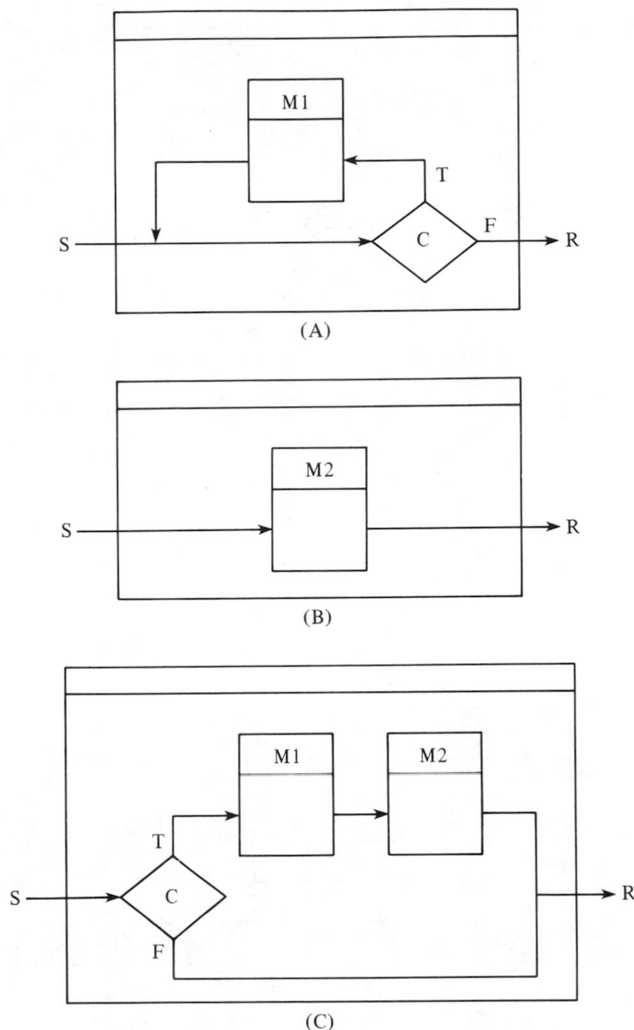

Figure 4.3-7. The Iteration Theorem. (A) Iteration Clear Box; (B) State Machine; and (C) Alternation Clear Box.

The Verification Theorem provides a means to verify the correctness of a state machine expansion into an iteration clear box. Rather than verifying by direct comparison of the state machine and iteration clear box behaviors, a difficult task, the Theorem permits verifying by comparison of the behaviors of the state machine and an alternation clear box, a simpler task.

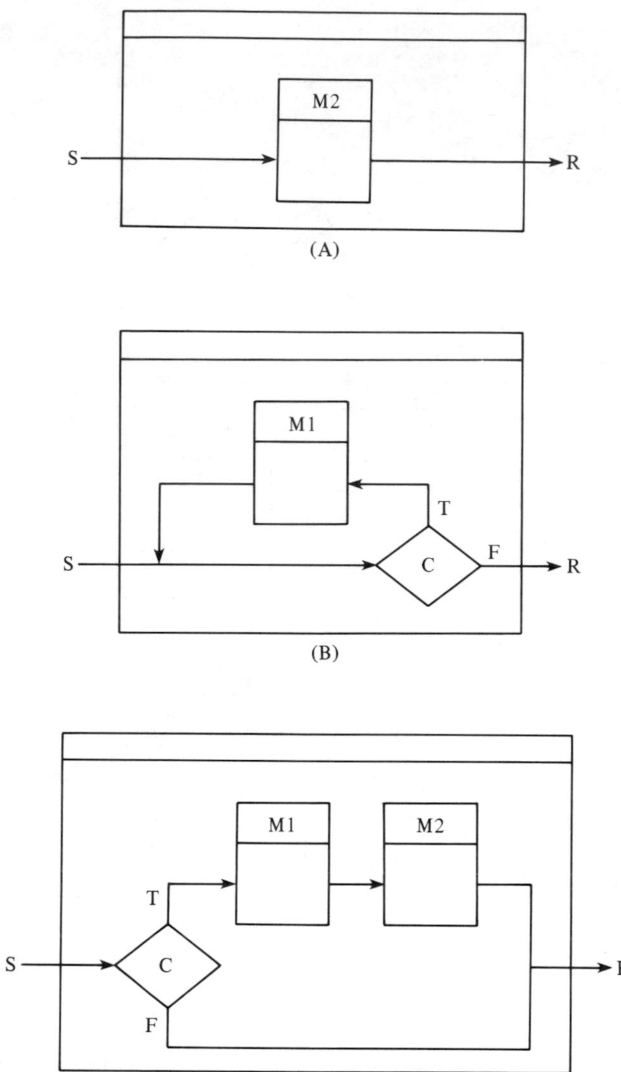

Figure 4.3-8. The Verification Theorem. (A) State Machine; (B) Iteration Clear Box; and (C) Alternation Clear Box.

In illustration, in the example above

C: S is odd

M1: R, NS := OS, OS + S

```
M2:  if S is even
     then
        R, NS := S, OS
     else (S is odd)
        if OS is even
        then
           R, NS := OS, OS + S
        else (OS is odd)
           R, NS := OS + S, 2 * OS + S
        fi
     fi
```

and M2 was discovered to describe the iterative effect of C and M1. To verify this discovery, we need only consider the alternation clear box of the Verification Theorem and compare its behavior with M2. We can analyze this alternation clear box in two steps, dealing first with the sequence M1, M2 as shown in Figure 4.3-9A. The unabbreviated sequence structure can be annotated as shown in Figure 4.3-9B and, by the definitions of M1 and M2

```
M1:  R1  := OS1
     NS1 := OS1 + S1
```

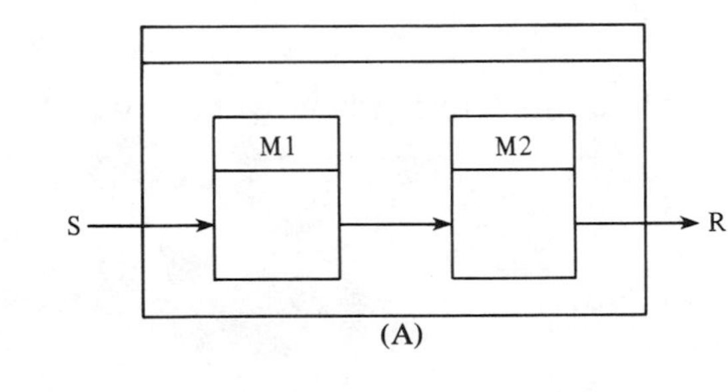

(A)

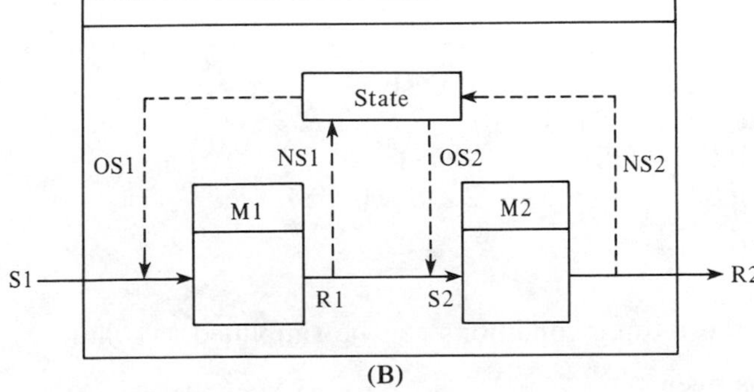

(B)

Figure 4.3-9. Abbreviated (A) and Equivalent Unabbreviated (B) Sequence Structures.

```
M2: if S2 is even
    then
      R2  := S2;
      NS2 := OS2
    else (S2 is odd)
      if OS2 is even
      then
        R2  := OS2;
        NS2 := OS2 + S2
      else (OS2 is odd)
        R2  := OS2 + S2;
        NS2 := 2 * OS2 + S2
      fi
    fi
```

Furthermore, by the sequence structure

```
 S2 = R1,
OS2 = NS1
```

In order to obtain the behavior of the sequence clear box, we need to determine R2, NS2 in terms of S1, OS1. Now

```
 S2 =  R1 = OS1,
OS2 = NS1 = OS1 + S1
```

so the M1;M2 sequence structure can be rewritten as

```
if OS1 is even
then
  R2  := OS1;
  NS2 := OS1 + S1
else (OS1 is odd)
  if OS1 + S1 is even
  then
    R2  := OS1 + S1;
    NS2 := OS1 + S1 + OS1
  else (OS1 + S1 is odd)
    R2  := OS1 + S1 + OS1;
    NS2 := 2 * OS1 + 2 * S1 + OS1
  fi
fi
```

Note that these three conditions can be simplified as follows

```
OS1 is even                      → OS1 is even
OS1 is odd, OS1 + S1 is even → S1 is odd, OS1 is odd
```

OS1 is odd, OS1 + S1 is odd → S1 is even, OS1 is odd

and the effect of this sequence structure (of Figure 4.3-9) is, with OS = OS1, S = S1, R = R2, and NS = NS2

```
if OS is even
then
   R, NS := OS, OS + S
else (OS is odd)
   if S is odd
   then
      R, NS := OS + S, 2 * OS + S
   else (S is even)
      R, NS := 2 * OS + S, 3 * OS + 2 * S
   fi
fi
```

Having worked out the sequence structure, we need to work out the alternation structure which contains this sequence structure, as follows:

```
if S is odd
then
   if OS is even
   then
      R, NS := OS, OS + S
   else (OS is odd)
      if S is odd
      then
         R, NS := OS + S, 2 * OS + S
      else (S is even)
         R, NS := 2 * OS + S, 3 * OS + 2 * S
      fi
   fi
else (S is even)
   R, NS := S, OS
fi
```

This can be simplified by recognizing that of the two innermost cases, one condition (S is odd) is redundant and the other condition (S is even) is a contradiction. Therefore, the alternation structure is, on rearranging:

```
if S is even
then
   R, NS := S, OS
else (S is odd)
   if OS is even
```

```
        then
           R, NS := OS, OS + S
        else (OS is odd)
           R, NS := OS + S, 2 * OS + S
        fi
     fi
```

which is identical with M2. Therefore, the form of M2 has been verified by the application of the Verification Theorem.

Summary: State machine abstractions of clear boxes define equivalent behavior while suppressing procedural details. The hypothesis that an iteration clear box exhibits behavior identical to its state machine specification can be verified by transforming the iteration clear box into an alternation clear box, which can be abstracted to a state machine and compared to the state machine specification.

4.4 DESIGN OF CLEAR BOXES FOR STATE MACHINE BEHAVIOR

Preview: Clear boxes are designed by expanding state machine transitions into equivalent BDL procedures. The state machine description of the hand calculator black box provides a basis for clear box design. Clear boxes can be organized into a hierarchy of smaller clear boxes by reusing the concept of a BDL procedure.

4.4.1 Clear Box Design Principles

The objective of clear box design is to express the transitions of a state machine in a procedure that accesses the same state objects, and possibly refers to working data and lower level black boxes.

The state machine transitions are a specification of the required clear box procedure. The initial expansion of any clear box procedure will be a sequence, alternation, iteration, or concurrent control structure. This control structure will reexpress the specification in terms of a sequence of two (or more) subspecifications, a choice between two (or more) subspecifications, a repetitive subspecification, or two (or more) concurrent subspecifications, respectively. Each subspecification will be smaller and simpler than the original specification, and can in turn be reexpressed in terms of new control structures and subsubspecifications. At any point in the process a subspecification may be regarded as a black box for which no further expansion is required.

In all but the simplest state machines, many different types of transitions based on stimulus and state may be specified, all of which must be recognized and carried out by the procedure of the clear box expansion. This observation leads to the following fundamental principle:

> **Fundamental Principle:** A clear box must determine which transition is required by the current stimulus and state, and then carry it out.

Thus, a useful strategy in clear box design is to begin the expansion process with a procedure that recognizes each stimulus and state, and directs control to the procedure part responsible for the corresponding transition. An alternation or case structure can be used to organize the tests of stimulus and state, with each thenpart, elsepart, or casepart carrying out a particular transition.

In some cases, individual transitions of a state machine may contain identical parts, or parts that differ only in the state data on which they operate, but are otherwise identical. Such commonality can be capitalized upon in clear box design, and leads to the following fundamental principle.

> **Fundamental Principle:** Any operations shared by state machine transitions should be expanded into clear box subprocedures and invoked by the clear box where necessary in carrying out those transitions.

For example, a file update state machine may define many possible file update transitions, all of which depend on the proper password stimulus for levels of file access and authorization. The shared password processing can be expanded as a common clear box subprocedure invoked by the various unique update transitions as required.

4.4.2 A Clear Box Design for the Hand Calculator

We use the hand calculator to illustrate an orderly top down step by step process of design expansion of a state machine into a clear box. Such an orderly expansion may not be easy to find without some analysis and insight. It may involve restating the form of the state machine, and possibly several attempts at an expansion. However, the final result provides an easy trail for the reader, and is well worth the designer's effort.

In Chapter 3 we developed a state machine design for the black box behavior of a simple hand calculator with stimulus keys,

C Clear Key
D Digit Keys (0–9)
F Arithmetic Function Keys (+, −, *, /)
= Result Key

for behavior not involving numerical overflow in either digit entry or result display. In that explanation, we assumed a state defined by four state registers:

BR: Begin Register (contains B for Begin or C for Continue)
VR: Visible Register (displays any number)
FR: Function Register (contains an arithmetic function)
HR: Hidden Register (contains any number)

The state machine transitions are defined in Table 3.4-2, repeated here in Table 4.4-1.

Table 4.4-1 provides the basis for working out a clear box design for the hand calculator. As noted, the state machine of Chapter 3 assumes that response R is always equal to the number in the visible register VR. Therefore, an initial clear box sequence structure can be designed as shown in Figure 4.4-1 in both diagram and BDL form. In order to save space in complex diagrams, black boxes are shown without internal lines. The state registers BR (begin register) and FR (function register) are defined as enumerated types, with permissible values listed, or enumerated, as (B,C) for Begin or Continue, and (+,−,*,/), respectively. Both

Table 4.4-1

Machine Transitions for a Hand Calculator

	(S ,	OS			)→	(R ,	NS			)
row		BR	VR	FR	HR		BR	VR	FR	HR
1	C					0	B	0		
2	any D	B		f	y	D	C	D	f	y
3	any D	C	x	f	y	D + 10x	C	D + 10x	f	y
4	any F		x	f	y	yfx	B	yfx	F	yfx
5	=		x	f	y	yfx		yfx		

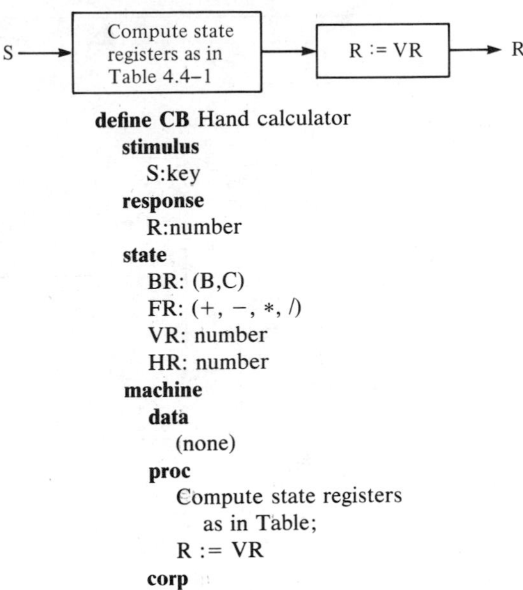

S
Compute state registers as in Table 4.4–1
R := VR
R
define CB Hand calculator
stimulus
S:key
response
R:number
state
BR: (B,C)
FR: (+, −, *, /)
VR: number
HR: number
machine
data
(none)
proc
Compute state registers
as in Table;
R := VR
corp

Figure 4.4-1. The Initial Hand Calculator Clear Box.

VR and HR are defined as type number. The machine has no local data, and the initial clear box decomposition is a sequence structure.

Next, we can translate Table 4.4-1 into clear box expansions line by line, step by step. Figure 4.4-2 shows an alternation expansion of the first black box in the sequence, to differentiate between line 1 of the table and all other lines, by testing for a stimulus of "C." Next, Figure 4.4-3 depicts an expansion of "Compute line 1 of Table" as a sequence of two black boxes.

Now that line 1 of the table has been expanded, we next expand lines 2-5 by checking for a digit stimulus to differentiate between the transitions on lines 2-3 and 4-5, as shown in the alternation expansion of Figure 4.4-4. Next, line 2 and line 3 of Table 4.4-1 can be distinguished by checking on SR, to get the case expansion of Figure 4.4-5. Continuing in this way, we finally arrive at the fully expanded clear box of Figure 4.4-6.

4.4.3 Segment Structured Clear Boxes

In systems of any size, clear box expansions of state machines can become quite large, so as a practical matter, a systematic means to break

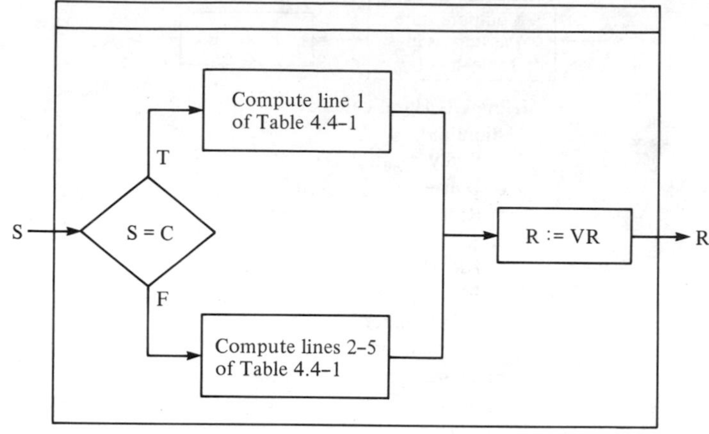


```
machine
   data
      (none)
   proc [Compute state registers as in Table]
      if S = C
      then
         Compute line 1 of Table
      else
         Compute lines 2-5 of Table
      fi;
      R := VR
   corp
```

Figure 4.4-2. An Alternation Expansion of "Compute state registers as in Table."

procedures into manageable parts is required. The concept of a clear box procedure can be reused for this purpose, by defining each part as a procedure with a name, in the form

```
proc <procedure name>
  data
    <procedure data>
  <procedure>
corp
```

then calling such a part into execution by a **run statement** of the form

```
run <procedure name>
```

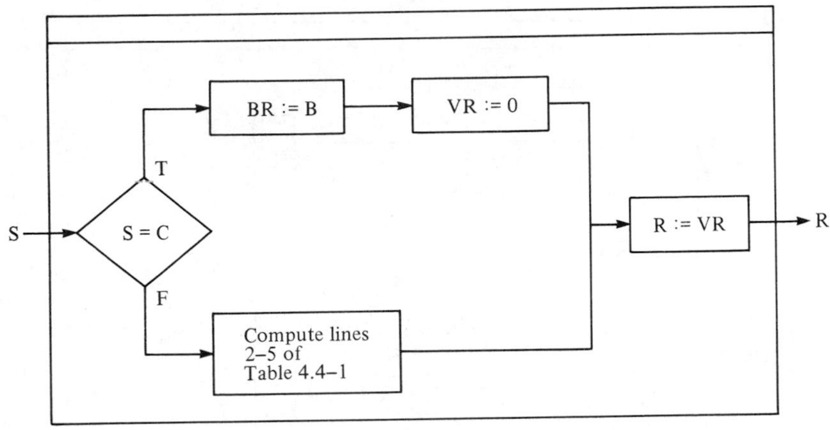


```
machine
   data
      (none)
   proc [Compute state registers as in Table]
      if S = C
      then [Compute line 1 of Table]
         BR := B;
         VR := 0
      else
         Compute lines 2-5 of Table
      fi;
      R := VR
   corp
```

Figure 4.4-3. A Sequence Expansion of "Compute line 1 of Table."

The <procedure data> defined in a procedure is available only within the procedure and not outside. Therefore data names can be reused without confusion. Such procedures can be listed as part of a clear box, or if often reused, as part of a library of procedures available to a clear box.

Procedures can be made even more reusable with the use of parameters. A parameter list of data can be defined with a procedure in the form

```
proc <procedure name> (<parameter list>)
  data
    <procedure date>
  <procedure>
corp
```

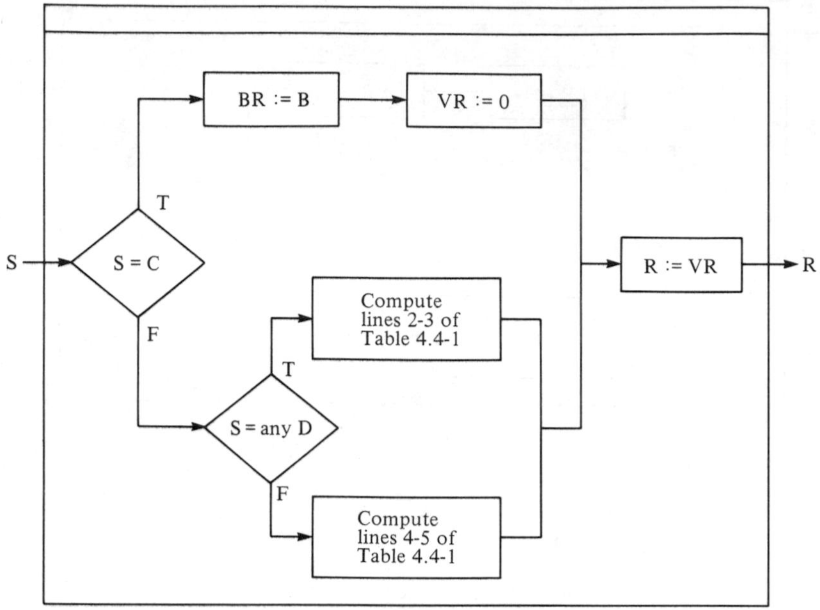


```
machine
   data
      (none)
   proc [Compute state registers as in Table]
      if S = C
      then [Compute line 1 of Table]
         BR := B;
         VR := 0
      else [Compute lines 2-5 of Table]
         if S = any D
         then
            Compute lines 2-3 of Table
         else
            Compute lines 4-5 of Table
         fi
      fi;
      R := VR
   corp
```

Figure 4.4-4. An Alternation Expansion of "Compute lines 2-5 of Table."

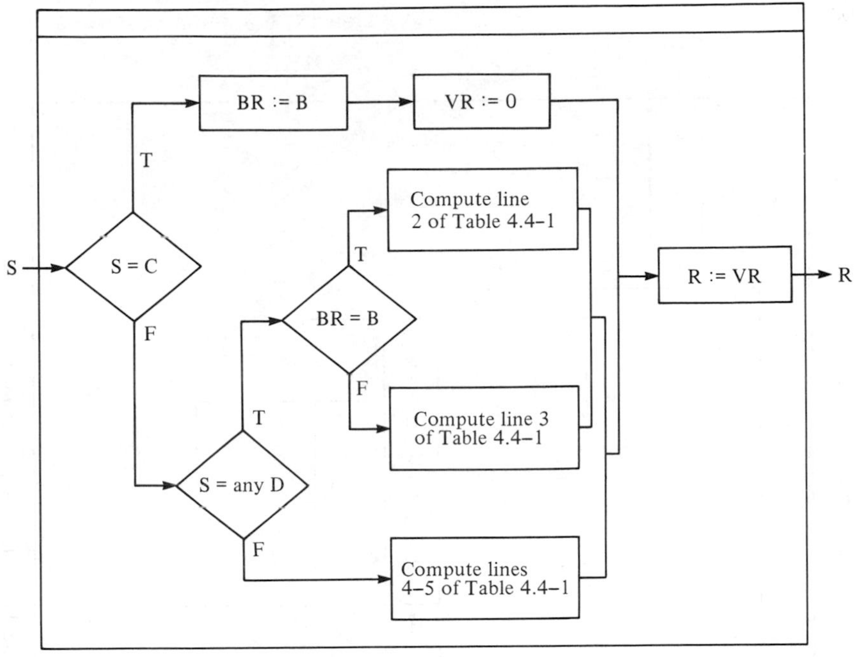


```
machine
   data
      (none)
   proc [Compute state registers as in Table]
      if S = C
      then [Compute line 1 of Table]
         BR := B;
         VR := 0
      else [Compute lines 2-5 of Table]
         if S = any D
         then [Compute lines 2-3 of Table]
            if BR = B
            then
               Compute line 2 of Table
            else
               Compute line 3 of Table
            fi
         else
            Compute lines 4-5 of Table
         fi
      fi;
      R := VR
   corp
```

Figure 4.4-5. An Alternation Expansion of "Compute lines 2-3 of Table."

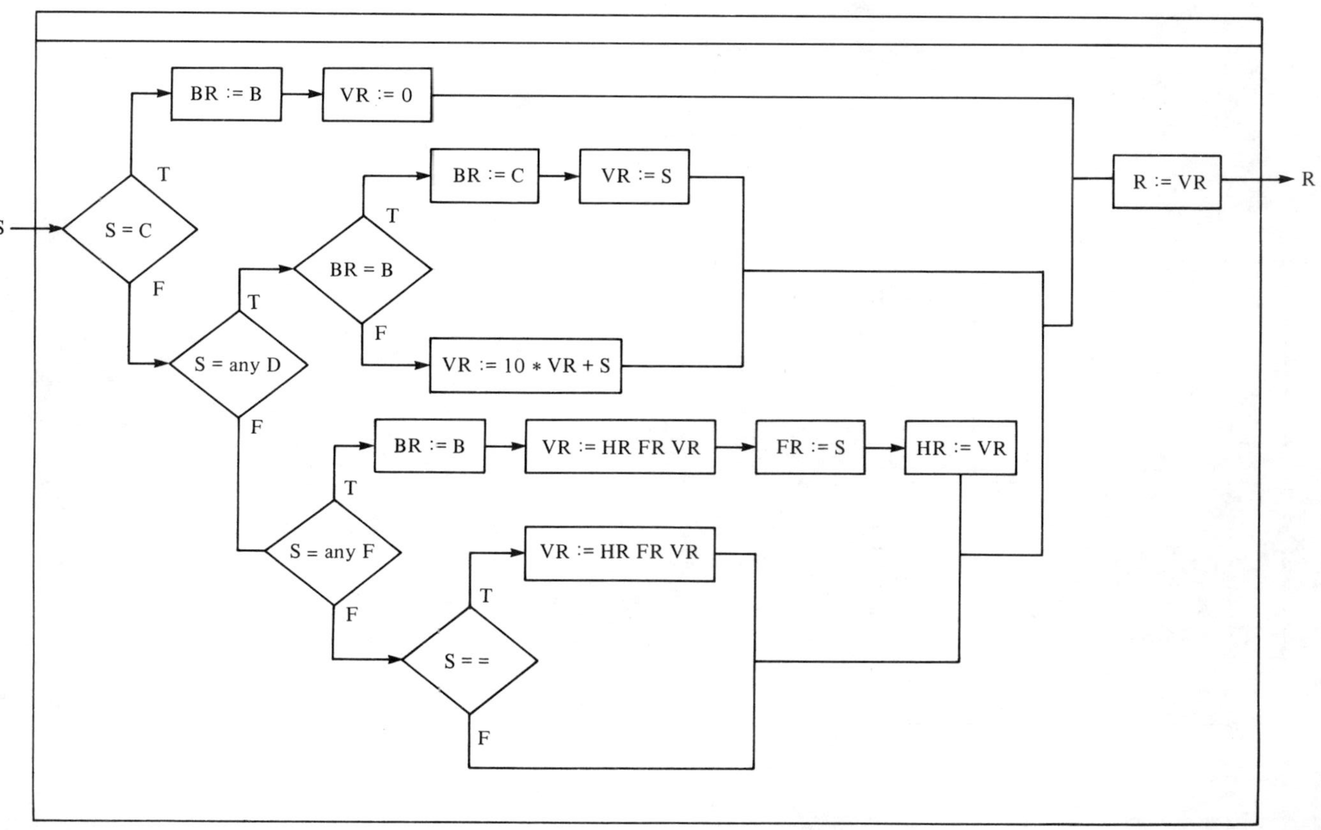

Figure 4.4-6. The Fully Expanded Hand Calculator Clear Box (*continued*).

```
define CB Hand calculator
  stimulus
    S:key
  response
    R:number
  state
    SR: (B,C)
    FR: (+,−,*,/)
    VR: number
    HR: number
  machine
    data
      (none)
    proc [Compute state registers as in Table]
      if S = C
      then [Compute line 1 of Table]
        BR := B;
        VR := 0
      else [Compute lines 2-5 of Table]
        if S = any D
        then [Compute lines 2-3 of Table]
          if BR = B
          then [Compute line 2 of Table]
            BR := C;
            VR := S
          else [Compute line 3 of Table]
            VR := 10 * VR + S
          fi
        else [Compute lines 4-5 of Table]
          if S = any F
          then [Compute line 4 of Table]
            BR := B;
            VR := HR FR VR;
            FR := S;
            HR := VR
          else [Compute line 5 of Table]
            if S = =
            then [Compute HR (function) VR]
              VR := HR FR VR
            fi
          fi
        fi
      fi;
      R := VR
    corp
```

Figure 4.4-6. The Fully Expanded Hand Calculator Clear Box.

and the procedure can access data in the parameter list as well as the state and working data. A call in the form

run <procedure name> (<parameter list>)

must define a new parameter list of data known to the procedure statement. This new parameter list must agree in length and data types with the parameter list in the procedure definition. For example, given a procedure

```
proc Add(x, y, z:number)
  x := y + z
corp
```

the statement

run Add (R, S, L)

will have the effect of the assignment

R := S + L

while the statement

run Add(L, S, L)

will have the effect of the assignment

L := S + L

In illustration, the final hand calculator clear box expansion of Figure 4.4-6 can be organized into segment structured form for better understandability, as shown in Figure 4.4-7. Two nested control structures have been removed and converted to named procedures invoked by outer syntax **run** statements. In this case no additional data is defined for the procedures so the **data** keyword is omitted. The result is a hierarchy of smaller procedures, called **segments,** each of which can be reviewed independently within its structural context. In a large system, the segment structuring process can be carried out repeatedly, to ensure that all segments are small and easy to read.

Summary: Clear box designs must identify and perform transitions specified by state machines. Similarities in transitions may permit shared subprocedures in clear box design. Successive clear box sequence and alternation expansions based on state machine transitions define a clear box design of a hand calculator. The BDL **run** statement permits segmentation of clear boxes for better readability.

```
define CB Hand calculator
   stimulus
      S:key
   response
      R:number
   state
      SR: (B, C)
      FR: (+,−,*,/)
      VR: number
      HR: number
   machine
      data
         (none)
      proc [Compute state registers as in Table]
         if S = C
         then [Compute line 1 of Table]
            BR := B;
            VR :=0
         else [Compute lines 2-5 of Table]
            if S = any D
            then [Compute lines 2-3 of Table]
               run Lines 2-3
            else [Compute lines 4-5 of Table]
               run Lines 4-5
            fi
         fi;
         R := VR
      corp
```

```
proc Lines 2-3
   if BR = B
   then [Compute line 2 of Table]
      BR := C;
      VR := S
   else [Compute line 3 of Table]
      VR := 10 * VR + S
   fi
corp
```

```
proc Lines 4-5
   if S = any F
   then [Compute line 4 of Table]
      BR := B;
      VR := HR FR VR;
      FR := S;
      HR := VR
   else [Compute line 5 of Table]
      if S = =
      then [Compute HR (function) VR]
         VR := HR FR VR
      fi
   fi
corp
```

Figure 4.4-7. The Hand Calculator Clear Box in Segment Structured Form.

EXERCISES

1. Derive statement functions for each control structure in the following clear box BDL segments:

(a)

```
do
   x := x + y + z;
   y := x − y − z;
   z := x − y − z;
   x := x − y − z
od
```

(b)

```
do
   x := 0;
   y := 0;
   k := 1;
   while k < n
   do
      x := x − 1;
      y := y + k;
      k := k + 1
   od
od
```

(c)

```
if s < t
then
   if u < v
   then
      x := t * v
   else
      x := t * u
   fi
else
   if u < v
   then
      x := s * v
   else
      x := s * u
   fi
fi
```

(d)

```
do
   x:= 0;
   y := n;
   while y ≥ d
   do
      x := x + 1;
      y := y − d
   od
od
```

2. Enumerate the processing paths of the clear box of Figure 4.2-18 and suggest improvements to the personnel procedures they define.
3. Design a BDL clear box that elaborates on the ''conduct interview'' operation.
4. Determine the state machine and black box behavior of a M1;M2 sequence clear box structure in which

   M1: R1, NS1 := OS1, S1 + OS1
   M2: R2, NS2 := OS2, S2 − OS2
5. Determine the state machine and black box behavior of an alternation clear box structure in which C tests if S is odd and M1, M2 are given as in Exercise 4.
6. Determine the state machine behavior of a iteration clear box when C tests if S > OS and M1 is defined as

   M1: R, NS := OS, S
7. Verify the result of Exercise 6 by use of the Verification Theorem.
8. Rewrite the description of M2 from Section 4.3.4 as a conditional assignment:

```
M2: if S is even
       then
          R, NS := S, OS
       else (S is odd)
          if OS is even
          then
             R, NS := OS, OS + S
          else (OS is odd)
             R, NS := OS + S, 2 * OS + S
          fi
       fi
```

9. Revise the hand calculator clear box design by changing the order in which cases of stimulus S are considered. Is there a better clear box possible?

10. Create a clear box for the state machine of Exercise 6 in Chapter 3, which introduces a clear entry (CE) key.

11. Create a clear box for the state machine of Exercise 7 in Chapter 3, which introduces a decimal point (DP) key and permits arithmetic to no more than 8 significant digits. The clear box must deal with overflow on key entry and arithmetic operations.

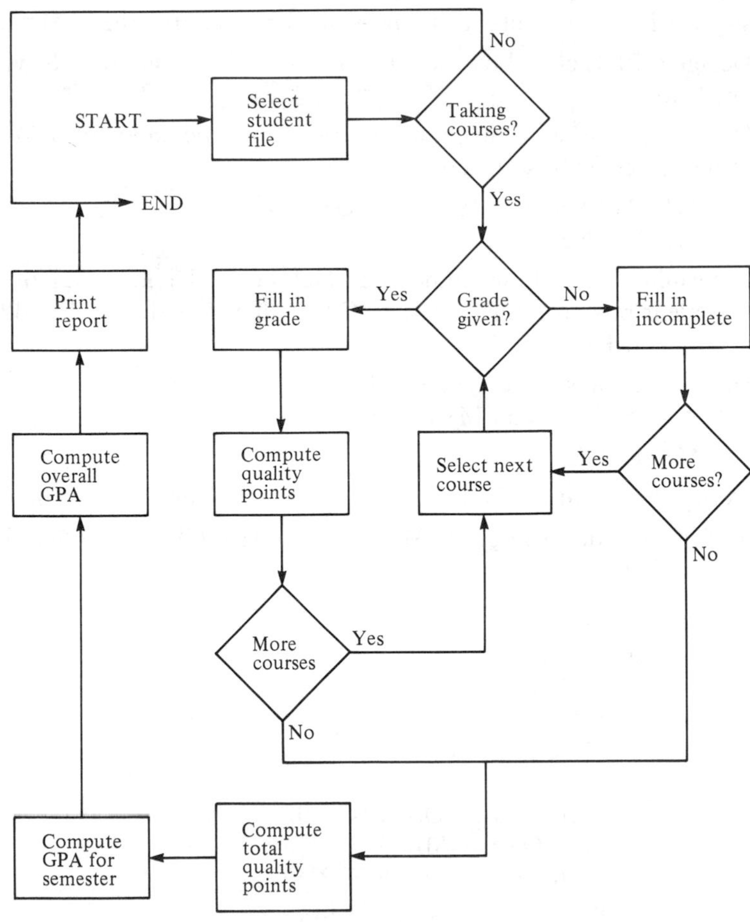

Figure E.4-1.

12. Discuss what is meant by the verification of a clear box design. How do you verify correct sequence, alternation, and iteration design expansions?

13. The flowchart in Figure E.4-1 produces a semester grade report form for a student. Transform this natural procedure into a BDL clear box using the techniques of Section 4.2.

14. Describe in a natural procedure (e.g., flowchart, natural language) your morning routine. Take this natural procedure and describe it in a structured BDL procedure.

Chapter 5 The Box Structures of Information Systems

5.1 THE CONCEPT OF BOX STRUCTURES

Preview: Box structures are hierarchies of black box/state machine/clear box expansions that limit complexity at each level of decomposition. Box structure hierarchies mirror hierarchies in business organizations. Box expansions can be limited in size and complexity as building blocks, but combined into larger and larger box structure hierarchies without limit, to deal with information systems of any size and complexity. The principle of transaction closure guides invention of the top level of the hierarchy. The work products of box structure analysis and design can be recorded in analysis and design libraries.

5.1.1 Box Structure Hierarchies

A box structure is a hierarchy of BB/SM/CB (Black Box/State Machine/Clear Box) structures, in which all black boxes used in each clear box head a BB/SM/CB structure at the next level, as depicted in Figure 5.1-1. That is, any black box in a clear box **use BB** statement will be identified at the next level. We call a BB/SM/CB structure a **box expansion,** depicted as shown in Fig. 5.1-2.

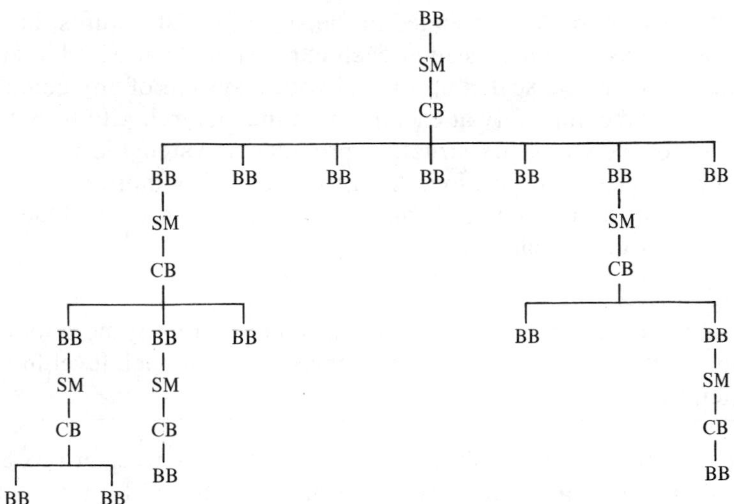

Figure 5.1-1. A Box Structure Hierarchy.

The black box at the top of a hierarchy or subhierarchy identifies the **what** of an information system or subsystem. But that **what** is usually too difficult to describe in one step for people to fully understand or to program for computers. Therefore, a box structure for that black box will be in order, beginning with a state machine, then a clear box design for the state machine. If the description can be completed (to understandability or programmability) with the clear box, using no unknown black boxes, the design is completed. If not, a set of one or more new black boxes will have been identified, and the description will proceed as above for each new black box. In turn, more new black boxes may be identified and described, until, after several levels, no more black boxes are required.

> **Fundamental Principle:** A black box identifies the what of information system or subsystem behavior, its clear box describes a how of that behavior.

Figure 5.1-2. A Box Expansion.

A box structure is open ended in the size of the structures, but it can use clear boxes of limited size at each expansion. That is, a hierarchical box structure can be scaled up to deal with a system of any complexity, yet use limited complexity at each point in the hierarchy. In turn, viewed from the top down, a box structure provides a systematic way to defer details of a system description. At each level, a few more details can be revealed, but the remaining details can be subsumed in the black boxes that remain to be expanded.

> **Fundamental Principle:** A box structure hierarchy permits the deferral of system details within black boxes at each level in the hierarchy.

The progression from black box to state machine to clear box at each expansion step in the hierarchy represents a canonical form for analysis and design. However, in some cases, problems can be solved more directly. For example, an expansion may be most effectively expressed directly as a state machine with no black box given. This case can arise where the stimulus history of a black box is especially complex, and its transitions can be better understood when defined in terms of stimulus and state. Or an expansion may be better defined by a clear box with no black box or state machine given. This case can arise when the transition rules of a state machine are especially complex, and are more easily expressed in procedural terms, possibly referring to black boxes at the next level. Note, however, that a decision to bypass a step in a box structure is reversible, but has the effects:

Black box omitted. State-free, procedure-free description unavailable

State machine omitted. State-defined, procedure-free description unavailable

Clear box omitted. State-defined, procedure-defined description unavailable; no new black boxes introduced

In these cases, there is no rule against thinking hard about the behavior of the box structure in terms of the foregone representations. And, if necessary, the other representations can always be derived or expanded for more detailed study and analysis.

5.1.2 Box Structures in Business Operations

Box structures are common in business operations, and mirror effective organizations in business. The box structure approach to information systems analysis and design makes use of a good deal of wisdom accumulated in successful business organizations.

We have already seen that black box, state machine, and clear box behavior is common in business operations. The correspondence is even more striking with hierarchical box structures and business operations. A major corporation will often be organized in product and service divisions, then divisions organized into major functions of marketing, finance, engineering, manufacturing, personnel, etc., the functions organized into departments, departments into smaller departments, and so on. This organization will be hierarchical, just as a box structure. Information will be stored at all levels, the more detailed information in lower level departments. Information will flow from one department to another, as outputs from the first and as inputs to the second. And information into a department will trigger information out to other departments. That is, each department will exhibit black box behavior in its information processing. In turn, each function, each division, and finally the whole corporation will exhibit black box behavior to its surroundings as it accepts, processes, stores, and produces information.

These box structured business organizations are no accident. They are due to no whims or aberrations of their executives. They have survived the natural selection of economic and business competition. There are no laws that require such organization. Corporations are free to organize internally in any way they choose. Small businesses may organize on some other basis than hierarchy, for example, on a communal or democratic basis, but no small business ever grows to even a medium sized, let alone large, business so organized.

Neither is it an accident that successful information systems are box structured as well. Box structures permit intellectual control in both building information systems and building business organizations. The analysis and design requirements are similar in both cases, identifying inputs and outputs, how information should be stored and processed. So a box structured approach to information systems analysis and design automatically draws on a good deal of accumulated wisdom of the business world.

5.1.3 The Top Level Black Box and Transaction Closure

It is one thing to describe a hierarchical box structure of an information system. It is quite another thing to develop it from scratch, to accumulate and assimilate the necessary information from the business organization, possibly by many people over many weeks or months, then to put it all together correctly. The top level black box is not the starting place of such an effort, although it is the principal objective at the start. The starting place is in the organization, to identify first hypotheses in the intellectual climb to this top level black box.

The search for transaction closure should guide this effort. What are the transitions and transactions required? Is that all, or are there secondary transitions and transactions required to make the primary ones possible? Is the top level state machine easier to describe to begin with than is the top level black box? Are there simplifying aspects from using the data in the description? Is the top level clear box easier to describe? There is no uniquely best starting place; instead, the search criteria are better focused on the objective of getting to the appropriate top level black box with due process, rather than leaping to a faulty top level black box prematurely.

A useful beginning of this search for a top level black box begins with the most obvious users of the system to be, but seldom ends there. These most obvious users often interact with the system daily, even minute by minute in entering and accessing data—for example a clerk in an airline reservations system. But usually, the data they use are provided in part by other users that enter and access data less frequently—for example those entering flight availability information. And other users even more distant from the obvious users enter and access data even less frequently—for example users who add route schedule information. All the while, an entirely different group, the operators of the system, is entering and accessing system control data that affects the users in terms of more or less access to the system because of limited capacity or availability.

The top level black box must accommodate all these users and operators, not just the most obvious ones. A cross check can be made between the top level black box and its top level state machine. Every item of data in the top level state must have been loaded with the original system or acquired by previous black box transactions. Are there any items not so loaded or acquired? It is easy, in concentrating on one set of transactions to assume the existence of data to carry them out. A close comprehensive scrutiny of these needed data items can discover such unwarranted assumptions early.

Another aspect of transaction closure arises in system integrity. The categories of integrity should be checked and rechecked, even in this search for the top level black box, for example:

Security. Need users be authorized; if so how and what transactions are needed to authorize them initially and allow them access subsequently?

Operability. What transactions permit system operation and dealing with unforeseen events?

Auditability. Are audits to be required; if so what transaction trails are needed and how are they to be accessed by audit transactions?

Reliability. What provisions are required for system checkpoints and recovery from unforeseen hardware or software errors and what are the transactions needed?

Capability. Are archives and restorations necessary for dealing with data in amounts not economical to keep on line?

In every case, the answers to these questions are to be found in the business organization in assessing questions of integrity and their impact on business performance.

5.1.4 Box Structure Analysis and Design

In the information system development process it is important to identify and distinguish between analysis and design. **Analysis** is a discovery process. The gathering of information and the forming of that information into descriptive box structures is a major part of analysis. The derivations performed in a box structure from clear box to state machine to black box are also discovery processes. **Design,** on the other hand, is a creative process. Given the information discovered during analysis, a box structure hierarchy for the new information system is created. Within a box structure, expansion from black box to state machine to clear box, then their rederivations, provide a rigorous method of verifying the correctness of the design.

The box structure diagrams of black boxes, state machines, and clear boxes provide general forms for generating and recording the results of information systems analysis. On the other hand, BDL provides a more formal and precise form of recording for information systems design. These two forms reflect the differences between analysis and design in information systems.

Box structure diagrams provide flexible, easily understood, graphical ways to discover and discuss ideas about information systems with managers and users. These diagrams can be annotated with terms and phrases of the business to facilitate information gathering and to ensure better accuracy in understanding ongoing operations and processes.

In contrast to the outward directed activities of analysis, information systems design is based on the results of analysis but is inward directed, dealing with inner consistency and tradeoffs in order to make good design choices. BDL provides precision and completeness, but at the price of foregoing easy and casual treatments.

The result of an analysis phase is an **analysis library,** a set of annotated diagrams and supporting documentation that covers the area of study.

This library of diagrams is not yet a complete and precise box structure. The diagrams are loosely compiled, and possibly overlapping, and with possible gaps. Almost all of the information in the analysis library will be useful in the design phase.

The result of a design phase is a **design library,** a set of BDL designs and supporting documentation that describes a complete and precise box structure. The analysis library is the raw material for the design library. The discovery and discussions that went into the analysis library are necessary ingredients for the construction of the design library.

Summary: A box structure hierarchy localizes and limits complexity by deferring details within black boxes at each level. Box structures permit intellectual control in building information systems and building business organizations. Transaction closure assures that top level black boxes will accommodate all possible users and uses. The analysis library contains raw material for the design library.

5.2 ANALYSIS OF BOX STRUCTURES

Preview: Transaction analysis, state analysis, and procedure analysis provide a basis for describing existing or intended information systems.

5.2.1 Deriving Box Structures from Business Operations

Any information system or part, real or intended, can be described in box structure form. The data interfaces between the part and its surroundings are described by black box stimuli and responses. The data stored in the part are given by the state of a state machine. The data processing is given by the machines of a clear box. The same data is often created and used in different ways for different purposes. The box structure approach places all these operations with the same data in a common box structure.

In illustration, consider an analysis of the charge account system of a department store. In this section, a preliminary box structure analysis is carried out to illustrate the need for a more thorough analysis to follow.

In department store operations, sales clerks may enter customer charges for merchandise purchased. Customers receive bills at the end of

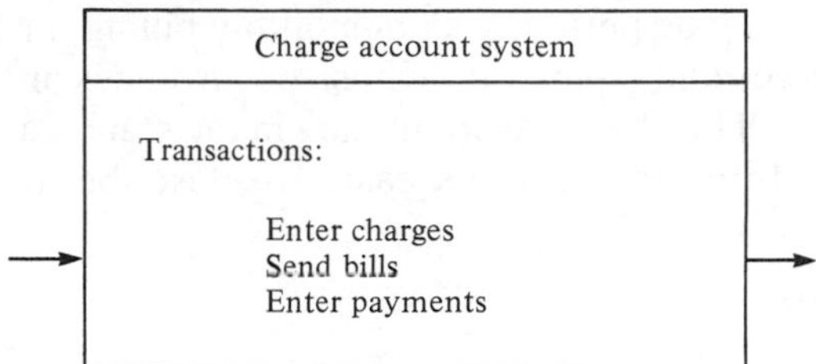

Figure 5.2-1. Charge Account System Black Box.

the month and return payments to their accounts. For clerks and customers alike the charge account system appears to be a black box. Clerks merely enter charges and get confirmations (that the customer has a charge account and has not exceeded a credit limit). Customers receive bills as black box outputs, and return payments as black box inputs. Figure 5.2-1 gives a box structure diagram in the form of a Charge Account System black box as a start.

The three types of transactions so identified can be described at some length. However, in order to describe the effects of these transactions, the data of the Charge Account System will be very useful. The charge account data of the department store is altered every time a customer charges an item. It is also altered every time a customer makes a payment. This charge account data is used to create customer bills, and to make credit limit checks. All of these ways of creating, altering, or using this charge account data can be reorganized into a state machine, as shown in Figure 5.2-2. Part of each input to this box structure will describe the way the data is to be created, altered, or used in this transaction.

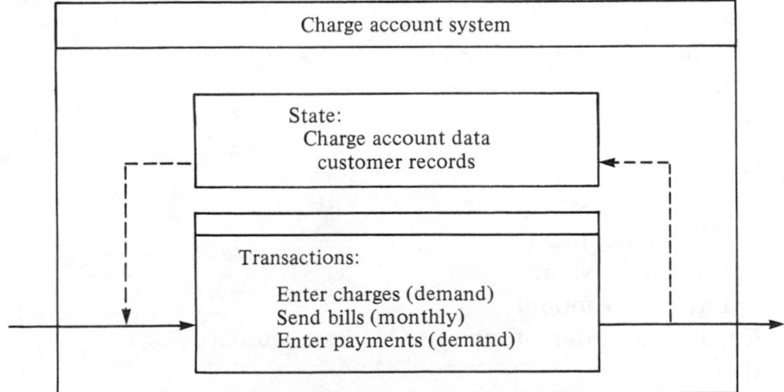

Figure 5.2-2. Charge Account System State Machine.

Transactions may be periodic, as in monthly billing, or on demand, as for entering charges and payments, whose occurrences are unpredictable within the system. The charge account data in the state machine of Figure 5.2-2 makes the three transactions easier to describe as more detailed procedures:

```
Enter Charges:
    Input Expected:
        "Enter Charge"
        Customer Name
        Charge Amount
    if Customer Name found in Customer Records
    then
        if Customer Balance + Charge Amount < Credit Limit
        then
            Increase Customer Balance by Charge Amount
            Confirm Charges to Clerk
        else
            Return Message "Credit Limit Exceeded"
        fi
    else
        Return Message "Customer Unknown"
    fi

Send Bills:
    Input Expected:
        "Send Bills"
    while more Customer Records exist
    do
        if Customer Balance > 0
        then
            Send Bill to Customer
        fi
    od

Enter Payment:
    Input Expected:
        "Enter Payment"
        Customer Name
        Payment Amount
    if Customer Name found in Customer Records
    then
        Decrease Customer Balance by Payment Amount
    else
        Return Message "Customer Unknown"
    fi
```

These descriptions, informal as they are, illustrate the necessity for the name of each type of transaction to be identified as part of each input, and to follow that part with input data.

Part of the value of such a description is its understandability by managers and users. The credit manager will immediately notice a deficiency in the above description. There is no provision to raise or lower credit limits, customer by customer, so another type of transaction will be called for. The financial manager may want a special report in order to anticipate future cash flows expected from current charges, and so on.

The foregoing example of a charge account analysis illustrates the kind of information that is needed, but several deficiencies are already visible. How can such deficiencies be avoided in an actual analysis? There are no foolproof methods. It takes good judgement, common sense, good listening, and an open mind. But a systematic approach can be very useful. Most systematic approaches to problems involve mastering a good deal of detailed, step by step procedures. And the systematic approach presented here is no exception. However, paradoxically, this systematic approach is aimed at preventing you from getting bogged down in step by step detail before you should.

The problem of systems analysis is how to discover the trees and leaves of a forest without losing sight of the forest itself. Eventually you will need to describe the trees and leaves, and for that you will need precise detailed descriptions. But unless you maintain your view of the forest, the trees and leaves don't mean too much.

Fundamental Principle: Systems analysis is a discovery process. It requires investigation, research, and insight into the system to be developed. The box structure methodology provides a framework for analysis. The results of systems analysis are diagrams and information that demonstrate a thorough understanding of the proposed system. These results are stored in an analysis library.

The next three sections deal with systems analysis more thoroughly in the transactions, states, and procedures of box structures. The addition of detail to an evolving box structure must be under good control and discipline as much as any other use of methodology. On occasions, you may need to go to considerable detail in one aspect of analysis before another. But the framework for dealing with detail must be there before going into it.

5.2.2 Transaction Analysis

The key to a disciplined approach to systems analysis is the use of **context.** Context plays a critical role in what we see and communicate with each other. Context is what we assume is common knowledge for a conversation. Context and precision are incompatible. The broader the context the lower the precision; the higher the precision the narrower the context. In information systems it is easy to be precise about the wrong things, and easy to be vague about the right things. The goal is to use context to be precise about the right things.

Box structure diagrams can be used to control context, by gradually narrowing context through increasingly precise diagrams. A **transaction analysis** provides a systematic way to narrow context and increase precision.

The objective of a transaction analysis is to identify the set of transaction types necessary for a black box. Simple black boxes may have only one transaction type, for example, a sales forecast black box does nothing but accept sales and issue forecasts. However, the charge account system black box has (at least) the three types of transactions dealing with billing, charging, and paying. In the previous example these three types of transactions were identified by direct intuition about a charge account system in a department store. The act of clerks entering charges comes immediately to mind, then the store sending out bills next, and finally customers sending in payments. But is that all? No. And how to find the rest?

The method of transaction analysis is to identify the information needs of the business, and the transactions that satisfy those needs, then other transactions needed to support the original transactions, and so on until no new transactions can be identified. Of the three transactions identified for the charge account system, one is primary, the other two are secondary from the viewpoint of the business. The primary transaction type is billing, to satisfy the information need of what to bill customers. But in order to bill customers, previous transactions of charging and paying are required. In other words, if billing was the only transaction type of the charge account black box, there wouldn't be any bills to send!

Another way to look at these three transactions is chronological, as done intuitively before. Charges are first, bills are second, payments third. That's a good cross check, to examine chronological sequences for completeness. But a hierarchy of information needs helps to organize a systematic completeness search for transaction types.

> **Fundamental Principle:** A transaction analysis begins with information needs of the business and the primary transaction types to satisfy those needs, then backtracks through additional transaction types needed to support these primary transactions, directly or indirectly.

Charge Account System Revisited

We begin again with the charge account system of a department store. In the broadest context, the black box merely labeled Charge Account System, as diagrammed in Figure 5.2-3, is known to clerks and customers. It is just the charge account system they see (and imagine) in action. This beginning is one name and the rest context! At least nothing has been left out so far!

Next we identify the information needs of the business and the transaction that satisfies it—Billing—in Figure 5.2-4. As already noted, Billing depends on the previous presence of two types of transactions—Charge and Payment—as shown in Figure 5.2-5.

Let's apply the analyses to Charge and Payment transactions. Are any prior transactions required? One has already been mentioned—Set Credit Limit. Are there any more? Both Charge and Payment require a customer name, which has to be acquired by the system previously. Let's call that transaction Open Account. This leads to the new version of Figure 5.2-6.

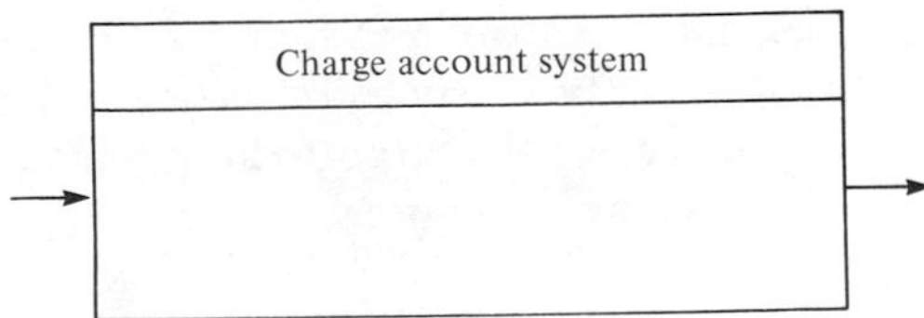

Figure 5.2-3. Charge Account System Black Box.

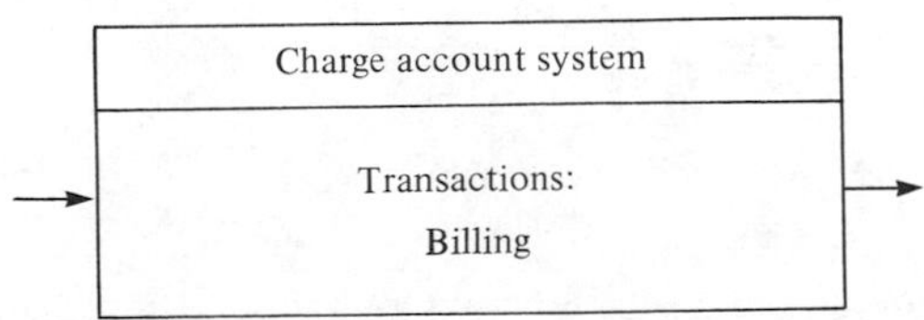

Figure 5.2-4. Charge Account System Black Box.

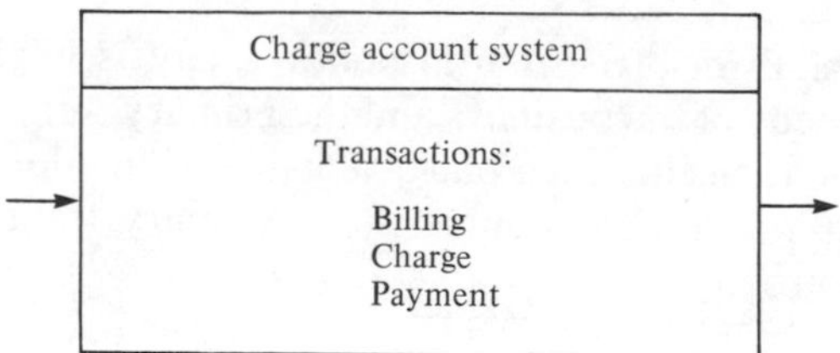

Figure 5.2-5. Charge Account System Black Box.

In turn, are prior transactions required for Open Account or Set Credit Limit? Open Account may be sufficient in itself, but the credit manager would probably require three more transactions, namely Credit Application, Credit Check, and Salary Check, which each provide the Charge Account black box with information for the Credit Limit transaction. The difference between Open Account and Credit Limit in generating new transactions is in the nature of information that might be required for the transaction. The store is willing to open an account for almost anyone with almost no information, but willing to grant credit only after prudent checks.

The credit application is required to make any checks, and a Credit Check (willingness to pay) and Salary Check (ability to pay) seem prudent. The result of these additions to transactions is given in Figure 5.2-7.

The Charge Account System may have a hierarchical structure that would simplify or better organize its description. The Credit Limit transaction may be an opportunity to define a new black box in such a hierarchical structure. The Charge Account System would still have a Credit Limit transaction, but its clear box may call on a Credit Limit black box to carry it out in a hierarchical box structure. In this case, a new black box of transactions can be defined, as in Figure 5.2-8.

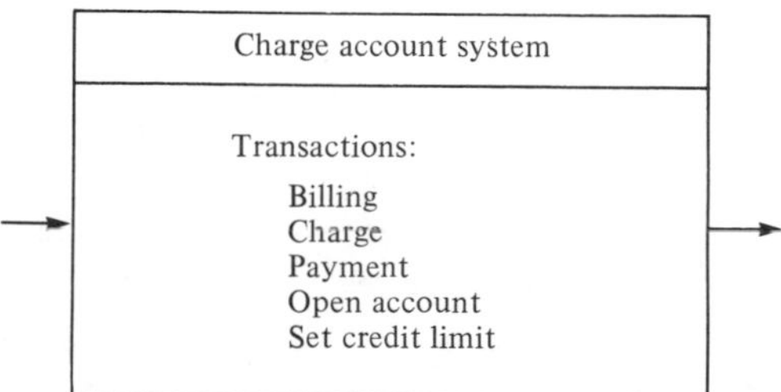

Figure 5.2-6. Charge Account System Black Box.

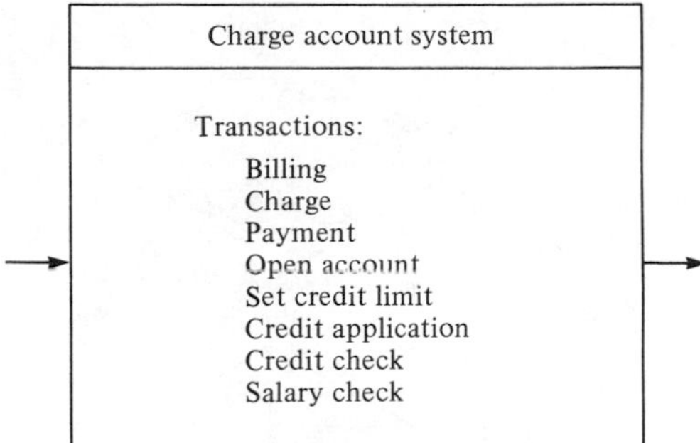

Figure 5.2-7. Charge Account System Black Box.

The Credit Limit black box supports the Charge Account System black box and provides a new starting point for its own transaction analysis. Each of its transactions can be examined in turn for necessary previous transactions as before. Eventually one of its transactions may be a candidate for a separate black box, and so on.

In summary, a transaction analysis begins with transactions that satisfy primary information needs of the business, then identifies necessary previous transactions, direct and indirect, until no new transactions can be identified. The final transactions identified obtain all information required of them from inputs to the black box. These transactions can be diagrammed in a **dependency tree,** beginning with the primary transaction. In the case of the Charge Account System black box, the dependency tree is as diagrammed in Figure 5.2-9. The dependency tree is suggestive of possible box structure and places to establish new black box transaction analyses.

> **Fundamental Principle:** Transaction analysis identifies transaction types of a black box, which can be diagrammed in a dependency tree and possibly organized into supporting black boxes.

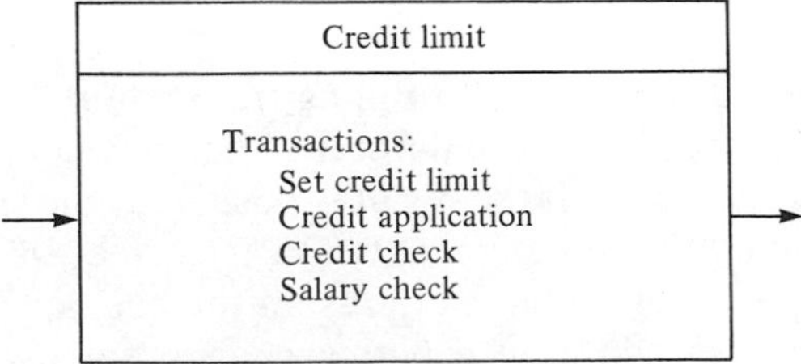

Figure 5.2-8. Credit Limit Black Box.

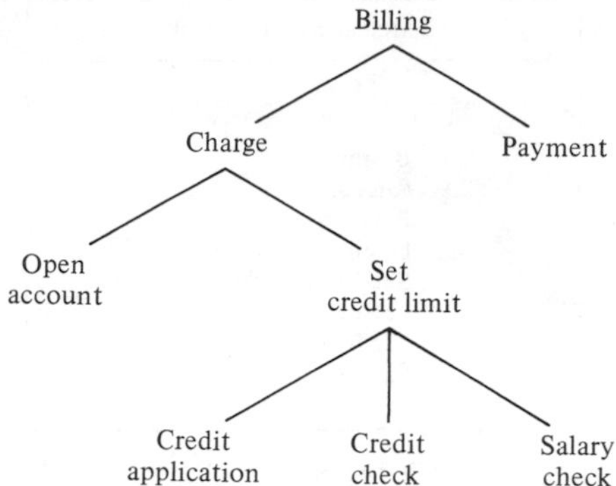

Figure 5.2-9. Charge Account Transaction Dependency Tree.

Once the transactions are identified the **inputs** and **outputs** of each transaction should be identified. The inputs to a transaction consist of stimuli from outside of the system and all required state information defined at a higher level in the system box structure hierarchy. Transaction outputs include responses to the external environment and updated state information internal to the system. The analysis of appropriate input and output formats is an important and necessary task. A user of the transaction should be prompted for essential input in a friendly manner condusive to efficient use of the system. Likewise, output interfaces and reports should promote effective system use.

In the Charge Account System, the inputs and outputs needed for each transaction are recorded in Table 5.2-1. Note that all inputs include a specific request to perform that transaction. This can be seen as a clerk entering the words 'ENTER CHARGE' on a terminal screen or simply pressing a single function key for entering charges. Such transaction request inputs are normally found at high levels in a system that contains user interfaces.

5.2.3 State Analysis

The black box description of a Billing transaction for a customer can be given as a computation of the difference between all previous charges and all previous payments. If the result is positive, send the amount to the customer as a bill. While the computation is correct, it is awkward and unnecessary. Instead, each bill can present the balance to the customer, which represents this difference between all previous charges and all previous payments, making any further reference to, or thought about, the latter unnecessary.

The customer balance is state data of a state machine that simplifies calculations for both the customer and the store. Each month the new balance is the old balance plus all charges for the month less all payments for the month.

The objective of **state analysis** is to identify state data that makes transactions easier to describe. As already discussed in Chapter 3, the value of state data is to reduce the dependence of transactions on entire black box stimulus or input histories. The customer balance reduces the billing transaction dependence on charges and payments to only those that occurred in the past month.

The transaction analysis provides a framework for the state analysis. For each transaction what data should be available from previous transactions? In illustration, the data required for Billing can be given as:

Customer Name:	The name of the customer account.
Balance:	The balance of the previous month.
Current Charges:	All charge transactions made in the current month.
Current Payments:	All payment transactions made in the current month.

These are the data items produced by previous transactions that Billing requires in its old state. They will be transformed in the new state as follows:

Balance:	Set to the old state Balance plus Current Charges minus Current Payments.
Current Charges:	Set to empty to begin a new month.
Current Payments:	Set to empty to begin a new month.

The output of the Billing transaction is the bills for all customer accounts that have had charge or payment activity during the month. The bills contain the customer name and the old and new balances, together with the month's charges and payments. On completion of the Billing transaction, the new state data will become the old state data for the next month's billing.

The input to the Charge transaction is a Name and an Amount to be charged. What state data is required to describe this transaction? With a little thought, it can be seen that four items are required, namely,

Customer Name:	The name of the customer account.
Credit Limit:	The maximum credit to be extended to the customer.
Current Charges:	All charge transactions made in the current month.
Balance:	The balance of the previous month.

If Name matches a Customer Name, the transaction will leave Customer Name, Credit Limit, and Balance unchanged in the new state, and Current Charges will be increased by Amount if Credit Limit is not exceeded. In this case, the output will be an approval if the limit is not exceeded, a rejection if it is. If Name is not recognized, the charge will be rejected.

The Payment transaction accepts Name and Amount as input and requires two state items from previous transactions:

Customer Name: The name of the customer account.
Current Payments: All payment transactions made in the current month.

If Name matches a Customer Name, the transaction leaves Customer Name unchanged in the new state and increases Current Payments by Amount. In this case, the transaction output is a confirmation of payment acceptance. If Name does not match a known name, the output is a rejection of the payment.

A similar state analysis can be carried out for the remaining transactions. The full analysis results are summarized in Table 5.2-1. Note that the Table reflects the transaction dependencies previously described. A transaction cannot be carried out if the state items it requires have not shown up previously in the input history of the Charge Account System black box. Thus, the Open Account transaction must precede Charge, Payment, and Billing transactions. Similarly, Charge depends on Set Credit which in turn depends on Credit Application, Credit Check, and Salary Check. These dependencies provide a natural structure for ex-

Table 5.2-1

Charge Account State Machine Transitions

Transaction	Input	Old State	New State	Output
Billing	Billing Request	Customer Name Balance Current Charges Current Payments	(same) New Balance No Current Charges No Current Payments	Bills
Charge	Charge Request Name Amount	Customer Name Credit Limit Current Charges	(same) (same) New Current Charges	Approval (or not)

Table 5.2-1 (*Continued*)

Transaction	Input	Old State	New State	Output
Payment	Payment Request Name Amount	Customer Name Current Payments	(same) New Current Payments	Confirmation (or not)
Open Account	Open Account Request Name		Customer Name No Balance No Current Charges No Current Payments No Credit Limit	Confirmation (or not)
Set Credit Limit	Credit Limit Request Name	Customer Name Application Credit Results Salary Results	(same) (same) (same) (same) Credit Limit	Credit Limit Confirmation (or not)
Credit Application	Credit App. Request Name Application Data		Application	Confirmation (or not)
Credit Check	Credit Check Request Name Credit Data	Application	(same) Credit Results	Confirmation (or not)
Salary Check	Salary Check Request Name Salary Data	Application	(same) Salary Results	Confirmation (or not)

plaining proper system operation to users, in instruction courses and user guides. They are also an important topic in design, in determining how much checking for proper transaction sequencing must be done by the system, and at what cost in efficiency.

5.2.4 Procedure Analysis

The state definitions of the previous section can be used to create procedural explanations of the transactions of the Charge Account System as shown below. These explanations make use of BDL structures, but are not in themselves clear boxes. They are intended to be used in a process of analysis and review with prospective owners and users of the Charge Account System, to arrive at understanding and agreement on system functions. Such procedures specify the function of the ultimate system design, but are not intended to substitute for, or prescribe the structure of, that design.

```
Billing:
   while more Customer Records exist
   do
      Set Balance to Balance + sum of Current Charges − sum of
        Current Payments
      Send Bill
      Set Current Charges to empty
      Set Current Payments to empty
   od

Charge:
   if Name is a Customer Name
   then
      if Balance + Current Charges + Amount does not exceed
           Credit Limit
      then
         Set Current Charges to Current Charges + Amount
         Return message "Charge approved"
      else
         Return message "Credit limit exceeded"
      fi
   else
      Return message "Name unknown"
   fi
```

```
Payment:
   if Name is a Customer Name
   then
      Set Current Payments to Current Payments + Amount
      Return message "Payment confirmed"
   else
      Return message "Name unknown"
   fi

Open Account:
   if Name is a Customer Name
   then
      Return message "Existing Account Open"
   else
      do
         Create Customer Name from Name
         Set Balance to zero
         Set Current Charges to empty
         Set Current Payments to empty
         Set Credit Limit to zero
         Return message "Account for Name confirmed"
      od
   fi

Set Credit Limit:
   if Name is a Customer Name
   then
      if Application available and Credit Results and Salary Results
            acceptable
      then
         Set Credit Limit to 0.05 * Salary
      else
         Return message "Credit refused"
      fi
   else
      Return message "Name unknown"
   fi

Credit Application:
   do
      Create Application for Name using Application Data
      Return message "Application confirmed"
   od
```

```
Credit Check:
   if Application available for Name
   then
      Create Credit Results from Credit Data
      Return message "Credit Results confirmed"
   else
      Return message "No Application for Name"
   fi

Salary Check:
   if Application available for Name
   then
      Create Salary Results from Salary Data
      Return message "Salary Results confirmed"
   else
      Return message "No Application for Name"
   fi
```

Review of this preliminary level of transaction, state, and procedure analysis with owners and users will likely result in still more transactions with corresponding additions to state and procedure definitions, for another iteration of analysis and review.

Summary: Annotated box structure diagrams facilitate manager and user discussion and discovery in information systems analysis. Box structure diagrams provide a basis for the effective control of the context of analysis by gradually narrowing context with increasingly precise diagrams. Transaction analysis, state analysis, and procedure analysis should proceed together in providing a framework for dealing with increasing detail without losing perspective.

5.3 DESIGN OF BOX STRUCTURES

Preview: The results of information systems analysis, recorded in the **analysis library,** are used as a basis for information systems design, which is recorded in a **design library.** Transaction design, state design, and procedure design provide a systematic basis for developing the box structure of an intended information system.

5.3.1 Designing Box Structures for Business Operations

The principal objective of system design is to create a system that satisfies the information needs discovered in the analysis of business operations. But other objectives must be satisfied as well. It does no good to create a system that satisfies information needs if it is delivered late, costs too much, is too difficult to use, or is error prone and unreliable. So the right design process is design to cost and schedule and quality. Such a process requires good intellectual control at all stages of design, to distinguish essential system functions from frivolous features, and to balance the work remaining with the schedule and budget remaining. The principal objective of the box structure design process is to maintain intellectual control in meeting the objective of an information system. Analysis is the art of the possible, but design must be the art of the practical.

Context is used to advantage in different ways in analysis and design. In bottom-up analysis, the natural context of business operations is used to simplify explanations, to promote understandability and broader participation by prospective system owners and users. In top-down design, precise context of box structures at each level defines the environment and function of box structures at the next level. A black box is the context for a state machine description, which becomes the context for a clear box expansion.

System design culminates in a top-down recording process. The box structure methodology enforces this process to permit good intellectual control at each stage of design. But the box structure methodology does not guarantee intellectual control. For example, it is foolhardy to attempt a top-down design process without knowing where you are going. The analysis which precedes design is intended to help you discover where the design must go. With this understanding, the design task is to organize the results of analysis into a coherent box structure.

Even with comprehensive analysis beforehand, design is still an iterative process. Design refinements may suggest simplification at higher levels, leading to redesign from the top down to take advantage of the new insight. Unanticipated complications or simplifications encountered in design can affect analysis results as well. Slight changes in user needs, often of little consequence to business operations, may result in significant reductions in design complexity. Conversely, more effective user function may be possible at little or no cost as a side effect of increased design simplicity.

5.3.2 Transaction Design

The foregoing analysis provides a basis for a Charge Account System design. The transactions identified can be designed one by one as required by the analysis. The input to the Charge Account System black box must identify the transaction type required for each transaction. This identification may be done automatically by the equipment (only charges from cash registers, etc.), but within the information system the external origin of inputs may be lost, so the transaction type must be identified, whether added automatically or not. In illustration, we suppose transaction inputs are self-labeled as to transaction types (so the clear box will contain a case statement which handles the various transaction types as cases). The BDL for the Charge Account System black box is given in Figure 5.3-1.

The input of this black box includes data of the transaction type, called Trans, and a generic type, called In Data. The key word **generic** denotes data of a type yet to be specified. In this case the type specification depends on the value of Trans, namely, the transaction type of the input, as given in Table 5.2-1.

The designs of input and output will make use of the information structures discovered during analysis. User languages for input and report designs for output can be designed using syntax structures. Data structures provide the design for input and output data. The system design may also include facilities for sophisticated user input/output interfaces such as natural language processing or even voice communication.

5.3.3 State Design

The previous analysis has already identified the need for state data in summarizing input history into such items as Balance, Credit Limit, etc., for ready access. Therefore, a design for the Charge Account System state machine can proceed as shown in Figure 5.3-2.

```
define BB Charge Account System
  input
    Trans:   (Billing, Charge, Payment, Open Account, Set Credit Limit)
    In Data: generic
  output
    Out Data: generic
  transaction
    As given by analysis
```

Figure 5.3-1. Charge Account System Black Box.

```
define SM Charge Account System
   input
      Trans:  (Billing, Charge, Payment, Open Account, Set Credit Limit)
      In Data: generic
   output
      Out Data: generic
   state
      customer file:
         file of record
            name:            character
            address:         character
            balance:         $ value
            charge file:     file of $ value
            payment file:    file of $ value
            limit:           $ value
         end record
      ...
   machine
      As given by analysis
```

Figure 5.3-2. Charge Account System State Machine.

5.3.4 Procedure Design

The foregoing analysis has already identified the procedures required for the various transaction types. As noted, analysis deals with the art of the possible, design with the art of the practical. In the Charge transaction, it is possible to compare the balance plus all current charges plus the present charge with the credit limit, even charges made ten minutes ago. However, this will require that customer files be available for update and access continuously in time. It may happen that the computer equipment envisioned for the system has no provision for continuous update of customer records, but provides for updating them at the end of the day in a batch execution. In this case, the design may require that charges be accumulated during the day, and added to customer records at night. A way to do this is to define a charge transaction file in the state, and to add a new transaction, called a merge transaction, to the black box. In turn, this forces a return to the analysis phase, analyzing the effect of this design with the store managers. Is it satisfactory to allow customers to charge all day against the previous day's running balance or not? It is a business question on tradeoffs between the information system and the needs of the business.

Let's suppose that the credit manager regards such a design unsatisfactory. Can anything else be done? The charges of the day can be retained on line and each new charge transaction can search the day's charges for that customer, in order to get a true credit check. It will take more processing for each charge transaction than originally planned, but it allows customer records to be updated only once a day. It is a tradeoff made between the business and the information system that is acceptable to all parties.

There are more ramifications for this new design decision, however. The payments of the day should be retained on line, as well, because the sales manager points out that the store could lose sales and even more good will if payments are not current in credit checks. Furthermore, new customer accounts will be accepted each day, but need to be added to the customer file at night. This causes the sales manager a small problem. Customers will not be able to make charges until the day following their application. But the sales manager decides that it is a tolerable solution. These analysis/design decisions are reflected in the clear box given next. Note that a Day Charge File, Day Payment File, and Day New Customer File have been added to the state, and a Merge Transactions procedure added to the black box.

The clear box, as already noted, will contain a case statement to handle each transaction type separately, and is shown in Figure 5.3-3.

The procedures for these cases are given in Figures 5.3-4 to 5.3-9. Note that Day Charge, Day Payment, and Day New Customer files are required in the state for the Charge, Payment, and Open Account procedure designs, respectively, to permit updating the Customer File just once a day. The **run BB** Credit statement in the Set Credit Limit procedure design invokes a nested box expansion which must have carried out the Credit Application, Credit Check, and Salary Check transactions, in order to have available state information required by this invoking procedure. The & notation means concatenation of data.

Is this design adequate for business needs? The design process can often reveal gaps in the analysis process that precedes it, and lead to further analysis and design activities. It is easy to see that analysis and design of a real charge account system could involve many additional transactions, with corresponding state data and procedures.

Summary: A top-down box structure development, based on a well organized analysis library, facilitates reviews and tradeoff studies for consistency, completeness, and practicality in information systems design. BDL box structures provide effective control of the context of design, by creating a precise context at each level for the next level of design. Transaction, state, and procedure design should proceed together in providing a framework for increasing detail without losing intellectual control of an evolving design.

```
define CB Charge Account System
  input
    Trans:  (Billing, Charge, Payment, Open Account, Set Credit Limit)
    In Data: generic
  output
    Out Data: generic
  state
    Customer File
    Day Charge File
    Day Payment File
    Day New Customer File
  machine
    data
      confirmation
    procedure
      case Trans
      part (Billing)
        run Billing (In Data; Out Data)
      part (Charge)
        run Charge (In Data; Out Data)
      part (Payment)
        run Payment (In Data; Out Data)
      part (Open Account)
        run Open Account (In Data; Out Data)
      part (Set Credit Limit)
        run Set Credit Limit (In Data; Out Data)
      part (Merge Transactions)
        run Merge Transactions (In Data; Out Data)
      esac
```

Figure 5.3-3. Charge Account System Clear Box.

```
data
  bill: record
  bills: file of record bill
proc Billing(In Data: none; Out Data: bills)
  while
    more records in customer file
  do
    bill := name & address & balance;
    balance := balance + sum of charges in charge file − sum
      of payments in payment file;
    bill := bill & charges in charge file & payments in
      payment file & balance;
    bills := bills & bill;
    charge file := empty;
    payment file := empty
  od
corp
```

Figure 5.3-4. Billing Procedure Design.

```
proc Charge(In Data: name, amount; Out Data: approval)
  if
    name is in customer file
  then
    if
      balance +
      sum of charges for name in day charge file −
      sum of payments for name in day payment file ≤ limit
    then
      day charge file := day charge file & name & amount;
      approval := "charge approved"
    else
      approval := "credit limit exceeded"
    fi
  else
    approval := "name unknown"
  fi
corp
```

Figure 5.3-5. Charge Procedure Design.

```
proc Payment(In Data: name, amount; Out Data: confirmation)
  if
    name is in customer file
  then
    day payment file := day payment file & name & amount;
    confirmation := "payment confirmed"
  else
    confirmation := "name unknown"
  fi
corp
```

Figure 5.3-6. Payment Procedure Design.

```
data
  customer record
proc Open Account(In Data: name, address; Out Data: confirmation)
  if
    name is in customer file
  then
    confirmation := "existing account"
  else
    customer record.name         := name;
    customer record.address      := address;
    customer record.balance      := 0;
    customer record.charge file  := empty;
    customer record.payment file := empty;
    customer record.limit        := 0;
    add customer record to day new customer file;
    confirmation := "account for name confirmed"
  fi
corp
```

Figure 5.3-7. Open Account Procedure Design.

```
data
   application data
   credit data
   salary data
proc Set Credit Limit(In Data: name; Out Data: confirmation, limit)
  if
    name is in customer file
  then
    run BB Credit (indata: name; outdata: application data,
                   credit data, salary data);
    if
      application and credit and salary results acceptable
    then
      limit := 0.05 * salary;
      confirmation := "credit is" & limit
    else
      confirmation := "credit denied"
    fi
  else
      confirmation := "name unknown"
  fi
corp
```

Figure 5.3-8. Set Credit Limit Procedure Design.

```
proc Merge Transactions(In Data: none; Out Data: confirmation)
  while
    more charges in day charge file
  do
    charge file (name) := charge file (name) & charge
  od;
  day charge file := empty;
  while
    more payments in day payment file
  do
    payment file (name) := payment file (name) & payment
  od;
  day payment file := empty;
  while
    more customer records in day new customer file
  do
    merge customer record into customer file
  od;
  day new customer file := empty;
  confirmation := "merge completed"
corp
```

Figure 5.3-9. Merge Transactions Procedure Design.

5.4 BOX STRUCTURE DESIGN PRINCIPLES

Preview: Box structures provide intellectual control in complex information systems development. State migration distributes state data to appropriate levels in box structure design. Common service box structures avoid duplication in state migration. Concurrency control in concurrent box structures must be explicitly designed.

5.4.1 Intellectual Control of Complex Designs

It is a Chinese proverb that a journey of a thousand miles begins with the first step. That is certainly true, but if the traveler meanders in circles in uncharted lands, the journey of a thousand miles may never be completed, no matter how many steps are taken. Complex information systems design can be subject to the same pitfalls and hardships. If the designers meander or circle they may never complete a design. However, it is easy to ensure progress in a journey with proper maps and a plan of travel. But it is not so easy to ensure progress in a design, because there are no maps.

Fundamental Principle: The function of box structures is to ensure progress in the design of complex information systems.

A box structure of a thousand black boxes must be traveled a step at a time. But box structures provide the designer with a mapmaking ability to measure progress and refine plans as unexpected obstacles (or easy tasks) are encountered. This theory begins with the fact that every system to be designed will have black box behavior. The usual problem is that this behavior is too extensive and complex to write down in one step in an understandable way. Instead, the knowledge of this behavior will be distributed among several or many people, and some of the behavior may not be yet worked out or explicitly known to anyone.

What is the first step then? It is to identify this top level defining black box and decide how to decompose and defer its description by a top level BB/SM/CB structure, depicted as follows, with several new black boxes identified at the next level.

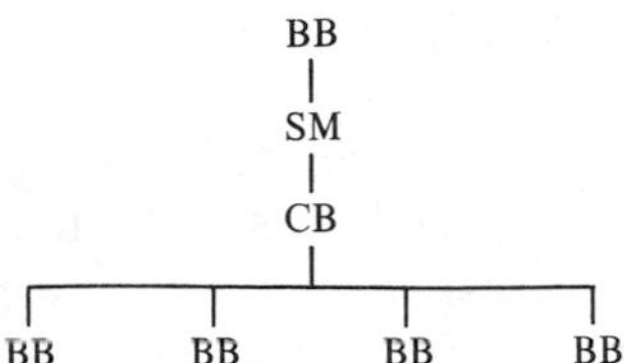

Obviously, this first step of design must have been preceded by considerable planning and analysis. Good choices of decomposition at the top level require in-depth understanding of the activities that will make up lower levels. So good design requires good analysis in advance, in order to decide what data is to be defined in the state machine, and what clear box of next level black boxes best decomposes the original black box.

Usually, each new black box will pose the same problem—it can be identified but not adequately described. Now there are two kinds of tasks: verification and expansion. In verification, it must be shown that if the next level black boxes have the identified behavior, then the top level black box, as designed in its clear box, will have the desired behavior. In fact, this verification task serves to ensure the correct identification of behavior for the next level black boxes. In the second task of expansion, any black box which cannot be completely and precisely described must be itself expanded into a BB/SM/CB structure, possibly generating new black boxes in the process. Everyone involved with this expansion should be thoroughly familiar with the verification step, in order to know what is expected of the black box being expanded. Thus, each new black box begins with its own first step. In short, there are nothing but first steps in this design journey. At any point in the journey, the box structure hierarchy provides a progress map, and the black boxes remaining represent the design work to be done. A word of caution is in order about this journey. It defines a top-down design process, but it requires enough bottom up analysis to ensure that each step is well chosen. This analysis should identify the best way to "divide and conquer" the complexity remaining in the black box identified. In fact the clear box forces another consideration, to "divide, reconnect, and conquer," so that the box structure is solid and well founded during its development.

Fundamental Principle: The basis for intellectual control of complex information systems design is top-down development of its box structure, in which black box behavior is structured into a hierarchy of box expansions, each expansion step a limited activity of analysis and design.

5.4.2 State Migration in Box Structures

A clear box expansion of an original black box and its state machine into several other black boxes and their state machines already implies a distribution of state parts across the hierarchy. The state of the entire box structure is made up of the states of state machines defined throughout the hierarchy. However first conceived, further analysis and study may reveal possible improvements to the distribution of state parts across the hierarchy of the box structure. This section discusses sound and safe ways to migrate state parts in box structures in order to maintain their external behaviors.

As discussed in Chapter 3, it is possible to distribute parts of an entire state of a hierarchy of state machines in an arbitrary way. When this hierarchy is generalized through clear boxes that reference several black boxes at each expansion, the distribution of state parts must be restricted in certain ways. For example, if a state part is accessed or altered in only one of several black boxes in a clear box expansion, then that state part can be migrated to the state machine of that lower level black box. However, if a state part is accessed or altered by more than one black box in a clear box expansion, that state part cannot be migrated downward, because the behavior of the clear box would be changed by such a migration.

A simple example of such an effect can be seen in the alternation structure of Odd:Add2|Add2. The Odd:Add2|Add2 clear box with shared state, depicted in Figure 5.4-1, exhibits behavior as follows, for stimulus history 3 6 1 9 6 and initial state value of 0:

S	S1	OS1	R1	NS1	S2	OS2	R2	NS2	R
3	3	0	3	3					3
6					6	3	9	6	9
1	1	6	7	1					7
9	9	1	10	9					10
6					6	9	15	6	15

In contrast, the Odd:(Add2)|(Add2) clear box with migrated (duplicated) state, shown in Figure 5.4-2, behaves as follows,

S	S1	OS1	R1	NS1	S2	OS2	R2	NS2	R
3	3	0	3	3					3
6					6	0	6	6	6
1	1	3	4	1					4
9	9	1	10	9					10
6					6	6	12	6	12

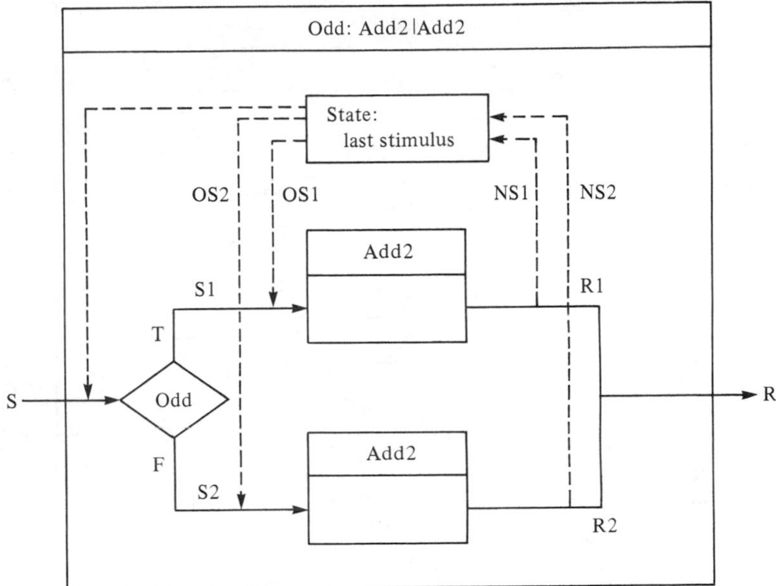

Figure 5.4-1. The Odd:Add2|Add2 Clear Box with Shared State.

which is not the behavior of the shared state version at all. That is, when a state part is migrated into two different state machines at a lower level, they simply behave as two different state machines, not one as before. There may be impelling reasons to migrate state parts downward in the hierarchy to more than one state machine, say for reasons of geography (a distributed database) or for security, however, the clear box must then be redesigned to keep the duplicated state parts always identical in content. For example, every invocation of one state machine (through the black box) must trigger an identical invocation of every other state machine in the duplicated set, say by a concurrent or sequential structure that contains all of them.

> **Fundamental Principle. State Migration:** State parts should be migrated as low as possible in the box structure hierarchy without requiring duplicated updating; if lower migration is necessary, the clear box should be redesigned to ensure the duplicated updating required.

At any point in the design of a box structure hierarchy, identification of new black boxes in a clear box expansion of a state machine provides a

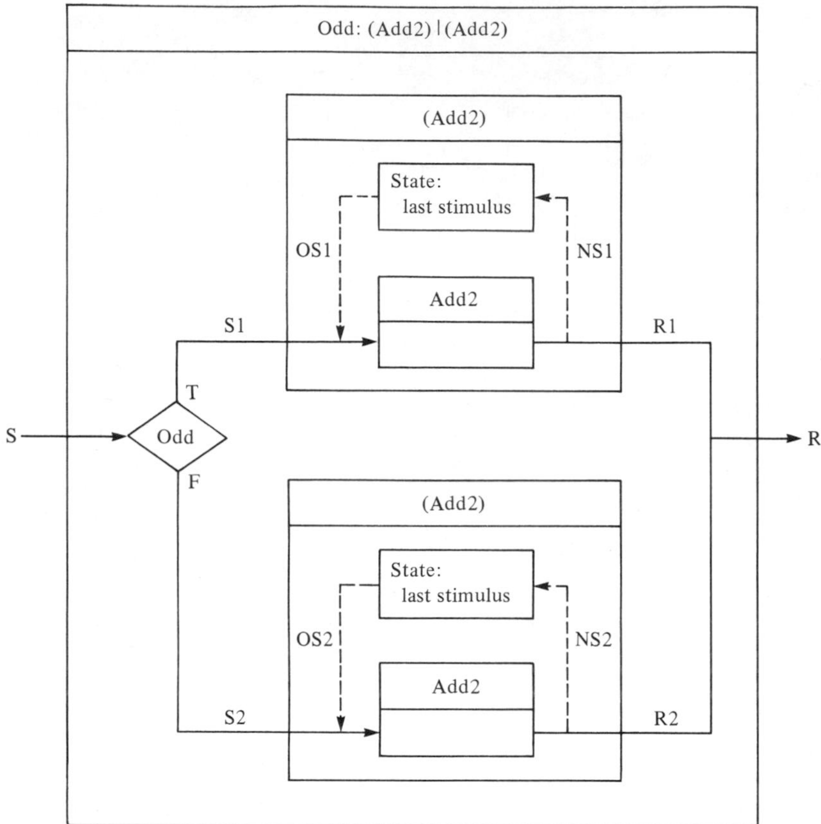

Figure 5.4-2. The Odd:(Add2)|(Add2) Clear Box with Migrated, Duplicated States.

potential opportunity for state migration. State migration permits simplification of the clear box state, and isolates the migrated state parts into new subhierarchies for better control of access and update operations. The value of state migration in limiting and organizing complexity in box structure design leads to the following fundamental principle.

> **Fundamental Principle. Clear Box Design:** Clear boxes should be designed with state migration possibilities in mind, by isolating operations on state parts into individual black boxes, whose state machine expansions become migration opportunities.

5.4.3 Common Services in Box Structures

When several black boxes of a clear box expansion access or alter a common state part, it is generally inadvisable to migrate the state part to those levels. But it may be advisable to define a new box structure to provide access to or alter this common state part for these several black boxes. (Of course, such a new box structure must be invoked in the clear box expansions of these black boxes.) This new box structure thereby provides a **common service** to these several black boxes. Such a common service box structure in effect encapsulates a state part, by providing the only means for accessing or altering it in the overall box structure.

State encapsulation requires defining a new box structure whose state will contain the common state part, and whose transitions will provide common access to that state part for multiple users. In essence, state encapsulation permits state migration to be carried out in another form, with the provision that the only possible access to the migrated state is by invoking transitions of the new box structure that encapsulates it.

Common service box structures are ubiquitous in information systems. For example, any database system behaves as a common service box structure to the people and programs that use it. In simple illustration, consider a clear box expansion of a master file update state machine. Such a clear box would contain a number of black boxes which operate on the master file, for example, to open, close, read, and write the file, as well as black boxes to access transaction files, directory and authorization information, etc. The master file of the clear box state cannot be migrated to the lower level black boxes without duplication. However, the master file can be encapsulated, without duplication, in a new box structure that provides the required transitions to open, close, read, and write the file. This box structure can be designed to ensure the integrity of the master file, and all access directed to it. In fact, when the master file is migrated to this common service, it is protected from faulty access by the box structure in an effective way.

Imagine the box structure hierarchy for a master file update clear box, as conceptually illustrated in Figure 5.4-3. Black boxes to open, close, read, and write the master file appear at various points in the hierarchy, all of which access the master file contained in the clear box state.

Figure 5.4-4 shows the same clear box, redesigned to invoke a new box structure named Master File, which encapsulates the master file and provides common services to open, close, read, and write the file. Note in the hierarchy that the previous black boxes to open, close, etc., have been replaced with invocations of the Master File box structure, where each invocation must now identify the particular transition requested. The

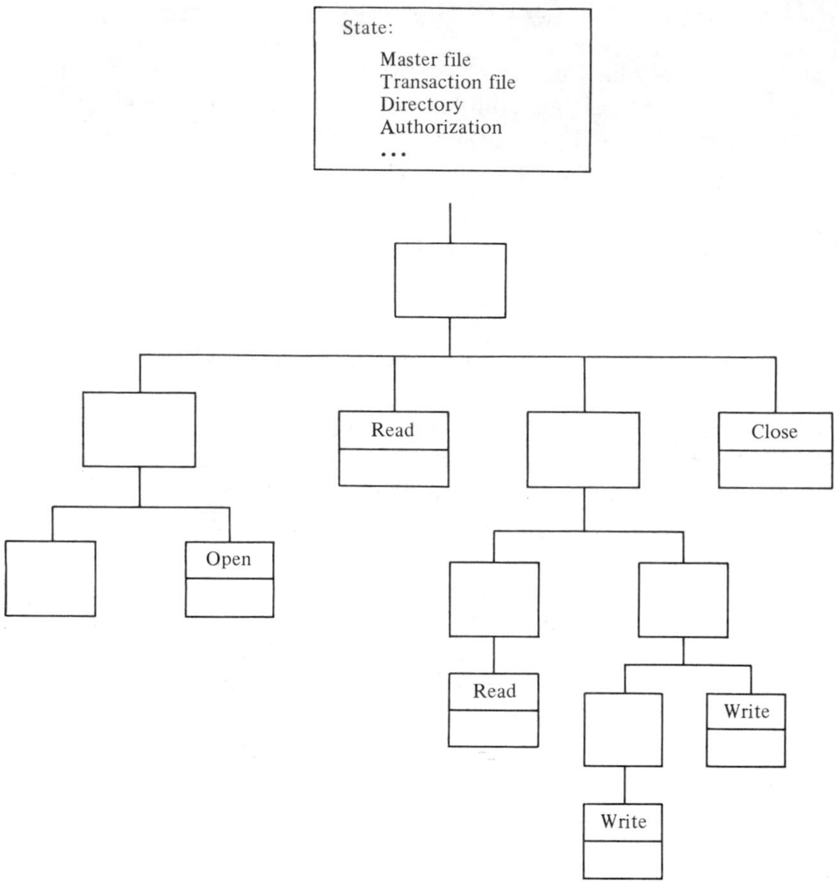

Figure 5.4-3. A Conceptual Box Structure Hierarchy for Master File Update.

Master File box structure is depicted in Figure 5.4-5 in clear box form, with the migrated master file as its state, and four possible transitions that can be requested by its users. Such an encapsulation offers a number of advantages. First, it permits state migration to proceed, to help simplify the original clear box and isolate the migrated state and its operations. Second, a clean interface between the new box structure and its black box users is created, to permit concurrent development of both.

Common service box structures often require definition of permissible sequences for correct use. Any other transition sequence is incorrect and would result in an error response to the invoking clear box.

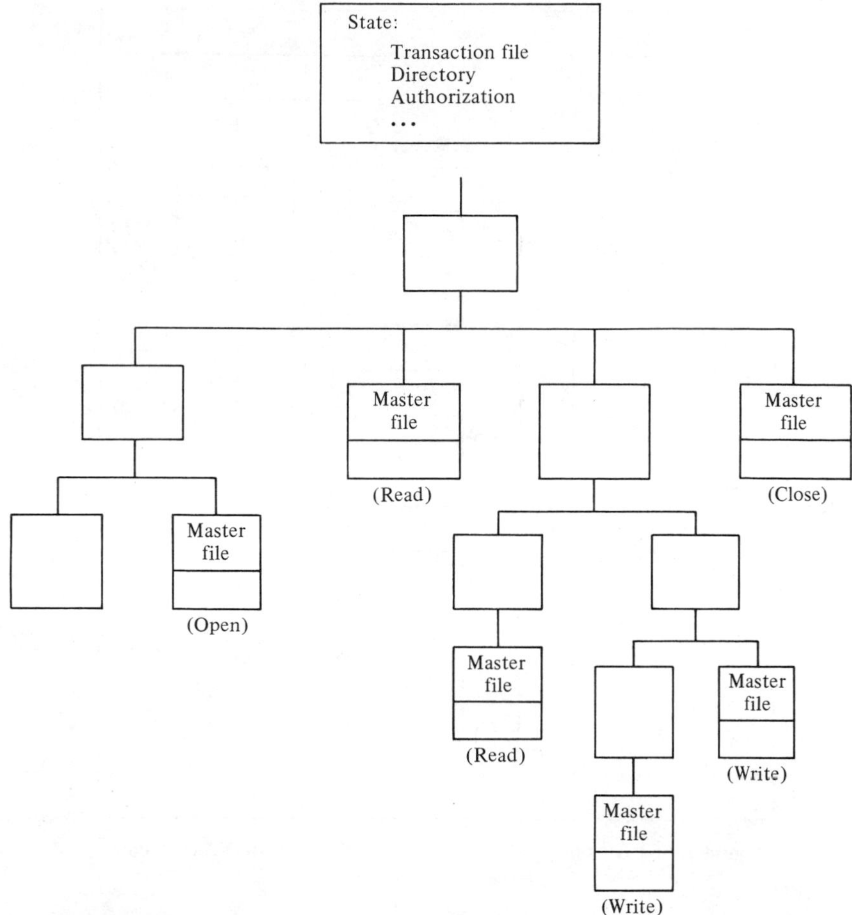

Figure 5.4-4. A Conceptual Box Structure Hierarchy for Master File Update Using a Common Service.

> **Fundamental Principle—Common Services:** When more than one expanded black box accesses or alters a state part, it is advisable to consider the encapsulation of that state part in a common service box structure to be used by these black boxes.

5.4.4 Black Box Replacement in Box Structures

A black box is a unit of design or description that can be isolated and treated on its own, independently of its surroundings in a system descrip-

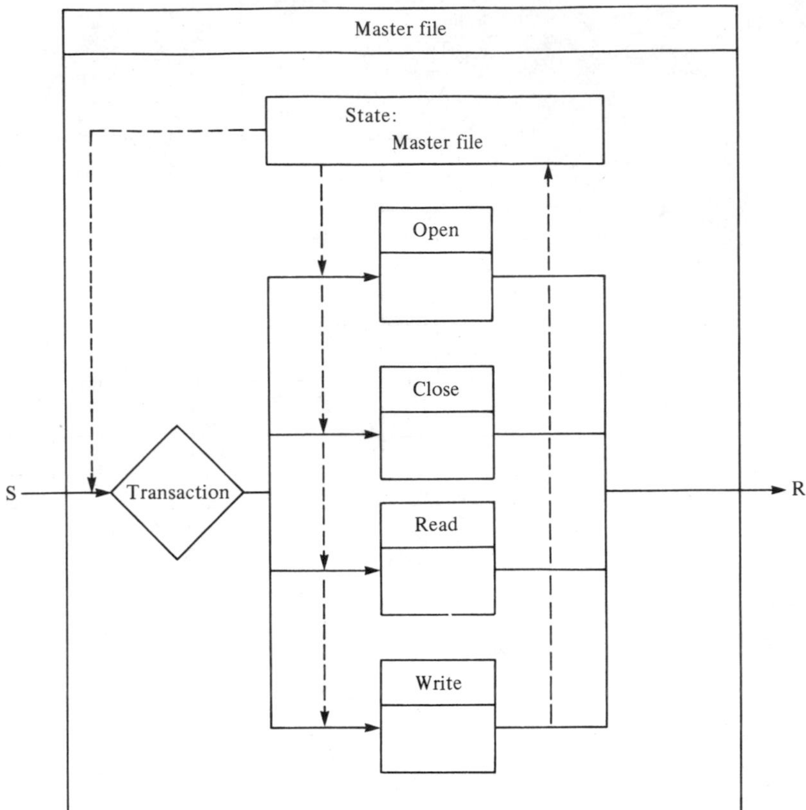

Figure 5.4-5. A Common Service Clear Box for Master File Update.

tion. In particular, a black box can be replaced by another black box of identical behavior and the rest of the system will operate exactly as before. Such black box replacement may be required or desirable for purposes of better efficiency, changing hardware, or even in changing from manual to automatic operations.

In some cases, however, it may be required or desirable to replace a black box by another black box of different, improved behavior. For example, consider the Inventory reorder rule clear box of Figure 5.4-6. The Sales forecast black box shows responses of S (Sales), and SF (Sales Forecast). If the sales forecast is a running average, this black box might be replaced with another black box more efficient than this one, or by another deemed more suitable for this particular item of inventory. Similarly, if the Inventory calculation is based on the months of supply reorder

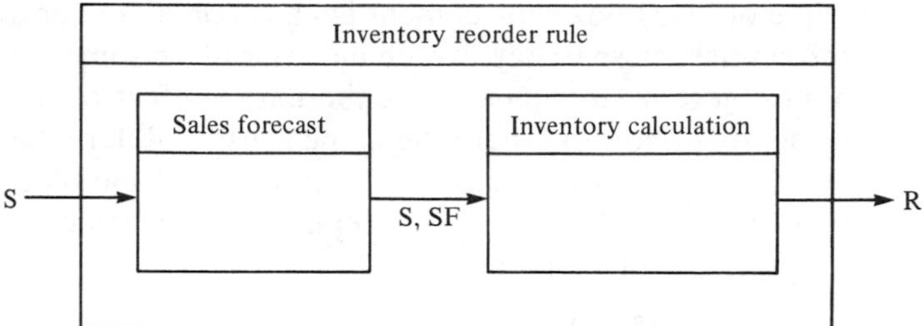

Figure 5.4-6. Inventory Reorder Rule Clear Box.

rule, which is known to have undesirable properties from the analysis of Chapter 1, it could be replaced with a more suitable black box, as well.

In illustration, consider a sales forecast for a seasonal item in which seasonal adjustments are based on a 5-year average and the forecast is based on the past year. That is, next month's sales are forecast as the fraction of that month's sales of total sales for the past 5 years times the total sales of the past year. In particular, given the past 5 years of sales (60 months) S1, S2, ..., S60, the sales forecast SF is:

$$SF := \frac{(S12 + S24 + S36 + S48 + S60)}{(S1 + S2 + \cdots + S59 + S60)} * (S1 + S2 + \cdots + S12)$$

Similarly, consider an inventory calculation of an (s,S) type in which L,H (Low, High) are two factors applied to sales forecast SF such that if inventory $I < L * SF$, then inventory reorder $R = H * SF - I$, otherwise $R = 0$. Such a sales forecast and inventory calculation will produce very different and improved behavior for a seasonal item than the original k months of supply inventory reorder rule.

Fundamental Principle: Black boxes can be freely replaced by other black boxes of identical behavior for improved responses or better box structures.

Note that the foregoing analysis not only suggests a simple but general black box for inventory reordering, but also identifies the requirement that both sales S and a sales forecast SF are required by the inventory calculation. That requirement is not itself difficult to invent independently, but in more complex, less familiar situations, such a systematic analysis ensures sufficient data for calculations in the box structure derived.

The principle of black box replacement is based on an important concept called **referential transparency,** which means that, in some context, a reference to an object by its name gives the entire effect of the object itself, independently of how or where the name is used. Referential transparency is widely used in mathematics. For example, if the object is an arithmetic expression, say 3 + 5, it can be replaced by its name, the value 8, in any larger expression, such as

$$(3 + 5)/4,\ (9 + 15)/(3 + 5),\ 3 + 5 + 7$$

to get new expressions

$$8/4,\ 24/8,\ 8 + 7$$

regardless of where it appears. Such arithmetic expressions can also be expressed in hierarchies, with operations denoting internal nodes and numbers denoting end nodes. For example the expression $(24 - (3 + 5))/(6 + (5 - 3))$ has the hierarchy shown that can be evaluated, by referential transparency, a step at a time (that creates a new hierarchy with each step).

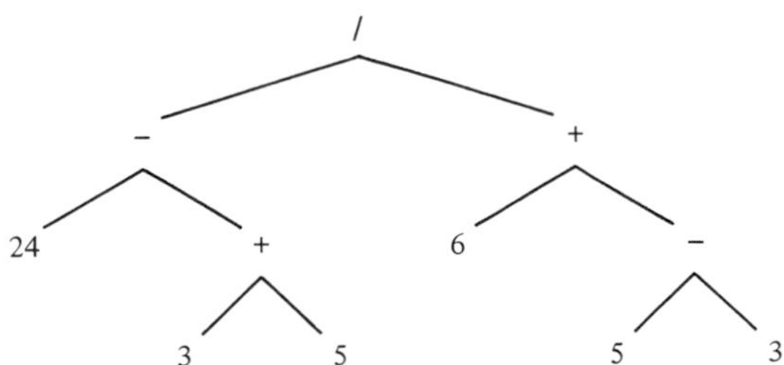

In box structures, a black box serves as a name (a description of complete behavior—a specification) and its clear box is an object with that name. Since the clear box may reference other black boxes by name, a hierarchical structure permits a divide, reconnect, and conquer strategy in box structure design.

But hierarchical structures in themselves do not ensure referential transparency. For example, it is possible to imagine hierarchies of data flow diagrams, but they do not provide referential transparency because data flow diagrams summarize certain aspects of system or subsystem behavior rather than specify or describe that behavior. Such data flow diagrams serve as artist sketches for these aspects, and are suggestive, but not definitive, of system behavior, whose final determination is provided by the implementation. While such ambiguity and freedom may improve the self esteem of implementors temporarily, it usually provides unpleasant surprises for managers, users, and operators, to the eventual frustration of implementors in having to rework the system into a satisfactory form. The lack of referential transparency is less demanding in discipline, but in the end, inhibits real creativity and productivity in system development.

5.4.5 Concurrency Control in Box Structures

The concurrent control structure provides a means of representing concurrency at all levels of a box structure hierarchy. The control of concurrent subsystems requires explicit analysis and design. In a clear box concurrent structure (Figure 4.1-8) each component machine accepts a stimulus and old state and produces a response and new state. The concurrent structure, then, produces a response that is a grouping of individual machine responses and a new state that is some resolution of the individual new states. The design of the Resolve black box will handle the details of generating the new state.

A primary concern in analysis and design of concurrency control is whether the concurrent machines are independent or dependent in terms of resource requirements. Machines are dependent upon on one another when their resource requirements overlap. Shared resources may include state data, input/output devices such as terminals, printers, or communication lines, or even computer processing cycles and memory.

The design of concurrency control when all component machines are independent is straightforward. All machines accept the same stimulus and old state and independently produce a response and new state. The overall response is a grouping of the component responses. The Resolve black box can be designed to form a new state by recognizing the changes in each machine's new state and merging these changes into a single new state. Note that the independence criteria requires that no two machines change the same items in the state data. Thus, no conflicts are possible in the resolution of the state.

Many examples of concurrent, independent subsystems can be found in business processes. For example, consider a large catering business that has separate departments, that specialize in preparation of entrees, salads and appetizers, desserts, and drinks. Figure 5.4-7 shows a concurrent clear box that describes the processing of a typical order for this business.

A catering order is accepted by the Take order black box. The detailed order is recorded in the state of the business and a stimulus is sent to a concurrent structure for controlling the food preparation to execute the order. Then, independently, each department prepares its portion of

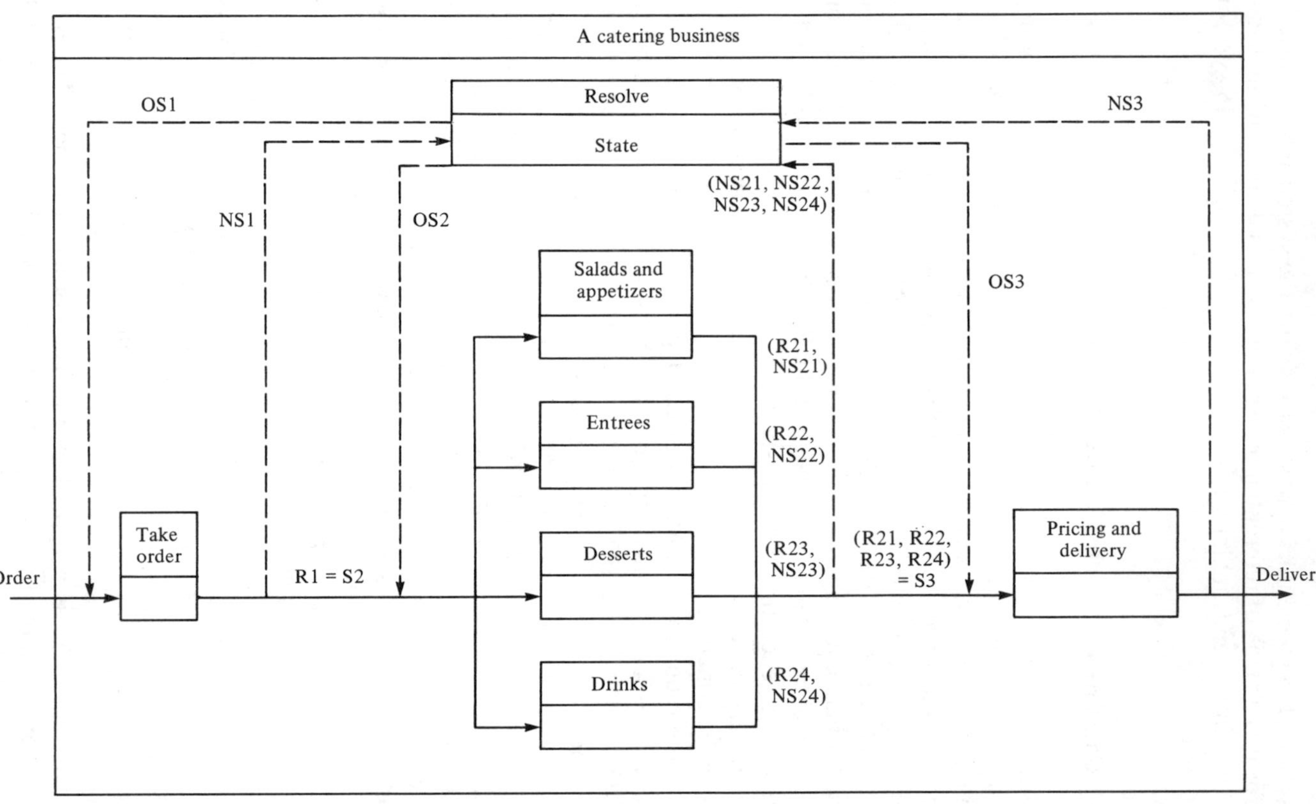

Figure 5.4-7. A Catering Business Clear Box with Concurrent Operations.

the order. From each department, the response is the availability of prepared food and the new state is an updated order status, plus pricing and inventory information for materials and labor. The Pricing and delivery black box prepares a final bill for the order for delivery to the customer.

The design of concurrency control for concurrent, dependent machines is necessarily more complex. As observed in Chapter 4, a common objective of concurrency control in this situation is **serializability.** That is, the behavior of a concurrent control structure must be equivalent to the behavior of one of the possible sequential orderings of its component machines. Note, however, that other concurrency control objectives may be required. For example, perhaps only a few of the possible sequential machine orderings result in acceptable behavior. In this case, additional controls in clear box structure must be designed to ensure that the desired behavior is produced by the concurrent structure.

The principle of transaction closure can be applied to concurrency control analysis and design. Once control objectives for a concurrent clear box have been determined and analyzed, additional subsystems may be required to supply the concurrency control processing. A well-known example is the design of locking methods for shared state data among concurrent machines. Locking methods would require the design of a subsystem to process request-to-lock transitions and analyze data dependencies among the concurrent machines. A lock table data structure would be required in the subsystem state. The response of the locking subsystem would identify permissible processing actions that each concurrent machine could perform. In most cases, this would require some form of iteration control structure to allow multiple transitions of the concurrent control structure. For example, in modern database systems with shared access among multiple users, a single user transaction will require literally thousands of transitions in a box structure several levels deep for locking, unlocking, and other checks for data integrity. In summary, a successful design and implementation of concurrency control in systems require a thorough understanding of the concurrency control objectives and methods.

Fundamental Principle—Concurrency Control: If component machines in a concurrent structure exhibit dependencies, for example, in state data or hardware resources, then a concurrency control subsystem, such as locking, is required to support a specified concurrency control objective, such as serializability.

Summary: Intellectual control in information system development depends on a "divide, reconnect, and conquer" strategy made possible by box structures. New black boxes identified in a clear box offer opportunities for state migration to their state machine expansions. Duplication of state data in migration can be avoided with common service box structures. Design of concurrency control depends on the degree of independence among concurrent machines.

5.5 THE BOX STRUCTURE OF THE NEW YORK TIMES INFORMATION BANK

Preview: The New York Times Information Bank was developed in 1969–1971 and set new standards of productivity and reliability. It was developed with box structures to permit reporters to access the Times reference morgue on-line. A top level system design decision helped define the top level box structure of the Information Bank, which included both batch and on line subsystems.

5.5.1 The New York Times Project

The New York Times Information Bank was developed in a two-year period in 1969–1971, by an IBM Chief Programmer Team headed by F. Terry Baker. It set new standards of programmer productivity and program reliability at the time, and represented an early demonstration of the value of top down structured programming. It was the single greatest spark in the "structured revolution" in program and information system development. The New York Times Information Bank was developed with the principles of box structures, which made top down structured programming natural and easy.

The New York Times maintains many years of articles and other material in a reference "morgue", for use by reporters in researching and writing new articles for the newspaper. In order to make the reference material more accessible in the limited time required of reporters in getting these articles out, the Times developed over many years an extensive set of abstracts of its reference material, and a large thesaurus of descriptors (words and phrases) that appear in those abstracts.

Although not visible to the ordinary reader, many people are involved every day in abstracting newsworthy material as it appears, and inserting their descriptors into the Thesaurus. The Thesaurus (literally "treasure") is extensive, too, over a thousand pages in length. The descriptors have many cross references, so that a reporter may enter it by looking up a descriptor of interest, find references to the abstracts and articles containing the descriptors, find cross references to other descriptors, look them up, and so on, in following out material for a new article.

As can be imagined, many years of articles and abstracts will occupy a large amount of cabinet space, and represent a considerable physical job of document retrieval, as well as logical challenge. But the value to the newspaper is also very great. The Times morgue is critical to the quality of its operation as a great newspaper.

Even though The New York Times Information Bank, and its remarkable productivity and reliability, was extensively reported, this is the first account of the box structure methodology used in its development. Rather than reporting in retrospect the design of The New York Times Information Bank which is itself proprietary, this account seeks to create an understanding of the problem, beginning with the ongoing human operations of using and maintaining the morgue, and how a top level solution, in the form of an information system, was conceived and begun.

5.5.2 Getting Started on The New York Times Project

In 1969, The New York Times decided to automate its morgue. Imagine yourself in charge of the project. What do you do? You know a lot about the computer systems available—about computers, operating systems, programming languages, and data management systems. You have two years, and you have a customer who knows the newspaper business, but not computers. But you do not know the newspaper business, and even if you did, you might not know exactly how your customer wants to conduct its business. The upshot is that, as much as you know, you've still got a lot to learn—not about computers, but about how your customer wants to conduct business with computers. The problem is that your customer can't tell you, because he does not know enough to visualize exactly how computers can help. The result is that you are going to have to learn a lot more about the newspaper business, and how your customer wants to conduct it, than your customer needs to learn about computers. You have to study all the alternatives that will effect the use of computers, but your customer only has to use the specific alternative that is eventually selected.

In those two years you must accomplish three things—investigation, specification, and implementation. You might consider starting out and just implementing the best system you can figure out. But that would be irresponsible and foolhardy, to put it mildly. (But it is surprising how many times that is tried!) The fact is, you'd better get your customer's agreement on what you're going to implement—that's a specification. And in order to make a sensible proposal in the form of a specification, you'd better find out what your customer does know (no sense in reinventing the wheel, especially when yours may not be round and your customer's is!)—that's an investigation. So, realizing that, you still need to allocate time to these three activities of investigation, specification, and implementation. The longer you spend on the first two activities, the less time you'll have left for the last. In fact, it would seem that the more time spent in investigation and specification, the longer it would take for implementation. So there is a balance to be achieved, which depends on the problem.

Let's say your allocation is 3 months investigation, 3 months specification, 18 months implementation. You now have these major milestones. First, in 3 months, you must explain in good detail how the morgue operation now works—how the morgue is used, how it is added to and maintained, and what other things you don't know, but need to know. Perhaps the most difficult thing to learn is what you know and whether that is all you need to know. Next, in 3 more months you must have an agreement with your customer on what you are to implement. You have to explain, in terms they can understand, what you propose to develop. (Incidentally, it must be something within your power to develop in the time you have left.)

Your experience in the investigation phase will be very valuable, in learning what language your customer speaks, and how to speak it yourself in the specification phase. Finally, in 18 more months, you must deliver an information system that meets the specifications you've agreed on. If you're in trouble in this activity, it's of your own making. Your trouble may be that you have promised too much. But it's much more likely to be that your box structures are too vague, too soft, with too little rigor in their parts.

You can expect your box structures to be your best mental assets in every phase. You are looking for solutions even in the investigation (not leaping to conclusions), perhaps right out of current operations, or perhaps from a solution to a related problem. You had better have your solutions in the specification—the 5% inspiration that will be followed by the 95% perspiration.

However, in order to make your box structures as crisp and precise as possible, you should look to their inputs and outputs, and the syntax and data structures (see Chapters 7, 8) that will help you define your box structures as clearly as you can.

In The New York Times morgue, you discover that there are three main classes of data. First, there are the periodicals and books of the morgue, themselves—past editions of The New York Times, but also many additional books, periodicals, and pamphlets not printed by the Times. Second, there are the abstracts of the articles, which reporters may use to decide whether to get the publications or not. Third, there is The New York Times Thesaurus of Descriptors, two loose leaf binders of over a thousand pages of terms (Descriptors) taken from abstracts and organized alphabetically, but with cross references and other structures of value for searching for abstracts and articles in the morgue.

It does not take long to discover that the Thesaurus is key to users, and key to automation. The users reach abstracts and articles through the Thesaurus. Further, the existing abstracts and articles are unchanged through time, there are just new abstracts and articles added daily. But the Thesaurus changes daily, first to accommodate new terms appearing in the abstracts, second to reference new abstracts and articles, and third, to correlate new and old terms appearing in new abstracts and articles with the use of terms already in the Thesaurus.

5.5.3 A Top Level System Design Decision

The first step in the development of a system is to identify its black box, and achieve transaction closure. For all intended users, list not only their transactions, but also any previous transactions required to allow their transactions to be carried out by the system, and so on. For example, in The New York Times system, the primary users are reporters accessing past articles and reference material through the Thesaurus. Therefore, previous transactions are required to get the Thesaurus, abstracts, and locations of full text into the system, and secondary users will need to be entering such data. Outside users will be charged for the service, so billing transactions will be required. Users must have proper authorization for the information they request, so authorization transactions are required. Thus, the black box for the New York Times system will contain at least four kinds of transactions:

Data Query and Retrieval
Data Entry
User Authorization
User Billing

In the case of The New York Times Information Bank, considering the operations of The Times and the computer hardware/software available, a top level system design decision was required. After considerable analysis, it was decided to move User Authorization and User Billing transactions off-line, and to break the Data Entry transactions into two parts, part on-line and part off-line. Data entered on-line during the day can be put into a transaction file; then the on-line database can be updated with the transaction file off-line overnight. Such a decision takes a joint analysis of the operations of the enterprise and computer performance/economics of appropriate depth. It would have been possible, but economically prohibitive, to put all transactions into a single on-line system. However, it was satisfactory, and economically feasible, to keep the on-line database current up to the day, not up to the last minute.

With a decision to move some transactions or parts off-line, the on-line system is better considered as an off-line, all day transaction itself along with the parts moved off-line. Furthermore, with further analysis, a new user class of system operations, and transactions for controlling and tuning the system, were identified.

Note that this top level box structure recognizes user input and output explicitly. The data processing techniques for storing and retrieving abstracts and Thesaurus entries are yet to be elaborated, even though they are critical and interesting. A natural tendency for information systems developers is to think about their own internal problems before they think about the users' problems. As a result, the final integration of solutions to their problems into a system often uncovers unsolved or unresolved user problems. In contrast, the box structure approach forces a system view from the very beginning that includes the users.

Another tendency of information systems developers is to focus prematurely on parts of a system and inadvertently to suboptimize the parts focused on. For example, the on-line reference system is the most visible and interesting part of The New York Times Information Bank. In contrast, the box structure approach forces the identification of the off-line operations of database update, authorization, billing, etc., that need joint consideration with the on-line system. In this way, the whole system gets top level scrutiny by analysts, designers, and managers with fewer afterthoughts and system patchups required.

5.5.4 A Top Level Box Structure for the Entire System

The conventional view of The New York Times Information Bank would be a system of several programs, for example, one for its on-line

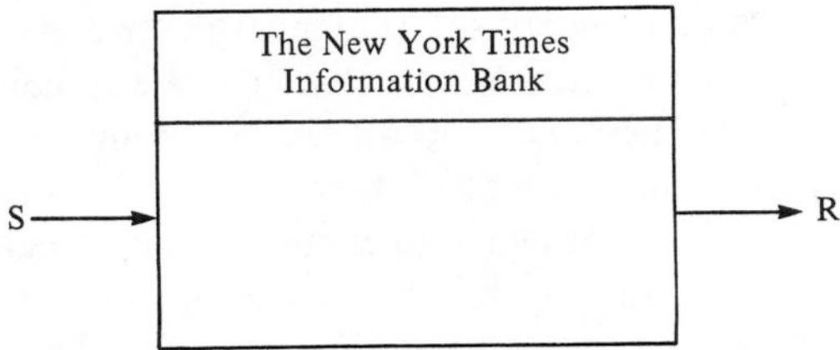

Figure 5.5-1. The New York Times Information Bank Black Box.

operations, one for incorporating the data entry transaction file into the on-line database, one for granting authorizations, one for analyzing database usage, and one for billing. In contrast, the box structure view begins with the New York Times Information Bank as one black box as shown in Figure 5.5-1 with stimuli and responses that accumulate into five types of input/output as shown in Table 5.5-1.

In this box structure view, the users of The New York Times Information Bank enter data for various purposes: as operators of the on-line system, as users of the information services, as data entry people, as database specialists, and as financial specialists. But the entire system behaves as one big black box for all of them. Each of the users individually sees a part of this black box behavior; each enters stimuli (e.g., keystrokes) that accumulate into inputs (e.g., lines of data) and receive

Table 5.5-1

The New York Times Information Bank
Input and Output

Input	Output
On-line	
Control data	Confirmation
Retrieval requests	Information
Entry data	Transaction file
Database Update	
Transaction file	Confirmation
Authorization Update	
User data	Confirmation
Billing	
Control data	User bills
Usage Statistics	
Control data	Usage statistics

responses (e.g., character echoes) that accumulate into outputs (e.g., messages on screens or printers). The on-line system handles many users concurrently, so it must interweave all these stimuli and responses in a split second way to behave as a black box as a whole, and also to provide seeming black box behavior to each user. For example, a hundred users at terminals may produce stimuli that reach and are recognized by the system in a specific sequence, even though many keystrokes are depressed in each second.

Such black box behavior is impossible to comprehend without a great deal of structure, but it is black box behavior nevertheless.

The next step to a box structure for The New York Times Information Bank begins with its overall state machine shown in Figure 5.5-2. The state data can be classified into various categories, as described in Table 5.5-2. This data is used in different ways by different transactions of the system. For example, a specific input (stimulus history) will serve to turn on the on-line system for the day's operation. Although the users are on-line, the on-line system itself is a batch job that takes all day to run. The stimuli following that will be interpreted as terminal messages from users signing on, entering data, retrieving data, or controlling the operation of the system until a specific input serves to shut down the on-line system.

This entire day's operation represents a single transaction at the system level. The stimuli of this transaction are the collection of all stimuli from all terminals that have occurred during the entire day. The responses of this transaction are the collection of all terminal screens and printed output produced during the entire day. These stimuli and responses, seg-

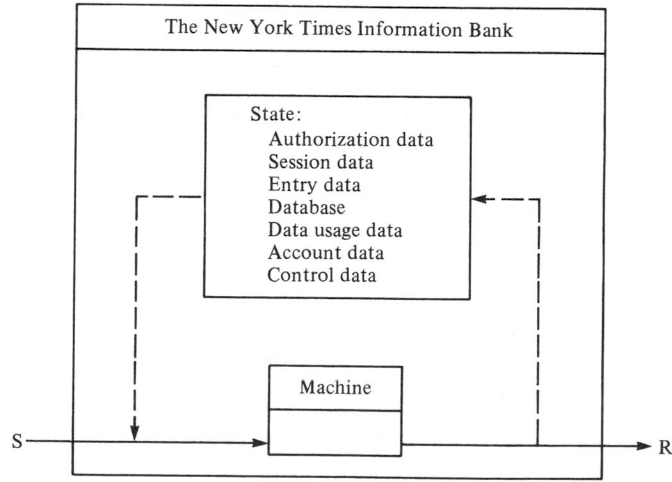

Figure 5.5-2. The New York Times Information Bank State Machine.

Table 5.5-2

The New York Times Information Bank State Data Categories

Authorization Data	Data used to grant authority for user sign ons and queries
Session Data	Data used to conduct on-line sessions with individual users, e.g., sign on data, terminal data, user data, current mode of interaction, etc.
Entry Data	Data accumulated in the transaction file today
Database	Thesaurus, abstracts, and locations of full text available today (note items generated today are in Entry Data)
Data Usage Data	Data used to analyze the use of data in the database
Account Data	Usage data posted to user accounts for billing and analysis
Control Data	Data used to control the system, allocate space to files and database, give priorities to individual users, etc.

regated by terminals and sequenced in time, are interpreted by individual users as the stimuli and responses for their individual transactions. The on-line state machine transaction requires all day, from the state at the beginning of the day to the state at the end of the day. It is a large, indeed gigantic, transaction. But it is just one transaction of The New York Times Information Bank.

In addition to the all day on-line transaction, another stimulus history will invoke a (batch) database update transaction, still another will invoke a (batch) user billing transaction, and so on. If authorizations are to be added or deleted, another transaction will be called for. A database usage analysis represents still another transaction for the system. In each case, on-line or batch, a transaction must be completed before the next transaction is begun, just as in any black box or state machine. For example, the database update can not be carried out concurrently with on-line operations. The type of state changes likewise vary with the transactions, as shown in Table 5.5-3.

The next step in the box structure of The New York Times Information Bank is its clear box, which can be described as in Figure 5.5-3. Each system transaction must first be recognized in a stimulus history that defines the system transaction required. Once identified, the chosen system transaction can be conducted on the basis of a continued stimulus history. In the case of the all day on-line transaction, a large additional stimulus history from many terminals is expected. In the other four batch cases of Database Update, Authorization Update, Usage Statistics, and Billing transactions, a relatively short additional stimulus history will follow, which defines any particular conditions in a single input, and the transaction will be completed in a single batch run of the computer.

Table 5.5-3

The New York Times Information Bank State Machine State Changes

Transaction	State Changes
Authorization Update	Authorization Data User authorities added/deleted
On-line	Session Data None—session data disappears at end of each day Entry Data Accumulates the day's work of data entry personnel Database None—used for retrieval only Data Usage Data Updated for day's usage Account Data Updated for day's usage Control Data Updated for day's operation
Database Update	Entry Data Emptied to database Database Updated with day's entry data
Billing	Account Data Reinitialized after user billing completed
Usage Statistics	Data Usage Data Reinitialized after analysis completed

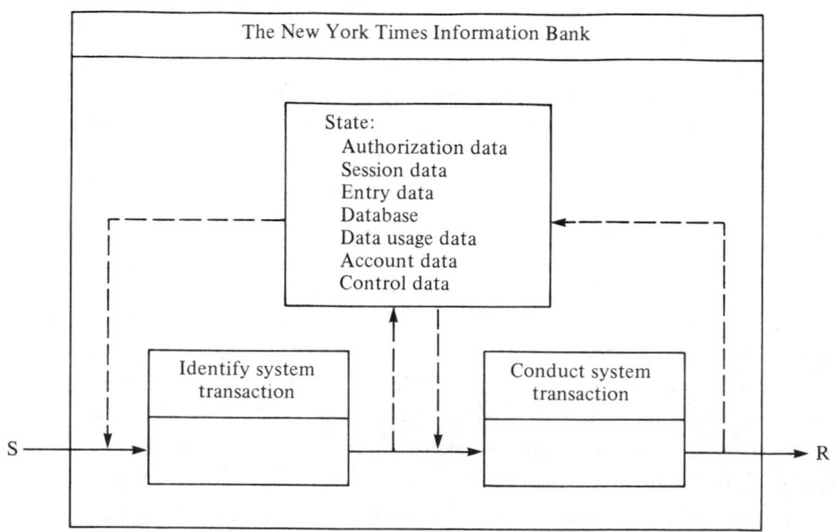

Figure 5.5-3. The New York Times Information Bank Clear Box.

5.5.5 A Top Level Box Structure for the On-Line System

The all day on-line system interacts with terminals all day, each engaged in many user sessions, each a sequence of interactions between a user and the terminal (and system). Each user session can be divided into subsessions such as sign on, sign off, browse Thesaurus, retrieve abstracts, data entry, etc. In turn, each of these subsessions is made up of line (of data) transactions and each line transaction is made up of char(acter) transactions (keystrokes/displays) of the system. Such a structure of transactions, including the batch transactions, of the entire New York Times Information Bank is depicted in Figure 5.5-4. Each day the system can expect hundreds, even thousands, of sessions, each session up to a dozen subsessions, each subsession from dozens to hundreds of lines, each line a few dozen characters.

The box structure of the on-line system mirrors this structure of transactions. The terminal interactions are concurrent, with no sequential requirements between them. But the transitions and transactions at each terminal are sequential. Thus, the on-line system can expect from each terminal:

A sequence of sessions, each
  a sequence of subsessions, each
    a sequence of lines, each
      a sequence of characters

In each case the sequence is determined by a user at a terminal in an unpredictable way, and the system must be prepared for any character stimulus at any time—even if illegal, in which case an error message is called for.

This structure of transitions gives a form for the box structure of the on-line system. At the top level, every stimulus, characters and lines, must be accepted and immediately identified by its originating terminal. Therefore, the state of the on-line system must contain a separate file of stimuli for each terminal. In fact, this state must contain a data area for each terminal to record the progress of sessions and subsessions of each terminal. Then, each terminal can be treated independently, and the box structure for each terminal can be developed independently of other terminal considerations.

The box structure of an all day terminal describes the behavior of the terminal throughout the day, session after session and user after user. However, to the system, the terminal behavior is defined by a large stimulus history, a character at a time. The same characters are used whether signing on, entering data, browsing in the Thesaurus, retrieving informa-

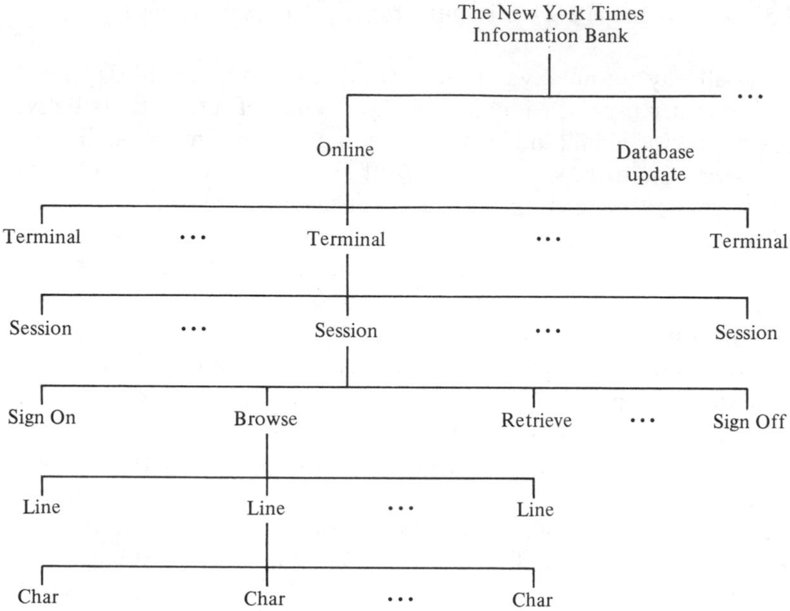

Figure 5.5-4. The New York Times Information Bank Transaction Structure.

tion, signing off, or whatever. Rather, it is not the character keystroked in isolation that determines the behavior of the on-line system for the user, but the sequence of characters keystroked up to any point.

The black box behavior of a terminal can be explained in a state machine with a general transition, Respond to Terminal, as shown in Figure 5.5-5. At first glance, such a general transition does not seem to explain much. But we can expand it in a clear box structure, as shown in Figure 5.5-6. The Respond to Terminal clear box shows the basic formation of inputs out of stimuli. With each keystroke, the user sees a new display (usually with a single character added). If the keystroke defines an input, then Respond to Input is invoked; otherwise the last stimulus is echoed in the display.

In turn, Respond to Input can be expanded, as in Figure 5.5-7. The components of this clear box are illustrative of the design of the Information Bank, but not intended to show exact dctails, which would require more explanation than space permits. Mode is a part of Session Data that specifies which mode of user interaction the terminal is engaged in. At the start of the day the mode is Sign On (waiting for first user). After Sign On, the mode may change (with proper authority) to one of the other modes

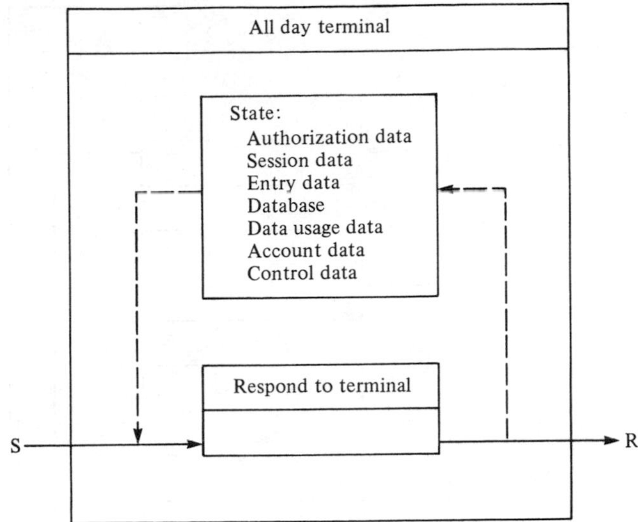

Figure 5.5-5. All Day Terminal State Machine.

by user request, then to other modes, and so on, and finally to Sign Off. Following Sign Off, the mode reverts back to Sign On (waiting for the next user).

At one more level of detail, a clear box expansion for Sign On is given in Figure 5.5-8. The first step of Sign On, namely Get and Check Name,

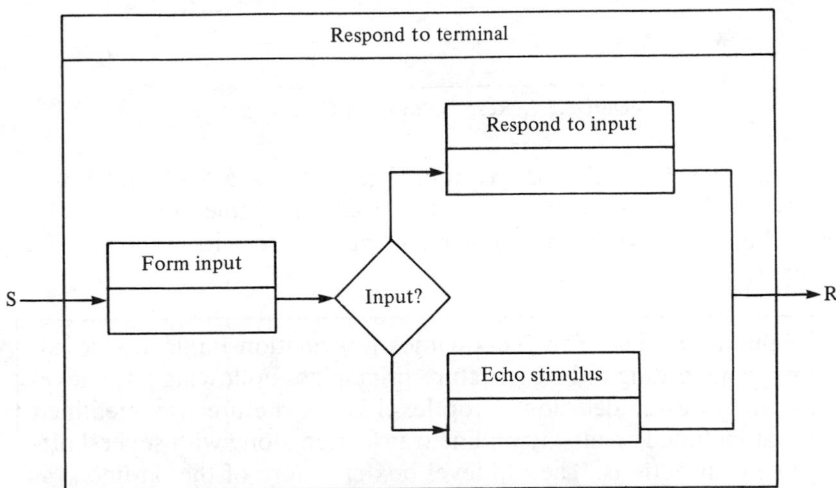

Figure 5.5-6. Respond to Terminal Clear Box.

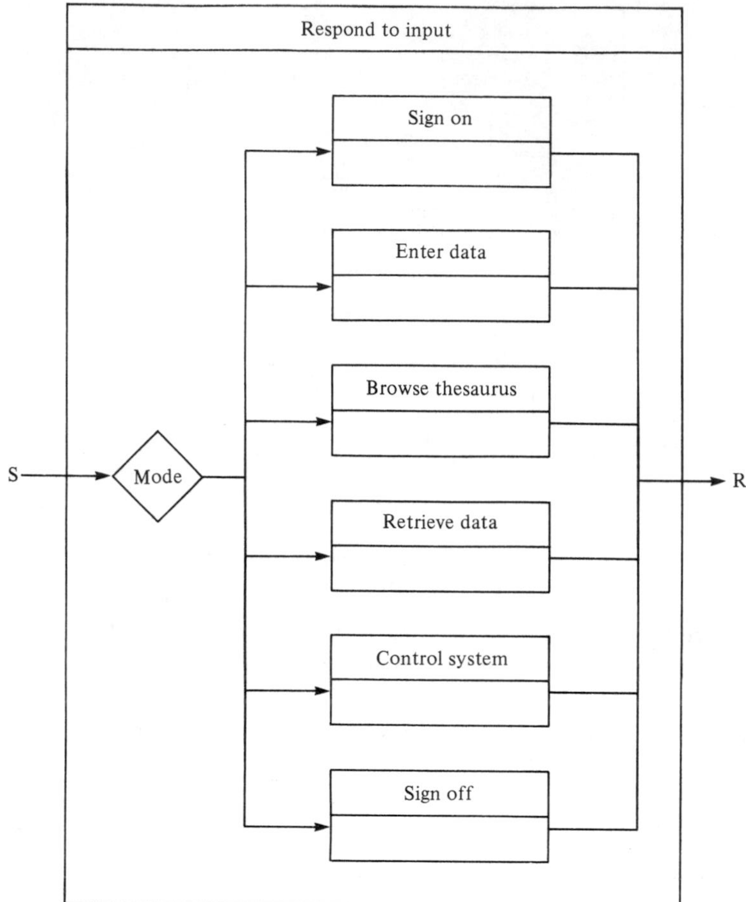

Figure 5.5-7. Respond to Input Clear Box.

can be further expanded, as given in Figures 5.5-9, 5.5-10, and 5.5-11, to Get Name from the terminal input and Check Name with a black box name file, which when given a name as input returns its authority (if any) as input.

Summary: The New York Times Information Bank was developed according to box structure principles. Following a top level system design decision, a top level box structure was identified that included an all day on-line transaction along with several off-line transactions. The top level box structure of the on-line system was also identified.

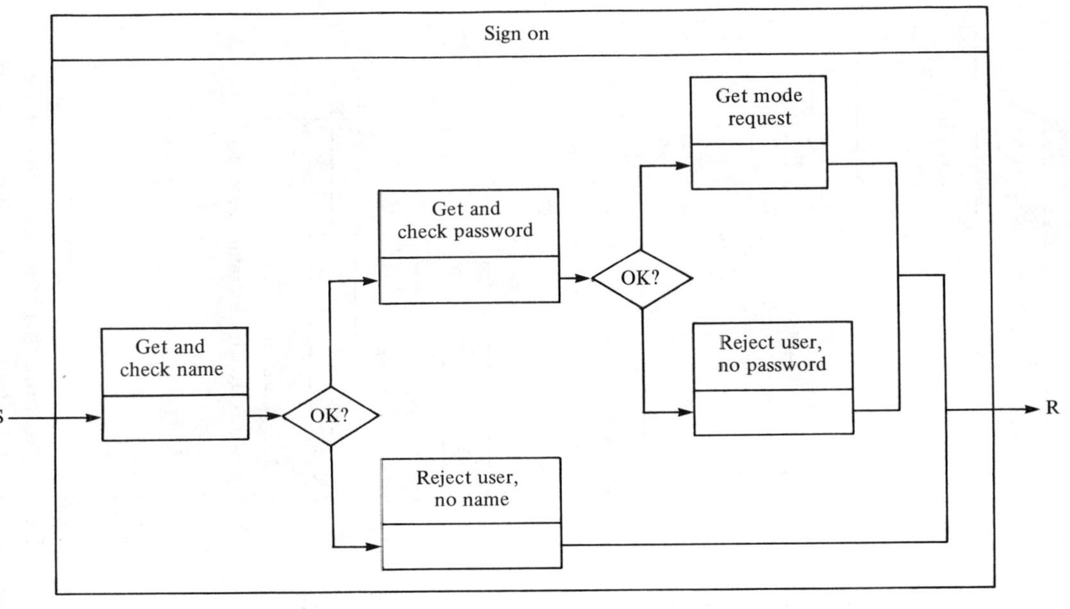

Figure 5.5-8. Sign On Clear Box.

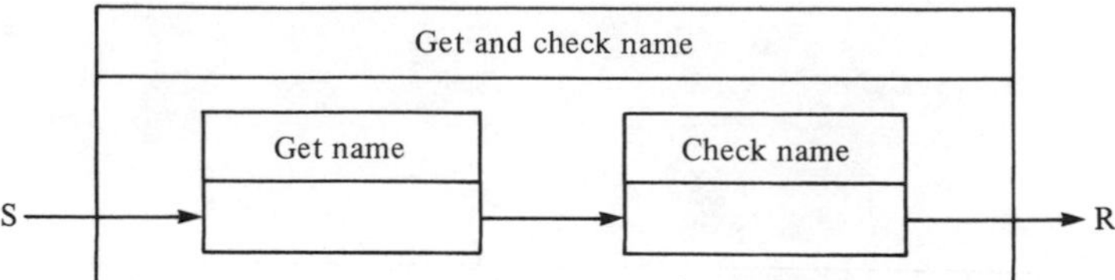

Figure 5.5-9. Get and Check Name Clear Box.

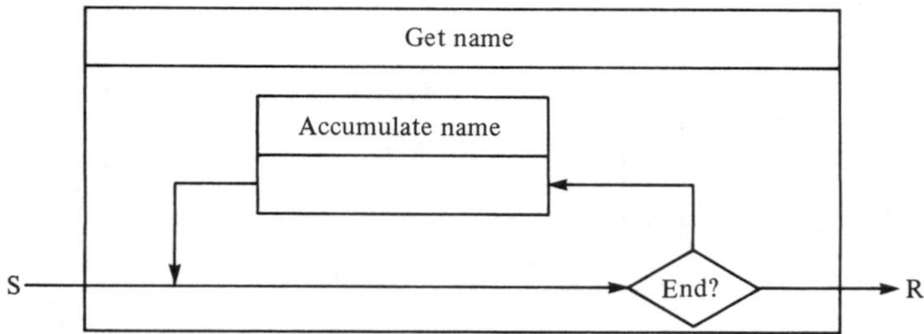

Figure 5.5-10. Get Name Clear Box.

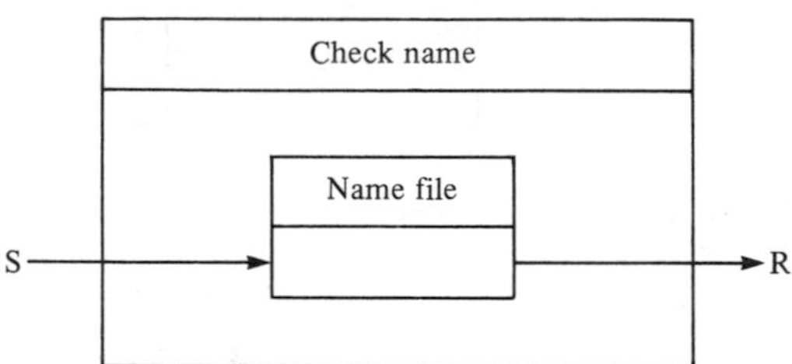

Figure 5.5-11. Check Name Clear Box.

5.6 EXERCISES

1. Develop a Delete Account transaction for the department store, together with any state data required and a procedural explanation.
2. Reanalyze the state data definitions and transactions of Table 5.2-1 for consistency and correctness. Can you find any transaction sequences that satisfy the dependency tree of Figure 5.2-9, but which could result in incorrect state data or output? If so, modify Table 5.2-1 to produce correct results.

3. Imagine extending the analysis to accommodate a financial management function that forecasts 12 months' cash flow for the department store management. Develop a sensible forecasting model, and the transactions, state data, and procedural explanation to specify it. Should the model be seasonal?

4. Develop a skeleton users guide based on the tree of transaction dependencies and Table 5.2-1. How should the users and managers of the system be organized in terms of responsibilities and accountabilities? What does the human side of the work flow look like? Incorporate these ideas in the users guide.

5. Develop an analysis of transactions, state data, and procedural explanations for the processes to be followed if the system breaks down for a day.

6. Develop an analysis of the archive and backup requirements of the system. That is, what information should be periodically purged from the system, but kept for unforeseen needs?

7. Work out a box structure for the black box Set Credit, using the analysis results of Section 5.2.

8. The sales manager of the department store wants a sales report to determine the role of credit limits in the size of purchases. Do customers with higher credit limits make larger individual purchases or just more purchases? How could the sales managers request be handled in the Charge Account System?

9. The finance manager of the department store wants a daily report on the Charge Account System to help in cash flow management of the store. What would you suggest in terms of data provided the finance manager, and how would you design such services into the Charge Account System?

10. Develop a sensible inventory reorder rule for a dairy product with a shelf life of 5 days. Describe its state machine and clear box behavior.

11. Develop a sensible inventory reorder rule for a set of 12 products in a store with limited storage space. In particular, the total inventory of the 12 products cannot be allowed to exceed a fixed amount defined by storage capacity. Describe the state machine and clear box behavior of such a reorder rule.

Chapter 6 Information Systems Management

6.1 MANAGING INFORMATION SYSTEM DEVELOPMENT

Preview: The system development process is a paradigm for generating time phased activities of investigation, specification, and implementation, according to a development plan that is updated for relevance to the business need at the completion of every activity. System development itself can be described by box structures whose transactions are the activities of investigation, specification, and implementation and whose state includes the development libraries. Work structuring and scheduling is an important aspect of activity management.

6.1.1 The System Development Process

The system development process is a paradigm for generating system development activities of investigation, specification, or implementation, based on:

1. A **development plan** consisting of a time phased set of planned activities to meet a specific business need.
2. At each completed activity, a **development plan update** to account for progress made, lessons learned, and changes in the business need.

Typically, a development plan is the joint product of business management and system development management. Frequently, the initial development plan is quite general, beginning with an investigation whose primary purpose is to recommend a more definitive development plan.

The time phased set of activities can be strictly sequential, or it may have concurrent activities. If a development is sequential, it can be pictured, in prospect or retrospect, as a **system development spiral** of activities, as shown in Figure 6.1-1. In this sample case, the activity sequence is a straightforward progression of

Investigation
Specification
Implementation

with a management approval to enter each activity and to end the entire development. Such a progression for developing a system is an ideal, but is not necessarily possible or even desirable. It may not be possible because the business problem is too complex and needs several investigation activities to arrive at a solution. It may not be possible because the system development problem is too complex and needs several specifica-

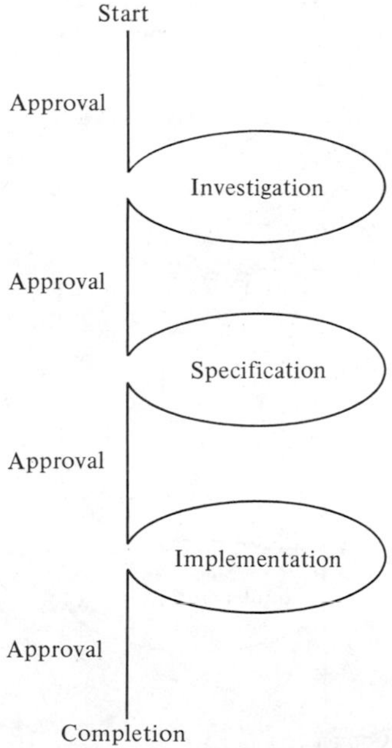

Figure 6.1-1. A Sample System Development Spiral.

tion/implementation activities in an incremental development. It may not be desirable because the business problem is too acute and a less than best implementation is called for as soon as possible. It may not be desirable because the happy outcome of the first investigation activity is the discovery of an existing implementation to meet the business need.

If a development is concurrent, it can be pictured in a network of spirals, as shown in Figure 6.1-2. In this network, activity dependencies are shown by the approval lines ("A" lines here). For example, Investigation 1 enables both Specification 1 and Investigation 2, while both Implementation 1 and Specification 2 must be completed before Implementation 2 can be started. The specific network pictured might, for example, represent the concurrent development of a database system (Implementation 1) and an application system (Implementation 2) that uses it.

The time phased activities of a development plan will be expressed in calendar time, often tied to business events. In fact, many times the system development itself will influence these events. For example, a system to improve customer service may be advertised, and so require

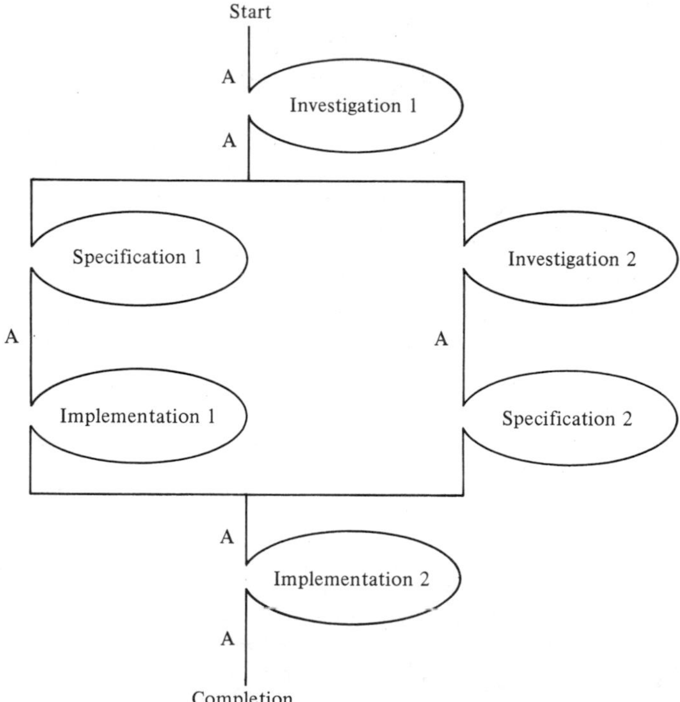

Figure 6.1-2. Sample System Development Spiral Network.

advertising copy and commercials to be developed and placed ahead of time, customer service personnel to be trained, equipment to be purchased, and so on. Needless to say, if the development is late, the business costs may be substantial and way out of proportion to the cost of development overrun. More and more, information systems are at the heart of businesses and their competitive positions, so the stakes for effective information systems development to calendar schedules can be very large.

> **Fundamental Principle:** The objective of information systems development is to improve business performance, not to develop information systems per se.

Information systems development can be successful even though no system is developed, and can be a failure even though a system is developed.

There are several ways a development can be successful without developing a system:

1. An investigation activity can discover an existing system to meet the business need, saving the cost of specification and implementation.
2. An investigation activity can discover how to improve the existing business process so much that a new system cannot be cost justified.
3. A specification activity can tailor the needs of the business to a form such that a specialized vendor can supply a system at greatly reduced cost.

An information system can be a technical success and still be a business failure in several ways:

1. The system addresses the wrong problem because of insufficient investigation and understanding of the real business process.
2. The system addresses the right problem, but is too hard to use because of insufficient investigation of user skills.
3. The system addresses the right problem and is easy to use, but cannot be kept on the air because of operator or integrity problems.

In short, there are any number of ways an information system development can succeed or fail. They are rooted in the business and the final judge of success is the business. For that reason the system development process must be flexible and responsive to the needs of the business. The box structure methodology provides management continuity between activities and management capability within the activities to better ensure progress. But it is finally up to management to focus and direct the development to the needs of the business.

6.1.2 System Development Illustrations

As noted, a system development spiral is an apt figure for tracking the system development process. The spiral demonstrates that activities are mixed throughout the development process. The next activity is not authorized until the previous activity is completed and its results evaluated. For every system a unique system development spiral will be constructed as the development progresses.

To illustrate more specific possibilities of a system development spiral, consider the following simple, yet realistic cases.

CASE 1. TERMINATED PAYROLL DEVELOPMENT

A business wants to determine if a new, automated personnel/payroll system would increase productivity and morale. A development team is established. The team begins the system development with several loops of investigation. Through interviews, system objectives are established and system requirements are analyzed. These analyses result in a final feasibility review report presented to the management. The review concludes that the proposed system would not be cost beneficial and recommends the system not be developed. However, the investigations also provide a much better understanding of the current payroll process, so much so that significant improvements can be recommended, as well. Management agrees, the project is terminated, and the improved payroll process is adopted.

The system development spiral for this case is seen in Figure 6.1-3. Even though no specification or implementation activities were performed, this is still a system development. In addition to the immediate benefits of an improved payroll process, the libraries built during the investigation contain useful information for future system developments.

CASE 2. A PHASED SEQUENTIAL DEVELOPMENT

A large supply company wants to automate its inventory and customer ordering systems. After an initial investigation activity based on feasibility concerns, a decision is made to break the development into two phases. First the inventory system is developed and then the customer ordering system is developed, as shown in Figure 6.1-4. A phased sequential development, such as this, has several advantages. Fewer resources

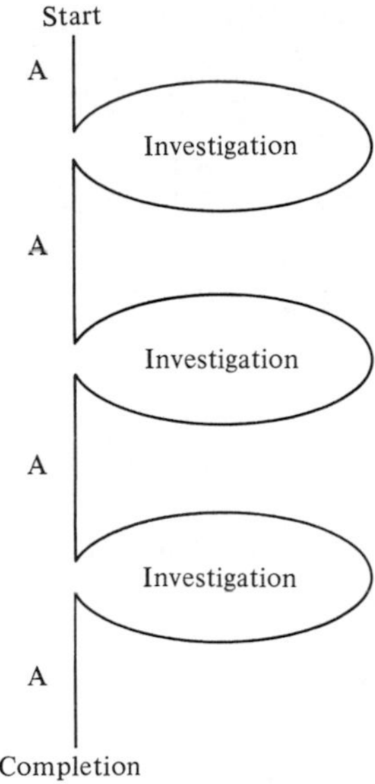

Figure 6.1-3. Terminated Development.

are needed at any one time; much of the work in the first development need not be repeated in the second; and the second development phase can learn from the experiences of the first.

CASE 3. A PHASED CONCURRENT DEVELOPMENT

The large supply company of Case 2 wants to shorten calendar time for bringing up the full system and decides to overlap the implementation of phase 1 with the investigation and specification of phase 2, as in Figure 6.1-5.

6.1.3 The Box Structure of System Development

The process of system development can be viewed as a box structured system itself. Figure 6.1-6 shows the black box of system development. The system development black box interacts continually with the business environment of the system. The stimulus history for system development is formed by information gathered from the environment. The primary sources of the information are business management, operators, and users. Additional information may come from customers or vendors of the business, business application experts, and numerous other sources.

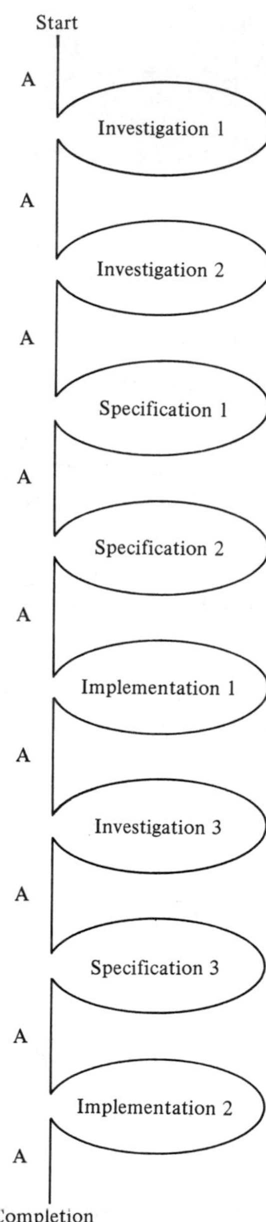

Figure 6.1-4. Phased Sequential Development.

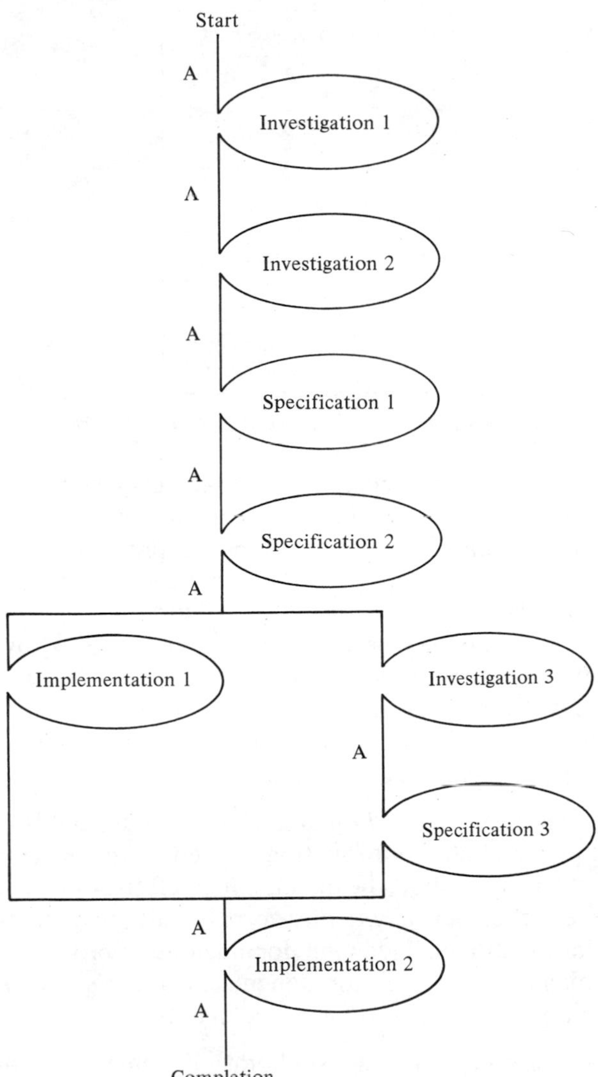

Figure 6.1-5. Phased Concurrent Development.

Each system development transaction produces a response that becomes a stimulus to the business environment. System development responses are requests for information or completed system components. In either case the environment black box accepts the stimulus and produces a response to the system development team in turn, and so on. For a

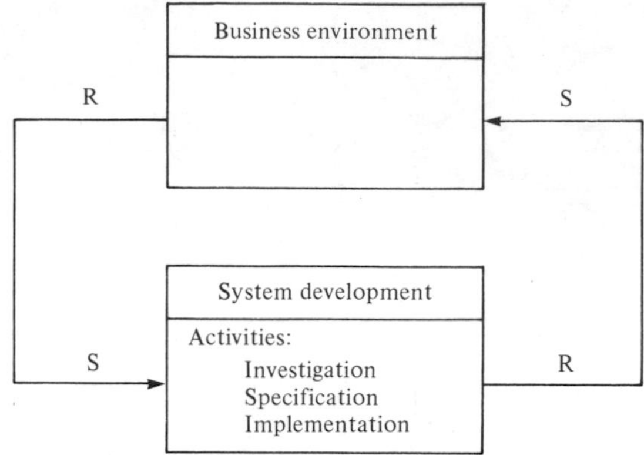

Figure 6.1-6. The System Development Black Box.

system component, or eventually the completed system, the response is an acceptance, a rejection, or a request for further development.

In this way the entire system development process can be viewed as providing a flexible ordering of investigation, specification, and implementation transactions.

The system development process is further described by the system development state machine, as shown in Figure 6.1-7. The state of the development process is held in its libraries that contain information used and generated by the development activities. The following four libraries are needed for development:

Management Library. The management library holds the development plan and other information needed by the development team to control and support the development effort. Control information includes schedules, day to day correspondence, and budgets. Support information includes the documentation provided to the business management, operators, and users, such as system proposals, feasibility studies, and review documents.

Analysis Library. The analysis library documents the analysis performed to create the new system. The box structure methodology emphasizes the use of box structure diagrams as a creative, flexible tool for analysis. This library is principally for communication among the development team and with the business environment during information system development.

Design Library. The developing system design is recorded in this library in a formal language such as BDL. The design library is used by implementation activities as the basis for the system. The

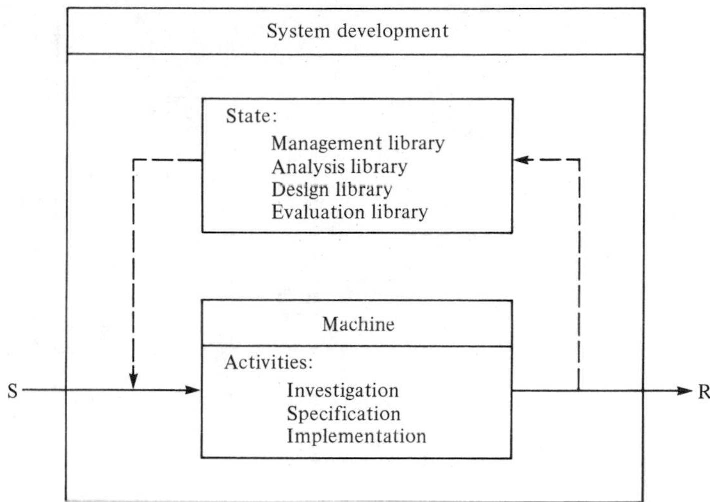

Figure 6.1-7. The System Development State Machine.

library is updated to contain any design changes that may occur during implementation.

Evaluation Library. Evaluations of development results are recorded in this library. Examples of evaluation include design verification through box structure analysis, software testing, and system testing. These results would serve as an information resource for proposals and review documents that are contained in the management library.

These four libraries are at the center of the system development process.

For sequential development, the system development clear box shown in Figure 6.1-8 illustrates the decision of which activity to perform during a development transaction. The stimulus history from the business environment provides the information to concurrently update the development plan and make the selection of an investigation, specification, or implementation activity. For concurrent development the clear box becomes a concurrent structure as defined by the development plan.

6.1.4 Work Structuring and Scheduling

Information systems are planned, managed, and reviewed as a set of time phased activities. But they are developed through their activities a

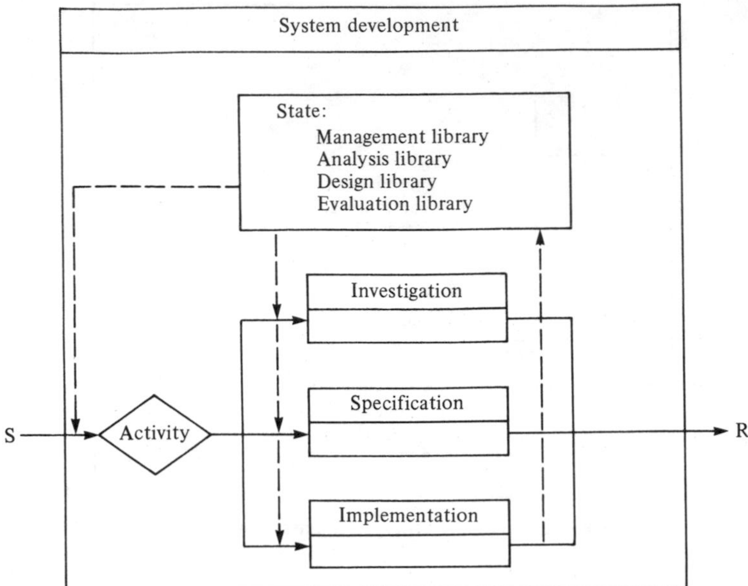

Figure 6.1-8. The System Development Clear Box.

person-day at a time, no matter how large the development may be. If every person every day works and worries about the entire system in a sizable development, the work will almost certainly founder for lack of effective progress and completion. In order to make effective progress, the work and its complexity must be divided and conquered in individual assignments. As already noted, the box structure methodology helps address the division of work. The assignment of this work to individual people is reflected in the work structures and schedules of development.

The effective management of information system development requires a precise decomposition of the work, and the evolving box structure of the system is an ideal basis for such work structuring. Work structuring should not only identify parts of the work to be addressed independently of each other, but also how these parts are to fit together when completed.

> **Fundamental Principle of Work Structuring:** Information systems development work should be structured by the principle of divide, reconnect, and conquer.

That is, the plan for fitting parts of completed work together should be developed before the parts are delegated for independent work. The expansion of a black box into a state machine, then into a clear box with new black boxes provides a plan for fitting completed work on the new lower level black boxes into the original black box expansion. So the box structure methodology provides a direct basis for structuring and managing work in systems development.

Work scheduling requires an additional dimension of management analysis and understanding in dealing with people in the development team. Work scheduling cannot be done in a vacuum, and people have different skills and abilities that must be accounted for in the scheduling. What is a three-week problem for one person may be a three-month problem for another and impossible for a third. Furthermore, the same person can take three weeks or three months to solve the same problem under different conditions.

First, for work scheduling both the work and the people must be there and be understood. It does no good to schedule work if nobody is available to do it. It does no good, and can do much harm, to schedule work which is beyond the capability of people assigned to it. It hurts the business, because false hopes are raised and counted on. It hurts the people who are then judged as failures. So work scheduling must take into account who is available and what their capabilities are. In some cases a sensible work assignment is a matter of enough time. In other cases the assignment may need to be changed to make it feasible at all.

Paradigm for Work Scheduling: With understanding and commitment of those who are to do the work,

1. Make a good schedule
2. Make the schedule good

A good schedule is one that can be made good and represents an effective response to the business need it addresses. Making the schedule good requires constant encouragement and monitoring of progress of work that is within the capabilities of the people assigned to it.

This paradigm for work scheduling may at first seem quite simple, but it sets in motion a set of secondary effects. The key is that a schedule is something to make good, not simply an estimate of how much time and resources an activity or task should take in the abstract. As such, the people doing the work are involved creatively as well in a work-to-schedule framework. In contrast, a pure estimating approach in system development, with no responsibility to meet the estimate, almost always leads to late and overrun performance. Creative work is always subject to self-criticism and rethinking, and the criteria for stopping used by people can vary greatly. If the objective is only to provide the very best system possible, better thoughts are always possible and there is no telling when the developers will come to believe the very best has been achieved. On the other hand, if the objective is to produce a good system to a realistic schedule, the schedule itself becomes as important as the system. That is, the schedule should act as a check and balance on the development work, not merely as an estimate. It is up to management to ensure that people know the cost and benefits of meeting schedules as well as of system operations. As already discussed, the business cost of missing schedules can be many times the development overrun.

6.1.5 Scheduling Mechanics

As noted, work structuring and scheduling requires deep understandings and sensitivities in management, but its results are very concrete, and schedule data can be handled mechanically. In large projects such schedule data can be usefully processed automatically by various project management software packages. As with any other application in which complex ideas are represented with concrete data, it is easy to generate GIGO (Garbage In, Garbage Out) if the concrete aspects obscure the complex meanings of the data. Two useful methods of presenting and processing schedule data are embodied in **Gantt charts** and **project network graphs,** presented next.

Gantt Charts

A Gantt chart is a simple bar chart that shows for each activity its start, duration, and end. A time scale is normally placed on the horizontal scale and the project activities are listed along the vertical scale. Figure 6.1-9 shows a Gantt chart for a small system development project with two people. It is possible to include additional information by having a separate Gantt chart for each member of the development team. Figure 6.1-10 shows a Gantt chart for the two individuals of Figure 6.1-9.

The graphic nature of the Gantt chart shows task responsibilities clearly to managers and non-system personnel. Scheduling can be done quickly for long-term projects and modifications to the schedule can be easily made. The major deficiency of using a Gantt chart is that task dependencies are not represented. This requires a project network graph.

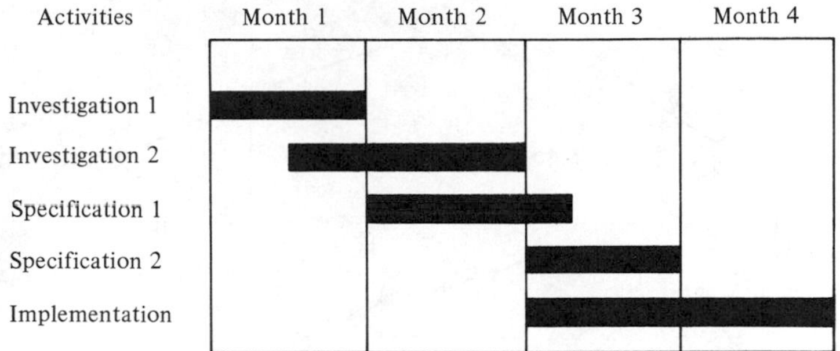

Figure 6.1-9. Project Gantt Chart.

Project Network Graphs

Project network graphs provide better methods for scheduling and tracking the progress of projects with concurrencies and dependencies.

Activities are represented in a graph by directed arcs between **events** (nodes) that signify the start and end of the activity. Each activity has a time duration (e.g., days, weeks, months). The connections in the network show the dependencies among the activities. Each activity arc is given a task name and each event is numbered in the network.

The project network in Figure 6.1-11 illustrates the notation. Activities A through L are connected by events 1 through 11. Each activity has the estimated duration placed beneath the arc. An activity cannot start until all activities coming into its starting event are completed. The dashed arcs are dummy activities with zero time duration. For example, the dummy activity from event 3 to event 4 states that activities E and F cannot start

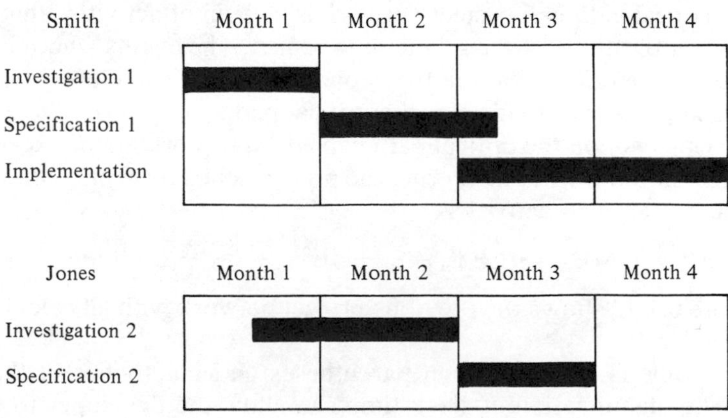

Figure 6.1-10. Individual Gantt Charts.

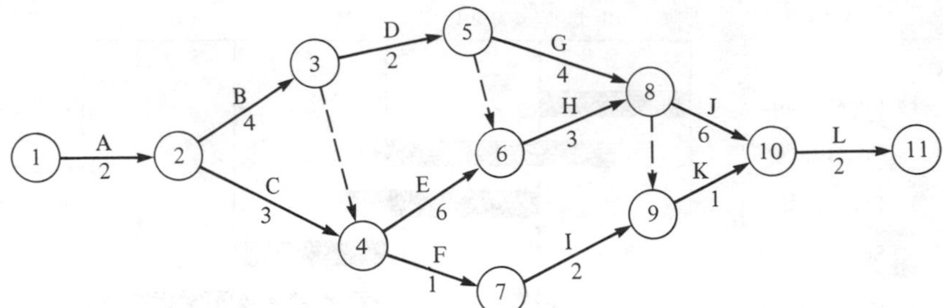

Figure 6.1-11. Example Project Network Activities and Events.

until both activities B and C are complete. This dummy activity is needed since activity D is only dependent upon the completion of activity B.

This project can be analyzed to determine the activities most critical to the schedule as follows.

For each event, we calculate an **earliest event time** (EET) and a **latest event time** (LET). The earliest event time is the earliest time that all outgoing activities could begin. It is calculated for all events, beginning at the start of the project as,

$$EET_E = maximum_A\ (EET_X + duration_A)$$

for each event E, and all incoming activities A from event X to E.

The latest event time is the latest time an outgoing activity can start without altering the schedule. LET is calculated for all events from the end of the project as,

$$LET_E = minimum_A\ (LET_X - duration_A)$$

for each event E, all outgoing activities A from E to event X.

The **critical path** in a project network is a chain of activities that must start and end on time for the schedule to be met. The events where EET = LET define the critical path in the project network. The closest tracking must be applied to activities on the critical path.

Activities not on the critical path can afford to start late or exceed the estimated duration without altering the project schedule. This is known as **slack time.** For each activity A,

$$SLACK\ TIME_A = (LET_{A\ END} - EET_{A\ START} - duration_A)$$

Figure 6.1-12 shows the previous project network with all calculations performed.

Slack time is represented in parentheses under activities in the network. The identification of slack time can allow the developer to better allocate resources. For example, a task that requires 150 person-hours, if separable into parts, can be done by 5 individuals in 30 hours. However, if

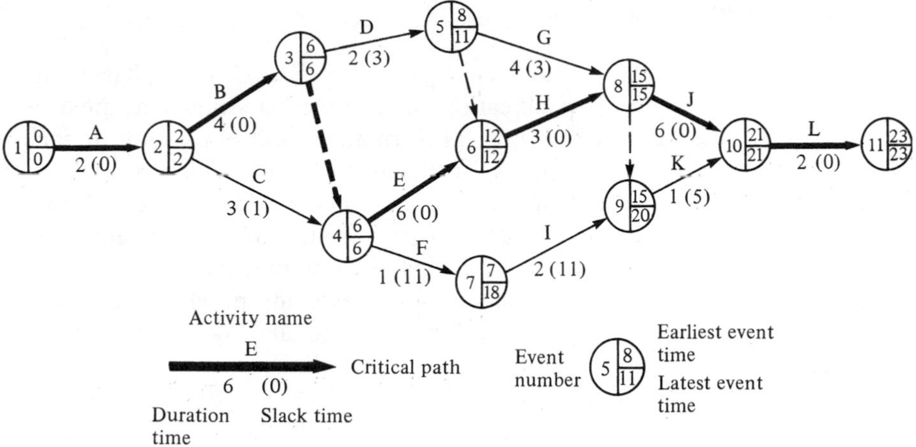

Figure 6.1-12. Complete Project Network Analysis.

sufficient slack is available, the developer may assign only 3 individuals over 50 hours to better utilize personnel. The other 2 individuals may be assigned to another activity on the critical path to improve the schedule.

The project network graph (with minor variations) is the basis for both PERT (Program Evaluation and Review Technique) and CPM (Critical Path Method) scheduling.

Summary: Managing system development requires a thorough understanding of the system development process. Every system has a unique system development spiral. Work structuring and scheduling should be done with understanding and commitment of people assigned to work. Gantt charts and project network graphs can be used to represent schedule data.

6.2 SYSTEM DEVELOPMENT ACTIVITIES

Preview: The activities of investigation, specification, and implementation have much in common and some differences. The commonality is exemplified by the applicability of the box structure methodology across these activities. The activities differ because of the state of development they represent, from fact finding and feasibility concerns in investigation, to system design and cost/benefit analysis in specification, on to completed systems and installations in implementation.

6.2.1 Activity Management

The system development process generates limited, time phased, activities of investigation, specification, and implementation that must be managed. The stages of planning, performance, and evaluation in each activity define an orderly process for this management. The box structure methodology provides a great deal of commonality across these activities for the analysis and design work that is required. The management problems are also very similar. As the names might imply, the most challenging stages for management are planning and evaluation, while the performance stage is most challenging for professionals. We discuss these stages next.

PLANNING

There are three basic results from the planning stage of any activity:

1. Activity Objective. A clear statement of what the activity is to produce.
2. Activity Statement of Work (SOW). A clear statement of how the activity will achieve its objective.
3. Activity Schedule. A clear assignment of work items in the SOW to professionals, together with completion dates which each of the professionals agree to.

With such a plan, the entire development team understands the objectives, statement of work, and the individual responsibilities for making the work objectives and schedule good. Such a plan not only requires the agreement of the professionals, but also requires their direct participation in the planning process. But the planning process must be led by managers to address the proper questions and problems for the activity in the overall development plan.

The outputs of all previous activity loops in the development spiral and the feedback from the business environment combine to help the management and the development team decide what type of activity is required next. The first task is to define an activity objective and to derive a plan for meeting that objective.

The activity loop can be scheduled with Gantt charts and project networks. The activity schedules are as detailed and specific as possible, since they are the primary means of management control. At the same time, the overall system development schedule is updated to reflect the resources allocated to this activity.

All of the planning information, objectives, SOW, and schedules are included in a formal **activity proposal** that is presented to the managers of the development. The proposal is accepted, modified, or rejected based upon management analysis. The eventual acceptance of an activity proposal marks the end of the planning stage and the beginning of the activity performance. The accepted proposal is stored in the management library for reference.

Performance

If plans are well made, performance is focused and predictable. The management job in performance is to assess and track progress against the SOW and schedules, to identify unexpected problems and help professionals decide how to meet them, and to identify unexpected windfalls in solutions that can free up people or other resources to help out with unexpected problems. It is here that good understandings and agreements on assignments and schedules pay off.

First, each team member understands his/her role in the activity, and the need for completing the work to schedule. In contrast, a common misunderstanding between managers and professionals pits the "manager's schedules" against the "professional's design". Such disagreements should be ironed out in planning, not late in performance. A common set of system values from the box structure methodology permits managers and professionals to communicate effectively and alleviate such disagreements. For example, a professional may be reluctant to adapt a less than the best system to a reservation system because a better one can be built; but if the professional understands that a quick and dirty system means the literal survival of the business, there will be little reluctance for a whole hearted effort to get the quick and dirty system on the air.

Second, good schedules and a common understanding of box structure methodology make progress assessments and tracking more accurate and more rewarding. When managers can recognize good and timely work, they can acknowledge it privately and publicly, and when warranted, arrange for awards for extra performance. A major morale problem among high performance professionals is just the fact that good work is often not recognized, and mediocre work by others is as well rewarded as their own. Good progress assessments and tracking of well understood assignments go a long way in recognizing good work.

Evaluation

Evaluation is both a closing out of one activity and a basis for selecting and commencing one or more following activities. The objectives and results of performance can be compared and related to the business and its situation. As illustrated in the terminated payroll development, the results may be surprising but still very useful for the business. Even if objectives are not met, the lessons learned may be useful. If the objectives are met, so much the better and the expected next activities can be initiated. In particular, the evaluation stage is the point where the development plan for future activities can be assessed and modified.

These activities and stages can be organized in table form, as shown in Table 6.2-1, that indicate typical activities in the system development matrix.

Table 6.2-1

Stages in Activities: Typical Tasks

	Stages		
Activities	Planning	Performance	Evaluation
Investigation	Activity Objective Statement of Work Scheduling	Business Process and Objectives Requirements Analysis System Prototype	Feasibility Assessment Review and Acceptance Development Plan Update
Specification	Activity Objective Statement of Work Scheduling	Systems Analysis and Design Operations Analysis and Design	Design Verification Review and Acceptance Development Plan Update
Implementation	Activity Objective Statement of Work Scheduling	Resource Acquisition Systems Integration Operations Education	System Testing Review and Acceptance Development Plan Update

6.2.2 Investigation Activities

The objectives of an investigation activity will be found in its activity plan within the development plan. In general, investigation objectives are to find facts and discover realistic opportunities for improving business processes, often by developing new systems. In order to illustrate potential investigation activities, we discuss the following kinds of tasks:

Describe a business process
Identify a system opportunity
Develop a system prototype
Assess system feasibility

Describe a Business Process

In order to improve on a business process it is imperative to understand and describe it. It does little good to produce a brilliant system with a fatal deficiency or flaw for lack of a real understanding of what goes on in the business. Usually a business process is targeted just because a system can be imagined that will improve the process. But in order to substantiate such an idea, considerable information will be needed about the business process.

The **sources** of this information are many and varied. The primary sources are the managers, users, and operators of the potential system. If current systems are in use, then documentation such as system manuals, user manuals, system logs, and current application programs may be useful sources of information.

Gathering business process information from individuals (managers, users, operators) requires good interpersonal communication skills. The two primary techniques are **interviews** and **questionnaires.**

In information systems development, the purpose of interviewing prospective system managers and users is to make explicit the information processing procedures, needs, and objectives of the business process. The current information processing system will be some combination of people and machines, and the planned system, some new combination of people and machines, all of which exhibit black box, state machine, and clear box behavior. Both the existing and planned systems will require explicit box structure descriptions. Thus, the knowledge gained in the interviewing process must eventually be represented in terms of box structures.

The interviewing process is intended to reveal box structure behavior to the interviewer. Often these box structures will emerge in fragmentary form, and will require corroboration and elaboration through additional interviews and feedback sessions. For this reason, interviews should not be regarded as solitary and stereotyped events, but rather as a continual process of discussions with individuals and groups to arrive at common understandings and objectives. Early box structure definitions, while often incomplete, can nevertheless be used to advantage in these discussions, as a means to focus on the correctness and completeness of planned system behavior.

The questions asked during a interview should focus on box structure behavior, but the words and phrases employed need not depend on box structure terminology. Even though a person being interviewed has no knowledge of box structure techniques, it is still possible to discuss box structure behavior in very precise terms. Consider, for example, the following questions, phrased in the everyday language of business, and their interpretation in terms of box structure concepts:

Question 1:

> "Do previous transactions against a credit account affect the processing of a current transaction?"

Box Structure Interpretation:

> "What is the black box behavior of credit account transaction processing?"

Question 2:

> "What information must credit account transaction processing have on hand in order to process a current transaction?"

Box Structure Interpretation:

> "What state information must be retained in the credit account processing state machine?"

Question 3:

> "How does credit account processing combine the information it has on hand with a current transaction to produce output and update the information on hand"?

Box Structure Interpretation:

> "What are the transactions of the credit account processing state machine?"

Question 4:

> "What steps are required to process a transaction in credit account processing?"

Box Structure Interpretation:

> "What is the clear box behavior of credit account processing?

Question 5:

"Is account verification performed before, during, or after credit account processing?"

Box Structure Interpretation:

"How are account verification and credit account processing related in terms of a box structure hierarchy?"

Interviews are an effective way to gather accurate and timely information. Questions can be asked and responses clarified on the spot. Participants in an interview feel actively involved in system development. However, interviews are time-consuming for both the developers and the persons interviewed, so the number of interviews should be kept to a minimum and each interview should be short and to the point.

The effectiveness of an interview is directly proportional to the preparation for it. The following guidelines support effective preparation.

1. Define the purpose of the interview. Fishing expeditions rarely result in quality information. Know the system area within which the person interviewed is expert, and stick to that area.
2. Only interview selected individuals. It is not necessary to interview everyone and time is limited. Start with upper management and work down the organization hierarchy. This provides the developers with an idea of how the system fits into the overall organization.
3. Be prepared with specific questions for each interview. Do homework and be ready to guide the direction of the interview.
4. Schedule the interview at the interviewee's convenience. However, set a strict deadline for all interviews to be completed. If an individual is not available in that time frame, schedule an acceptable substitute.

An interview has three phases, namely **opening, body,** and **closing.** In the **opening,** state the purpose of the interview, establish the legitimacy of the interview, and achieve a rapport with the person interviewed.

In the **body,** move from general, open questions to specific, detailed questions. Vague answers should be clarified. Take clear notes or ask to record the discussion. The notes should be summarized after the interview.

The **closing** should leave both parties satisfied with the interview. If questions remain, schedule another interview. Offer to send a copy of the interview summary to the person interviewed. Prompt feedback allows any misunderstandings to be discovered and rectified.

Questionnaires are used when information is needed from a large group of individuals, usually system users. The design of the questionnaire is very important. The questionnaire must be clearly written and easy to complete or it will be ignored by most users. It should be checked for clarity and misunderstandings as carefully as a computer program, and test cases run to verify its effectiveness. Analysis of the completed questionnaires is facilitated by the use of computers to tally and statistically analyze responses.

Another important source of information is the current system, manual or automatic, that handles the work of the proposed system. A complete understanding of the details in the existing system is important for the development of a system that will replace it. Box structures can be used to describe and analyze current system operations. Transactions, input, output, data states, and processing in the existing system can be identified and placed into an overall system context. This information can be found in system manuals, users manuals, program code, or even the system specifications left by the development team of the current system.

The results of the analysis of a current system are to be used as requirements information, not as a starting point for the box structure design of the new system. Additional system requirements should dictate an original design for the new system; otherwise why replace the current system?

Identify a System Opportunity

This is the most obvious task of an investigation—invent or check out a good systems idea for the business. Such good systems ideas depend on a good knowledge of the business and an equally good knowledge of information systems possibilities. There is little value in thinking up systems that cannot be developed, just because the business has a difficult problem. And there is little value in thinking up systems that solve problems the business doesn't have. Usually, a proposed information system surfaces early in a development, often in its first investigation activity.

It is critical for a development team to identify the objectives of the proposed information system. It is important to document these objectives in order to obtain a consensus among all parties. Obtaining agreement on system objectives may be difficult because different groups of people may have different expectations. Figure 6.2-1 lists sample objectives of management, users, and operators.

Reconciling these objectives may be difficult, but it must be done before further system development can proceed. A clearly worded document should state the objectives of the proposed system in as much detail as possible. The developers should define priorities to force individuals to make decisions on system objectives. Management, users, and operators should review the document and agree on its content. Figure 6.2-2 shows

Management Objectives

Increase Productivity
Reduce Costs
Improve Quality
Improve Work Environment

User Objectives

Easy to Learn
Easy to Use
Produces Correct Results
Efficient
Improves Working Conditions
Improves Personal Productivity

Operator Objectives

Easy to Understand
Easy to Operate
Reliable
Secure
Adaptable to Changes

Figure 6.2-1. System Objectives.

the outline of a System Opportunity Document that reconciles such objectives with development realities.

Develop a System Prototype

For complex or especially innovative systems, it may be difficult or impossible to find the information needed to perform an adequate requirements or operations analysis. A prototype is a limited version of the desired system built to provide requirements and operations information. Prototypes can range in scope from a simple study to see if two software packages can interchange data correctly to a large scale prototype of the complete system to answer feasibility questions before resources are allocated for detailed system development.

Once the decision is made to prototype a portion of the system, the prototype development takes on a life of its own. The prototype has its own system development process with investigation, specification, and

1. System Function: A description of what functions the system is to perform.
2. System Objectives: A prioritized list of objectives gathered from management, users, and operators.
3. System Realization: A description of how the system can be created.

Figure 6.2-2. A System Opportunity Document.

implementation activities. Separate development libraries are constructed. When the prototype is completed, objective information is gathered from its operation and used in the analysis of requirements and operations for the planned system.

Opportunities may exist to use portions of the prototype development in the system development. The best possible outcome is to evolve the prototype into the desired system. If the prototype, as a limited study, does not incorporate the full functionality of the real system, reusability may not be practical. However, the prototype development libraries will contain much helpful experience and information. These libraries should be integrated into the system development libraries.

Assess System Feasibility

One goal of investigation is to answer questions of feasibility that arise during system development. The feasibility answers determine the scope of the proposed system.

It is important to deal with the feasibility questions in order to define the scope of the system right at the beginning of development. There are several types of feasibility that must be considered:

Economic Feasibility. What costs are sensible for the development and the continued operation of the system? Can an adequate system be developed for the costs?

Equipment Feasibility. Do adequate hardware and software exist in the organization or must new hardware and software be purchased?

Personnel Feasibility. Does the organization employ qualified personnel to support the operation of the new system or must new personnel be hired?

Time Feasibility. Can the system be developed within the time that an organization requires?

Organizational Feasibility. Considering the organizational effect of implementing the new system, will managers, users, operators, and others effectively use the system?

Technical Feasibility. Does adequate, state-of-the-art technology exist in order to accomplish the system objectives? Must the development include a component of research to discover new and innovative hardware and software techniques?

Development Feasibility. Does the development team have the knowledge and skills to develop the required system? What consultant expertise is needed? Are more development personnel needed?

6.2.3 Specification Activities

The objectives of a specification activity will be found in its activity plan within the development plan. In general, specification objectives are to define information systems that can be used for improving business processes. In order to illustrate potential specification activities, we discuss the following kinds of tasks:

Systems Analysis and Design
Operations Analysis and Design
Cost/Benefit Analyses

Systems Analysis and Design

System analysis takes as input the requirements information found in the analysis library and forms a box structure hierarchy for the system. The analyses of transactions, state, and processing throughout the hierarchy provide a rigorous method of system discovery. The principles of state migration, transaction closure, common service black boxes, and black box replacement guide the final structure of the new system.

The findings of system analysis are recorded in the analysis library. Analysis tools, such as box structure diagrams, can be used to structure the results of the analysis activity.

System design uses the results of system analysis and builds top-down design specifications for the system. A structured specification language, such as Box Description Language, is used for this purpose.

For each box structure in the system hierarchy, design decisions must be made in how to specify the implementation of that subsystem. Design skill and creativity play a significant role in building a good system design. Examples of design decisions for a given box structure include:

1. Should this portion of the system be automated or manual?
2. If the subsystem is automated, what hardware is needed? Should commercial software be purchased or should the software be designed and programmed as part of development?
3. If commercial software is used, which package is best?

4. What common service box structures are possible in the subsystem?

5. What integrity issues are inherent in the subsystem design and what measures should be used to handle them?

The final design specifications for the system, along with all of the design decision analyses, are included in the design library for use during system implementation.

Operations Analysis and Design

The information gathered on system operations in investigation is used during specification to analyze and design operations for the new system. Issues that should be studied include:

System Administration. What individuals are needed to administer the system? What qualifications are needed? How does the system administration fit into the business organization?

Staffing. What are the requirements for system operators, application programmers, maintenance personnel, and other system support staff? How is the staff scheduled to provide the most effective use of the system?

Facilities. What renovation or new construction is needed to provide adequate facilities for the system? Are air conditioning, power, heat, etc., adequate?

User Access. Are user access capabilities appropriate? What are the best procedures for user identification and authentication?

Work Flow. Does work flow efficiently into the system, through the system, and out of the system to other parts of the business?

The analysis of these and other operational issues should lead to a specific operations design for a system. The result of an operations design should be placed in the design library along with the analysis that supports it.

Cost/Benefit Analyses

With the system and operations design available, the costs and benefits of the proposed system can be analyzed in more depth. Costs occur in two ways, as shown in Figure 6.2-3.

The primary justification for each activity commitment by managers is a proposal (and credible past performances) that must contain or imply a cost/benefit analysis. Such a cost/benefit analysis, as it is assessed by the

One-Time, Development Costs
- Personnel
  - Development Team
  - Expert Consultants
- Equipment
  - New Hardware and Software
  - Installation
- Facilities Renovation
- Materials and Supplies
- System Conversion and Training

Continuing, Operational Costs
- Personnel
  - Operators, Application Programmers
- Hardware/Software Maintenance
- Overhead—Heat, Power
- Materials and Supplies

Figure 6.2-3. System Costs.

managers for credibility and accuracy, is the principal basis for the commitment. The size of the activity in time and resources is often specified by the managers, but should represent a good balance between significant goals and predictability in meeting those goals.

The cost/benefit analysis usually compares estimates of costs to the completion of the development with the benefits of the development. The closer to the end of development, the more credible the cost estimates; successive activities should produce improved credibility in the cost estimates.

As with schedules, there are two parts to an effective cost estimate; first, make a good estimate; second, make the estimate good. Stated another way, a good cost estimate is one you can make good. The box structure hierarchy can help you make estimates good, because you can measure progress and the work remaining against the hierarchy. Both time and resources can be allocated top down through the hierarchy, and then managed to the allocations as work progresses.

Each black box in the hierarchy represents an opportunity to delegate responsibility for and to allocate resources to, in analysis or design. As progress is made in top-down specification or implementation, resources can be reallocated to black boxes remaining to be expanded, or even responsibilities redelegated. The benefits of black box replacement in the system apply to its development by people, as well.

The best way to establish credibility with managers is to meet your estimates and milestones of progress scrupulously. To do this, you need to keep your eye on the activity while in it, and let the remainder of the development take care of itself in due time. At every activity, the remaining development costs and benefits must be defined anyway. So, when properly planned, a sequence of properly executed activities will produce the desired system. The alternative temptation is to spend undue time thinking (and dreaming) about the system and the remainder of the development to the detriment of the activity. But it pays to concentrate on the task at hand, even though it requires discipline and diligence, compared to great thoughts about the future.

While the costs and cost estimates are always tangible, the benefits are not always so tangible. In some cases, direct cost savings or additional profits can be deduced as benefits. But in many cases, tangible benefits are harder to deduce.

Even so, intangible benefits can be stated in ways to permit more direct human judgment. There is a profound difference between emotional judgments and business judgments, and there are many ways to present intangible benefits to make business judgments with more confidence. For example, consider the question of preserving files from electrical power outages as discussed in Chapter 1. An auxiliary power supply will have a cost and a benefit, but what benefit? It can be proposed because it is the best thing to do, simply calling for an emotional judgment. A better way is to determine the actual costs of recovering files lost from a power outage and compare those with the cost of the auxiliary power supply. Even better is to include the cost of business lost while recovering files lost. These benefit costs may be difficult to determine and take some work, so there is a trade-off between the cost of the analysis and the accuracy of the results. That is, there are also trade-offs in making trade-off analyses.

But even with the effort put into trade-off analyses, some thought on how to present them can help managers make better business judgments. For example, simply determining break-even points can help in a judgment. In this case, the break-even time between electrical power outages to make the auxiliary power supply worthwhile will be a single number—say eight months. Now, people can make judgments more directly—even consult power outage history data (which should be brought in by the analyst)—taking into account improvements planned in power availability by the utility company, etc., in a business judgment that has been helped by an intelligent and perceptive analysis of an intangible benefit.

6.2.4 Implementation Activities

The objectives of an implementation activity will be found in the activity plan within the development plans. In general, implementation objectives are to create specified systems and prepare users and operators for operation. In order to illustrate potential implementation activities, we discuss the following kinds of tasks:

Resource acquisition
Software development
System integration and testing
Operations education

Resource Acquisition

The acquisition of system resources, such as hardware, software, and site renovation materials, requires exact information and planning. Acquisition tasks should occur as early as possible once system design decisions have been made. Figure 6.2-4 provides a brief listing of system resources that may be needed in building and operating an information system. The development team may or may not be responsible for all types of acquisition. For example, the business organization may perform the site renovations and the personnel hiring. However, such activities must be closely coordinated and should appear in the activity schedule.

Early and effective coordination with hardware and software vendors and utility companies is essential. Delivery and installation delays can ruin a carefully planned implementation schedule. Vendor obligations and warranties must be clearly understood and documented. The development team may become involved in the negotiation of hardware and software maintenance contracts if so desired.

Hardware
- Computer Systems
- Peripheral Devices—Disks, Printers
- Maintenance Contracts

Software
- Operating Systems
- Application Software

Physical Facilities

Communications
- Networks
- Phone Lines

Personnel
- Operators
- Application Programmers

Figure 6.2-4. System Resources.

Software Development

The development of software may be a large or small part of the implementation. The advances in software engineering over the past decade provide excellent guidance for writing software from design specifications. The use of the Box Description Language for the system design specifications greatly facilitates software development. The BDL can be translated into high-level structured programming languages, such as Ada, C, Pascal, Structured FORTRAN, and Structured COBOL. Through rigorous software engineering, the program coding can be performed to standards of clean modularity and high reliability.

System Integration and Testing

System integration is the process of bringing together all of the system components and making them work as an integrated whole. The box structure hierarchy in the design library is used as a guide for this process. The system top should be integrated and tested first, using "program stubs" to represent the subsystems it invokes. Individual subsystem tops can then be integrated and tested in the system environment, using program stubs of their own, which will in turn be replaced by their tops. With each integration step, the evolving system acquires new functionality, which can be tested in a stepwise manner as well. Each integration step verifies that the interfaces between the evolving system and the new increment are correct, and introduces new interfaces to be verified in subsequent steps.

System testing should test the capabilities of the total, integrated system. The system testing plan should include the following types of testing:

System Objective Test. The system objectives, developed in investigation, are recalled. In the order of priority, each objective should be tested on the system. Operators, managers, users, and developers should be satisfied that the objectives are fulfilled on the system.

Volume and Stress Test. The limits of the system, in terms of performance and capacity, should be tested. The system can be loaded with computer-generated test data to and beyond its normal operating capacity. Jobs can be submitted to drive the system to maximum utilization. Performance objectives can be studied with these limiting conditions.

Operational Test. A system environment that approximates normal operations should be tested. All managers, users, and operators should "walk through" a normal day's (week's) activities. This test studies normal system performance as well as such important factors as site planning, adequate personnel, and operational procedures.

Documentation Test. The documentation developed for the system (System's Manual, User's Manual, and Manager's Manual) should be tested for completeness and accuracy by selected personnel in the organization.

A comprehensive system test plan should be developed before the system is complete. The system specification provides adequate information upon which to construct the test plan. The system test plan should be reviewed and agreed upon by the organization before testing begins.

During the actual system testing a detailed record of all tests and results must be maintained. Any unsatisfactory results must be addressed before the system testing is complete. This may require a reiteration of previous activities. Additional requirements gathering, system analysis, and system design may be needed to rectify system problems.

Upon completion of a comprehensive system test, all testing documentation is added to the evaluation library. When the system is accepted as meeting its objectives, the implementation is complete.

Operations Education

Building the information system is not the only implementation responsibility of the development team. Equally important is educating the people in the business in how to operate and use the system to the greatest effectiveness. Management, operator, and user groups should be identified along with their different education needs.

The objective of operations education is to instruct the business organization in implementing the operations design specifications. As part of the educational process, system case studies and tutorials can be developed and retained for future educational activities in the organization.

Seminars and classes can be scheduled throughout the development period. Hands-on system training is best scheduled when the system is operational at the end of the development process.

Operators require an in-depth period of training in order to become completely familiar with system operations. Based upon the complexity of the system, this training may require from a few hours to several months of intensive study.

The system development team may provide this training or skilled instructors may be hired to construct training materials and course outlines. A standard procedure for ongoing operator training must be studied and planned.

Users require education on how to execute the system and retrieve and interpret results. User **tutorials** can be an effective method for user instruction. Tutorials may take the form of a documented walk-through of applications using sample data. Interactive tutorials, walking through a simulated application on the system, are conducive to rapid user learning.

Managers must be educated in the objectives of the system. They should understand what the system does and how it works. The use of the system as a management information system and as a decision support system is necessary knowledge for middle and upper managers.

6.2.5 Information System Operations

As the system development progresses to completion, plans are needed to bring the new system into operation. The new system is brought into operation by being placed into interaction with the business environment. The system development team remains active until the business organization has fully accepted the operational responsibility for the system. While the system is being brought to full operation, the development team has responsibilities in managing the system conversion, establishing operations assurance, and planning for system evolution and obsolescence.

System Conversion

When a system currently exists that is to be replaced by the new system, then a method of conversion from the old system to the new system is required. Three conversion methods have been used with different advantages and disadvantages.

Parallel Conversion. As illustrated in Figure 6.2-5, the old and new systems are run side-by-side until the new system has been satisfactorily tested and accepted by the organization. The advantage of this method is that the old system is still on-line during the conversion procedure. Operations can be maintained normally while the new system is being tested and corrected. This provides a nonthreatening conversion. The disadvantages are the expense of operating two parallel systems and the fact that the two systems may not be comparable in terms of results.

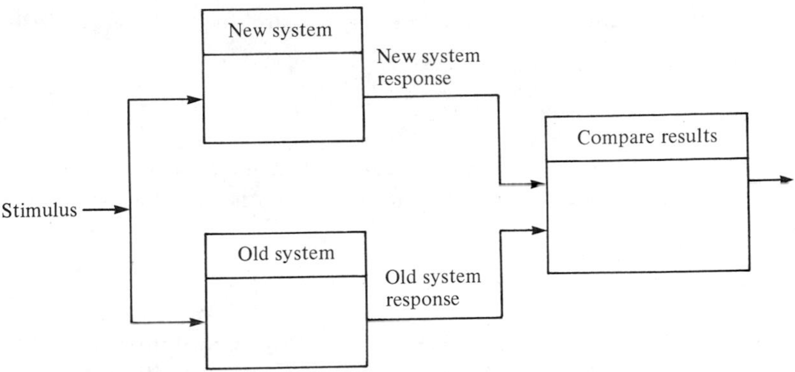

Figure 6.2-5. Parallel Conversion.

Phased Conversion. The new system is gradually phased in a subsystem at a time. When each new subsystem is tested and accepted, it takes the place of the old subsystem. The concept of phased conversion is critically important for large and complex systems. In fact, it may be decided during initial planning in investigation to structure the entire system project into phases. Each phase would be developed and implemented on its own schedule.

Direct Conversion. This is a one-time changeover from the old system to the new system. This technique also applies when no previous system exists. The advantage of direct conversion is that it is clean, quick, and relatively inexpensive. However, the risk is higher that the new system may develop problems—and no backup system will be available. This fact stresses the importance of thorough system testing before a direct conversion.

Operations Assurance

Operations assurance concerns the correct and efficient handling of the system integrity issues during operation. Standard Operating Procedures (SOPs) are established to utilize the devices and techniques for integrity control in the system. For example, the following system topics are important components of operations assurance:

Privacy and Security Controls

System Backup and Recovery Facilities

Performance Monitoring and Tuning Capabilities

Preventive Maintenance

Reviews and Audits

During operations, managers, operators, and users will have responsibilities for maintaining the integrity and effectiveness of the system.

System Evolution

Evolution and change are inevitable during a system's operational lifetime. New requirements and objectives will require alterations to the operational system.

System modifications will be managed in the same manner as the original system development. Each modification will go through the activities of investigation, specification, and implementation, in the context of the operational system. The analysis details and the design specifications will be added to the analysis and design libraries, respectively, of the original system.

The box structure diagrams and the BDL designs for the original system will be modified based upon the system changes.

System Obsolescence

Obsolescence occurs when the current system can no longer fulfill the organizational objectives demanded of it. Additional modifications will not substantially enhance the system. The system is then declared obsolete and investigation must begin on a new system to replace it.

The new system begins its own system development process. The obsolete system will continue to run until the replacement system is accepted as adequate. Note that requirements gathering on the old system will provide useful documentation in the form of the system development libraries to support the investigation of the new system.

Summary: Performing system development well requires knowledge, skill, and experience in the activities of investigation, specification, and implementation. Effective management control can be exercised by identifying the stages of planning, performance, and evaluation in each activity.

6.3 SYSTEM DEVELOPMENT LIBRARIES

Preview: System development libraries contain the information discovered through analysis (analysis library) and created through design (design library). The management library holds information pertinent to the business environment. The evaluation library records the results of system verification tests. The contents of the development libraries are used to create the system documentation.

The system development libraries are the central repositories for information in the system development process. The two principal libraries are the **analysis library** and the **design library.** The analysis library holds information discovered during investigation activities and specification analysis. The **design library** contains the information created during specification, design, and implementation activities. The **management library** and the **evaluation library** play special roles during system development as discussed later in this section.

It is a major responsibility of the development team to establish and maintain the libraries. All of the materials added during development should be dated to provide a historical record of the system development.

In a small development project the libraries can be handled manually. For example, each library can be maintained as a looseleaf notebook. However, even moderately sized projects will require computerized support for storage, retrieval, and updating of documents in the libraries. A majority of the information will be stored as text or in specialized form, such as box structure diagrams, project schedule networks, and Box Description Language.

6.3.1 The Analysis and Design Libraries

These two libraries contain the information central to the system development process. Throughout the discussion of the system development activities, we have seen how information is gathered and used in analysis and how information is created in design. The management of this information takes place through the respective analysis and design libraries.

The analysis library is used extensively during investigation activities. All of the system requirements information and the system operations information are structured and stored. The results of interviews, questionnaires, and prototypes, along with details of previous systems are maintained in the analysis library. Considerable storage capacity may be needed to hold all of the raw information gathered for system analysis.

System analysis and operations analysis tasks then analyze this information to discover new methods of satisfying the system objectives. The details of these analyses are included in the analysis library. Box structure derivations could be stored in terms of box structure diagrams. The box structure hierarchies that include transaction, state, and procedure analyses would be managed in the analysis library. Further analysis tasks performed during specification or implementation activities would add to or update the information in the analysis library.

The design library holds the design of the new system. The system design specifications and the system operations design are included in this library. The analysis library is used to create the designs. For large systems, the design library should offer designers automated support for the structured specification language (e.g., BDL). Design decisions should be documented together with the resulting design.

The result of system design will be a complete, top-down specification of the system to be implemented. During implementation, any design alterations must be recorded in the design library before they are implemented. The result of system operations design will be detailed standard operating procedures for the system.

6.3.2 The Management Library

The management library contains information for the management and control of the system development. It serves as the interface between the development team and the business management, operators, and users during the system development process.

At the start of development the system opportunity document and the development plan are stored in the library. Then with each activity loop in the system development spiral the following information is added:

Activity Evaluation. The results of the activity and their evaluation, with management acceptance, are placed in the management library.

Development Plan Update. The development plan is updated and, with management acceptance, is added to the library.

Activity Planning. The next activity plans, after management acceptance, are added to the management library.

The management library, thus, records the chronological development of the system from the business perspective. Decisions based on agreements between the business and the development team, such as system objectives, schedules, resources, budget, and scope are recorded here. In addition, the day-to-day management information for the development, such as activity schedules and SOW, are maintained in the management library.

6.3.3 The Evaluation Library

The evaluation library has the special role of documenting the evaluation stages of all activity loops in the system development. This evaluation information provides the data on which the recommendations of the activity management review are based. This library also documents the adequacy of the evaluations performed during development. The following types of evaluation information are stored.

Feasibility Assessment. During investigation activities the results of the system feasibility studies are recorded in the evaluation library. The feasibility results are presented in the review documents from which the system scope is defined.

Design Verification. As the design is created, verification is used to guarantee that the specifications meet the objectives of the system. The verification procedures and the results are placed in the evaluation library.

Software Verification. Programs written during implementation should be verified for correctness based on their specifications. Test data is generated and the software is tested under controlled conditions. The testing results, along with the test data, are included in the evaluation library.

System Testing. System testing should be comprehensive and rigorous, as discussed earlier in this chapter. The type of tests performed and the results of the tests are part of the evaluation library.

6.3.4 System Documentation

While the system development libraries are built and used almost exclusively by the development team, the libraries serve as a solid basis for required system documentation. The preparation of system documentation is an ongoing activity throughout all of the system development process. However, during the system implementation, a formal documentation activity must exist. This activity must have well-defined objectives and assigned resources. Based upon the documentation requirements of the organization, the following manuals should be produced by the documentation activity.

System Manual

The System Manual will be drawn from information in the analysis library, design library, and evaluation library. It is to be used by system operators and system programmers who require a technical understanding of the system and the design decisions. The goals of the manual are:

Enable operators and programmers other than the developers, to use, enhance, and maintain the system.

Record technical information that enables system changes to be made quickly and effectively.

Facilitate auditing and verification of the system.

Reduce dependency on individual developers and operators, to in turn reduce the effects of personnel turnover.

A suggested outline for the System Manual is:

1. Introduction
   - Background
   - Underlying Business Process
   - System Objectives
2. System Description
   - Box Structure Hierarchy
   - Box Structures
   - Analysis and Design Considerations
3. System Transactions
   - Input/Output
   - Transaction Use
4. System Data
   - Syntax Structures
   - Data Structures
5. System Integrity

6. System Evaluation
   - Design Verification
   - Software Verification
   - System Testing
7. System Evolution
   - Modification
8. Appendices
   - Glossary
   - Error Messages
   - Source Listings
   - Bibliography

User Manual

The purpose of the User Manual is to provide users with an understanding of the system's purposes, capabilities, and limitations. The details of analysis and design need not be presented to the user. The user should be presented with the overall structure and logic of the system, input requirements, output formats, and the use of system results. The goals of a User Manual are:

Assist the user in understanding what the system does, that is, its black box behavior.

Provide all information necessary to use the system accurately and effectively.

Enable potential users to determine whether the system will serve their needs.

Present system tutorials for users for self-learning.

A suggested outline for a User Manual is:

1. Introduction
   - Background
   - Underlying Business Process
   - System Objectives
2. Description of System
   - Transactions
   - Input/Output
   - Data

3. Use of System

    Input Requirements
    Output Reports
    Sample Executions
    Tutorials

4. Appendices

    Glossary
    Error Messages

MANAGER MANUAL

This manual provides managers and executives with an appreciation and understanding of how the system is used in the organization. Managers will be required to use and interpret the results of the system and to support its continued use and maintenance. The role of the system in organizational decision making must be clearly discussed. Portions of this information can be drawn from the management library and the design library. In particular, the system operations design should be clearly stated in this manual. The goals of the Manager Manual are to

Provide managers with information so they may determine whether system objectives have been met.

Provide plans and schedules for system operations, training, and evolution.

Demonstrate the system's role in organizational decision making.

A suggested outline for the Manager Manual is:

1. Introduction

    Background
    Underlying Business Process
    System Objectives

2. Description of System

    Box Structure Hierarchy
    System's Role in Organization
    Transactions
    Decision-Supporting Results
    Manager's Use of System

3. Development and Operation of System

    System Development Summary

System Operations Design
Standard Operating Procedures

4. Evolution of System
   Future Applications
5. Appendices
   Glossary
   Bibliography

Summary: The system development libraries are the central repository of all development information. The libraries are designed to support the goals of completeness, consistency, and efficiency during system development. System documentation is developed from the information in the libraries.

6.4 WORKING WITH PEOPLE IN SYSTEMS DEVELOPMENT

Preview: While box structures help with logic problems in information systems development, the people problems are still the deepest and most persistent, and require the same interpersonal skills as for any other field. Managers, users, operators, and developers each pose different problems because of their roles in the business, but the box structure methodology provides an effective basis for addressing these problems.

As noted in Chapter 1, the deepest and most persistent problems of information systems development are people problems, not logic problems. However, faulty solutions to logic problems can create unnecessary people problems. This book has concentrated on the logic problems because they can be treated comprehensively and gotten out of the way by systematic use of logical principles. Every person deals with people problems continuously, at home, in school, at work. The same skills and sensitivities for interpersonal communication and cooperation apply in information systems development. It is up to you to use all you know about dealing with people in information systems development, and you will have to work at it during your whole career. Getting good logic solutions is not enough without a sensitive understanding of everyone you deal with. But this is nothing new as a requirement for success in any field.

6.4.1 Working with Managers

The managers of a business have an obligation to make the business succeed, namely, to survive and prosper, as well as to meet additional objectives set down by its owners or directors. More and more, businesses voluntarily accept social and national responsibilities, but without first surviving and second prospering, such social and national objectives are irrelevant. In particular, the managers of a business have no obligation to use information systems, or to have information systems developed at the expense of the business, except as secondary requirements to meet their primary objectives of business survival and success.

In a small business, managers can be in on most decisions and make them on the spot. However, in large businesses, the upper level managers cannot be everywhere the business is, so they cannot be in on most decisions. There are two general ways to try to cope with size in business operations. One way is for each level of management to make as many decisions as possible, then leave the rest to the lower levels. The second way is for each level of management to spend part of its time defining business processes at lower levels rather than in direct decision making. The higher the level, the higher the fraction of time that goes into defining business processes. Almost every large successful business copes with size in this second way. In a large corporation, the top levels of management will spend 80% or more of their time on business processes rather than in direct decisions.

Information systems represent a way to implement and reinforce business processes, so it is no wonder that businesses use them extensively. However, it is worth noting that the reason for information systems is to implement business processes, whose reason, in turn, is business performance. As a result, discussions about potential information systems should begin with business performance and business processes, not with data processing systems of hardware and software, no matter how exciting they may be.

Although there is an ideal development plan, with successive activities of investigation, specification, and implementation, development plans and commitments are usually tied into a larger business process, so that managers of the business seldom commit to the entire development without milestones and options at several points in time. There are two main reasons:

1. It makes good business sense to check progress and make Go/Nogo decisions at frequent intervals. Over time, progress may be slower or faster than expected, and lessons learned can be incorporated into

future planning. One lesson may be to halt the development entirely; another may be to speed it up by increasing the level of resources applied to it.

2. The business exists in a real world that creates its own set of problems and opportunities that may be addressed by information systems development. For example, a competitor may announce a new on-line reservation system and an immediate competitive response is urgent. Or a change in banking laws may permit a better use of financial assets if an information system can provide more rapid, more current financial data immediately.

For these reasons, even though an ideal development plan exists, it may not always be followed, indeed, will not usually be followed in changing times for the business. For example, faced with a competitive reservation system, a business will not have the time for an investigation, specification, and implementation stretched out and interspersed with Go/Nogo decision periods. Instead, these activities must be bypassed or abbreviated to the greatest extent possible, possibly by choosing a less than best solution in existing hardware and software packages to get on the air as soon as possible. Once on the air, a more orderly transition can be planned to get to a more satisfactory long-term system solution. But by this time, reservation clerks will have been trained, customers grown used to, and considerable costs sunk into the initial quick and dirty system. So the planning must take these factors into account, as well as the ideal questions of information systems development.

In short, the reasons for information systems are in their businesses, and their development takes place under business conditions, not ideal conditions.

Even so, the principles of box structures are as valid as before. They simply need to be applied more flexibly than in the absence of business forces and pressures. For example, the quick and dirty reservation system can be described from the beginning in box structure terms, even though the people who created the hardware and software packages did not think of them that way. The situation is similar to discovering a clear box for a human sales forecaster who does not know of box structures. The box structures are there to discover and use, whether put there consciously or not.

6.4.2 Working with Users and Operators

Even though a system may provide all the capability required of it, it may be hard to use. When it is, its effective capability is reduced, perhaps in drastic ways. It is of little value to have the internals all worked out if people are not able to make use of them. Frequently, these internals represent the most interesting and challenging problems of analysis and design for the developers. But if the developers become so absorbed in these problems that the user interfaces are worked out as afterthoughts, the usual result is poor usability and a system with seriously reduced effective capability, even though it is all there inside.

It is relatively easy, even as afterthoughts, to add input/output support to make systems user friendly in entering commands and data and observing the results of such entries. And such user friendly facilities can help increase the effective capabilities of the system. But if that must be done as an afterthought, the major opportunities for providing system usability have already been missed.

Inherent System Rationality

The major opportunity for providing system usability is in the inherent rationality of the system itself—not the rationality of the developer, but the rationality of the user. This inherent rationality permits the user to be in charge of the situation, whatever it is, to use the system intelligently and to best advantage. Without this inherent rationality, the user is forced to use the system by rote, often without realizing better possible solutions for the problem at hand. Even if the input/output is user friendly, if input/output data are being conceived by rote, the potential capability of the system is not being used.

It is a job of perceptive analysis from the very beginning of development (in investigation and specification) to discover the rationality of the user and to embody that rationality in the design. The internals of the system may be challenging, but if they are not firmly rooted in the user rationality, and that rationality is not clearly visible to the user in the completed system, then the internals will not be used effectively.

For example, many problems of decision analysis can be formulated either in terms of mathematical optimization or in terms of spreadsheets. If the users are accustomed to spreadsheets and use them daily for other purposes, it is dangerous to just solve such problems in terms of mathematical optimization models, then specify inputs/outputs to suit these models, say as a matrix of coefficients. The system may solve the problem, all right, but the users will need to provide inputs in a form they do not well understand and have little intuition about. No matter how user friendly the input task can be made, the input data must be prepared by rote. Occasionally, in such situations, anomalies in input data not recognized because of the lack of intuition can lead to ridiculous results, even though they are mathematically correct. Hopefully, common sense will prevail and the user will regard the system as failed; even worse, common sense does not prevail and it is the business or part of it that fails instead of just the system.

In summary, to provide effective system visibility, the first requirement is to provide inherent system rationality for the user, a rationality that must be discovered and built into the system from the very start of development. Once inherent system rationality is achieved, the system should be designed as user friendly as possible for input/output, but not before inherent system rationality has been accounted for.

Effective User Manuals

The foremost tangible item for user support in a system is the user manual. It should contain all the information a user needs to know to use the system. User manuals should be written before the implementation, not afterwards. The user manual should provide guidance for the implementation, not the other way around.

In many cases, early user experience will lead to better understandings of the best user interfaces. Even so, this is no reason not to write user manuals before the implementation. In such cases, a system and user manual revision should be planned at the outset, as an immediate evolutionary step, possibly before full use of the system. The user manual, in such a case, represents the best working hypothesis of the required user services and interfaces available. Without such a crisp, definitive working hypothesis, the implementation can easily degenerate into an undisciplined trial and error process in which accountability for effective performance is lost between developers and users.

The specification and implementation activities of information systems development have good analogies in medical diagnosis and surgery. When internists diagnose a medical condition and prescribe surgery, it is up to them to specify the exact surgery required, for example, to remove an appendix or to remove gall stones. The surgery is required to be quick and clean, not to continue the diagnosis. For example, if an appendectomy is required, out comes the appendix, good or bad. The alternative temptation is to slice a little bit, diagnose a little bit, and so on, only to discover that the appendix is alright after all. The problem may be that the patient is dead by this time, but who is accountable, the internist or the surgeon? In information systems development, such an—implement a little bit, specify a little bit—process may discover a very good user interface, only to find the system has been terminated because of a lack of confidence by the managers due to missed schedules and budgets.

There is another fundamental reason why user manuals should be written before implementation, even when it is virtually certain that the user interfaces will need to be changed. It is to capture the inherent system rationality.

As already noted, the first priority in ensuring system usability is to create an inherent system rationality from the viewpoint of the user. Only then should the actual user interfaces be considered in detail. This means that a user manual for a system with good inherent rationality will be at risk primarily in its details; if the user interfaces must be changed, it will be the details of the user manual that must be changed. Its overall structure and rationality should survive such a change. Therefore, writing the user manual before implementation is not as risky or foolish as it may seem at first glance. Most of the work, particularly the conceptual work, will already be done, and the rework of the details will be sharpened considerably by the initial user experience.

The alternative temptation to wait for implementation and experimentation to write the user manual will be more expensive and far less effective in the long run, for several reasons:

1. After implementation, there will be pressure to use the system without adequate time and effort to get the user manuals into the best possible shape.
2. The implementation process itself produces an amnesia (for developers) in the rationality of the system caused by dealing with all the details and possible surprises of implementation itself.
3. Late user manuals mean skimpy attention to user training plans and facilities.
4. The process of writing user manuals should tie into the box structure descriptions available at specification and ensure that the inherent user rationality is reflected in the implementation.

Training for Effective Use

User training begins with a good system and a good user manual. Its objective is to help users be effective in their business purpose, not just to use the system well. If the system is well conceived, effective business performance and the good use of the system is usually synonymous, but not necessarily is this always the case. For example, if the system breaks down or produces nonsensical results, the user should use common sense on behalf of the business. That is, the system should be the servant of the users, not their master.

The analysis library may be useful in training or in development of training materials for users. The box structures in the analysis library will be described in user terms and should document the inherent rationality of the users. As a result, it may be useful to better ensure that all users see their own activities in a uniform, documented form, rather than depending on word of mouth or example to achieve standard practices among users.

Consulting and Help

A development team that has discovered the inherent user rationality of the system before implementing it will be in a good position to help and consult in its usage. Conversely, a development team that has made up its own rationality from the computer point of view rather than the business point of view will not be in a good position to help and consult in the use of the system to achieve business objectives. Yet, it is not always easy to see the difference in practice. Users are frequently intimidated or impressed with pure technology and may blame themselves (or be blamed by the managers) for ineffective use of a system. Indeed, users may make good use of a poorly conceived system without realizing the missed opportunity that the system could have been even more useful if better conceived. In short, the very principle of business first, system second, that produces superior business systems automatically produces the best ability to help and consult in the usage of the system.

Systems of People and Computers

We have already noted that people and computers both store and process information, at greatly different rates but, nevertheless, they provide the same function. For that reason, it is most effective to conceive of systems and subsystems implementation in a flexible way to best use people and computers in cooperative ways. While computers and software must be specified and realized, so also must people and training be specified and realized as well.

The joint implementation of systems by people and computers is easier said than done. A common mistake is to concentrate on the computer side, and treat the people side—the users—as after thoughts, so that people merely fill in those parts the developers were unable to implement in computers. Such a mistake leads to poor interfaces, but even more importantly, to the wrong systems rationality—designed for the computer rather than people.

When a system is conceived as a flexible arrangement between people and computers, the system functions should be divided so that computers do as much of the clerical data processing as reasonable, and people do critical decision making and recognition functions. For example, an airline reservation clerk serves as a voice to keyboard translator and a digital display to voice translator among other functions. The technology is here to move from digital display to voice automatically because little recognition is involved. For example, many telephone exchange information services use automatic digital display to voice translation. But the translation from voice to keyboard will come much later.

The airline reservation clerk also provides critical decision making in helping customers decide on flights or to answer other questions on fares, connections, and seat assignments. However, for the customer on a telephone, the airline reservation clerk is part of a system of people and computers whose immediate interface is the clerk. A good airline reservation system makes best use of people and computers with no preconceived notions of what either should be doing.

In advanced systems planning and development it is especially important to keep both people and computer developments in mind. A common mistake in advanced systems is to think very hard about what computers can do, but to take people for granted—to conceive of "future computers and present people". The reality is that people are usually the most advanced part of any new system. It is relatively easy to predict the future abilities of computers—their speed and storage capabilities, their input/output capabilities. But it is much harder to predict the future abilities of people.

As a simple example in another technology, seventy years ago it was easy for experts to predict that production automobiles would one day go seventy miles an hour. But how many of these experts could have predicted that seventy-year-old grandmothers would be driving them? In conceiving systems it is even more important and difficult to visualize what people can do than what computers can do.

Another example, the advent of networks of microcomputers is leading to entirely new methods of doing business in large organizations, where most communications become instantaneous and paperless. The ability to organize microcomputers into such networks has long been known, but the use of such networks by people is developing on a scale and in ways never predicted.

6.4.3 Working with Developers

Development Organizations

The product of information systems development is largely invisible. Unlike the computers that they use, the system design and software that implements an information system is a sea of details that only the developers see and understand. The user guides and manuals reflect the system rationality, but the ultimate result is the dynamic operation of the system, not in any static structure, such as a cathedral or a building. As a result, progress in system development is much harder to understand than for a static product. This difficulty must be dealt with in the development organization.

In a static structure, it may be possible to divide the structure itself for assignment to various suborganizations. For example, one might divide a building into four parts, and for each part use masons, carpenters, and painters in the same proportions. The interfacing walls between the parts could be worked out, and anyone not working on a common wall need not know anything of the other parts.

Unfortunately, things are not so simple with an information system. The box structures allow a good deal of isolation and delegation at every point in a box structure hierarchy. But even so, one cannot divide an information system up as simply as a building.

The guiding principle in organizing information system development work is to divide up the work, not the systems to be developed. The surgical analogy already used is a good one to start with here. A surgical team could be imagined in which a set of co-equal surgeons were all operating together, each sterilizing their own tools, providing their own sponges, checking the status of the patient, and so on. Of course, we know that is not the right way.

A surgical team does not divide up a patient, it divides up the work. One person does the surgery, another provides anesthesia, another radiology, another sterilizes scalpels and sponges, and so on. The rationality of the surgery is held in one mind, that of the surgeon. The remainder of the team helps guarantee the integrity of the operations. As with information systems, the integrity of the surgery is largely independent of the particular function of the surgery. Anesthesia, scalpel and sponge sterilization, and so on proceed similarly for a wide class of surgical procedures. The radiology and other services may be more closely related to the surgical procedure itself.

In a similar way, the rationality of an information system must be as consistent and thereby as closely held among a few minds as possible. To divide up the rationality, as one might divide up a building, is a serious mistake. But how can one divide up the work without dividing up the system? The surgical team gives some clues. Find supporting tasks that are not so specific to the rationality of a particular system.

A major division is to divide the work between conceptual and clerical activities. The clerical tasks will not be specific to the system, but will require general skills for entering and verifying designs and data, just as general skills for sterilizing scalpels and sponges are required for surgery. Both conceptual and clerical tasks can be divided further. Clerical skills for entering and verifying designs will be different than those for data. Again, data entry skills may be divided according to specific system requirements.

The conceptual skills can be divided along lines discussed in this book. The box structures provide a distinction between analysis and design activities—between working with managers, users, and operators, and working with procedures and computers. Each black box identified provides an opportunity to delegate a task of design and analysis to a group or a person. Upper level black boxes can be analyzed for consistency and appropriateness by several groups or persons when desirable. In fact, the box structure methodology can be regarded as a work structuring methodology as well, and has been largely motivated by the management problem from the outset.

Information Systems Engineering Practices

The basis for consistent high quality information systems development work is a set of systems engineering practices to provide uniformity and repeatability in analyses and designs. The box structure methodology of this book represents a set of practices for analysis and design. Additional business specific practices are needed for the activities of investigation, specification, and implementation. These additional practices should be augmented with management practices to ensure good use of personnel and close cooperation with the managers, users, and operators in the business.

Engineering practices can be viewed as straight jackets or as liberating disciplines. Well conceived, they provide a basis for a more creative, more productive work by highly capable information systems personnel. Without good engineering practices, groups face much rework, many misunderstandings at interfaces, and many frustrations and disappointments in going from idea to implementation. The marks of good practices are high morale and well deserved self-esteem from the effective information systems development they help make possible.

Information systems development is a challenging and creative activity. But variability should not be confused with creativity in system analysis and design. If a definite system is called for in a given business situation, then any creative person should invent that same system. And just inventing a different system does not make it a better, or even an adequate system. The box structure methodology begins with business processes and ends with better business processes, very possibly supported by new information systems. The challenge and creativity is to keep the business processes in mind even while navigating in the deepest waters of system development and the logical problems of computer hardware and software it represents.

Summary: Box structures provide a framework for addressing the needs of managers, users, operators, and developers. Information systems exist to support management-defined business processes, and their development takes place under business conditions, not ideal conditions. The usability of an information system depends on the extent to which it embodies the inherent rationality of its users and operators. Information system developers should divide up the work, not the system.

6.5 EXERCISES

1. You are hired by a small business to perform a systems development for a computerized word processing system. Currently a single secretary does all typing manually. Business is expanding and a second secretary may be hired. Three officers of the business produce the work to be typed. They expect the developed word processing system to cost less than $10,000 (including your fees) and to be ready for full use in one month.
   (a) Hypothesize and illustrate a typical system development plan for this project. What tasks would be performed in each activity? What planning and evaluation would be needed in each activity?
   (b) Discuss the feasibility issues found in the investigation activity.
   (c) Describe the contents of the four libraries for this project.
2. A system development is planned for a bank to provide first, branch teller services, then automatic teller machines in grocery stores. However, a network system is needed to support these teller facilities. Construct a development plan for this situation.

3. Construct a project network for the following activity information:

Activity	Immediate predecessor	Duration (weeks)
A	—	4
B	A	20
C	A	10
D	B,C	5
E	D	10
F	D	7
G	E,F	12
H	F	10
I	G,H	1

For each event calculate the EET and LET. Find the critical path of the project and indicate slack times for non-critical activities.

4. Develop scenarios in which each of the three types of system conversion would be the most advantageous.

Chapter 7 | Syntax Structures in Information Systems

7.1 SYNTAX STRUCTURES

Preview: Syntax structures of stimulus histories and responses lead to simpler and clearer descriptions of black box, state machine, and clear box behavior. A formal grammar is given by rules of syntax which describe all possible stimulus and response histories of a given class.

The common property of black boxes, state machines, and clear boxes is that they all accept stimulus histories and produce responses. The responses of simple black boxes, state machines, or clear boxes are easy to describe in terms of stimulus histories. But for more realistic ones, the responses are not so easy to describe. In these more realistic problems, more powerful methods are needed for describing stimulus histories, in order to specify what responses are required. This chapter introduces methods of syntax—how stimulus histories can be structured by their syntax for better understanding of the black box, state machine, or clear box behavior required.

As seen in the hand calculator, while many black boxes and state machines will accept any stimulus history, only certain histories make sense and provide intelligible responses. These histories need to be described for user guidance and information—without user knowledge of what information systems do, what they do is useless.

Fortunately, there is a simple, but powerful, method for describing intelligible user input. It is a method of long human history, based on the syntax of natural language and speech. But, whereas syntax methods can only deal with natural language to some approximate degree, they can deal with explicitly designed user languages in information systems completely and exactly.

7.1.1 The Syntax of Hand Calculator Inputs

The hand calculator provides a simple example of syntax methods. The inputs which will provide dependable outputs are arithmetic expressions preceded by a clear key and followed by an = key. There is no limit to the number of stimulus histories which satisfy this requirement—literally, an infinite number of such histories. But even so, we can define a finite set of syntax rules among syntax parts (as parts of speech) which all these stimulus histories must satisfy. Such a set of syntax rules is called a **formal grammar.**

These rules are written using special symbols, known as **metasymbols,** to distinguish the parts of the rules. The metasymbols < and > are called angle brackets, and are used to bracket names of syntax parts. Two additional metasymbols required to define syntax rules are ::= (a three character symbol), and | (a vertical bar). The ::= metasymbol means "is an instance of" and the | metasymbol means "or". Thus, for example, the syntax rule

<HC Input> ::= C = | C <expression> =

means that a syntax part named "HC Input" (HC for hand calculator) is an instance of "C =" (the sequence of a Clear key and = key), "or" (an instance of) "C <expression> =" (a sequence where a new syntax part named "expression" appears between C and =). In this case, <expression> must be defined elsewhere, but C and = are literals that stand for themselves. Note that the answer provided by the hand calculator for the stimulus history C = is always 0, but that is an intelligible response to the user nevertheless.

The stimulus history for <expression> (an arithmetic expression), must be of the form

<number> <operator> <number> ··· <operator> <number>

where each <number> is a sequence of digits, and each <operator> is one of +, −, ∗, /.

Another way to describe an <expression> stimulus history is as a

<number> followed by zero or more <operator> <number> pairs. The repetition metasymbol * which precedes a syntax part means "zero or more" such parts. (Note the distinction between * as metasymbol and as a literal multiplication operator, which will be clear from their context.) Thus, if we first define the syntax rule (<operator number> ::= <operator> <number>), we can then define <expression> in the syntax rule

<expression> ::= <number> *<operator number>

which means, expanding the repetition part,

<expression> ::= <number> |
<number> <operator number> |
<number> <operator number> <operator number> |
...

which means, in turn, expanding the <operator number> parts

<expression> ::= <number> |
<number> <operator> <number> |
<number> <operator> <number> <operator> <number> |
...

as was intended for <expression>.

Numbers are sequences of digits 0 through 9. Thus, all of the syntax parts needed to define <HC Input> can be expressed as follows:

1. <HC Input> ::= C = | C <expression> =
2. <expression> ::= <number> *<operator number>
3. <operator number> ::= <operator> <number>
4. <number> ::= <digit> *<digit>
5. <operator> ::= + | − | * | /
6. <digit> ::= 0 | 1 | 2 | 3 | 4 | 5 | 6 | 7 | 8 | 9

These six syntax rules constitute a grammar that defines all possible proper stimulus histories to our example hand calculator. Every syntax rule in the grammar has a left side and one or more possible right side expansions. Each alternative is a possible right side. Thus, for example, rule 5 has four alternative expansions, as +, −, *, or /, which we can label 5.1, 5.2, 5.3, and 5.4, respectively.

Every left side is a syntax part (named with angle brackets) that is ultimately expressed in keystrokes to the hand calculator. For example, the following expansion steps produce a sequence of keystrokes to solve the problem C 7 + 29 =:

<HC Input> ::= C <expression> = (by rule 1.2)
<HC Input> ::= C <number> <operator number> = (by rule 2)
<HC Input> ::= C <number> <operator> <number> = (by rule 3)
<HC Input> ::= C <digit> <operator> <number> = (by rule 4)
<HC Input> ::= C <digit> <operator> <digit> <digit> = (by rule 4)
<HC Input> ::= C <digit> + <digit> <digit> = (by rule 5.1)
<HC Input> ::= C 7 + 29 = (by rules 6.8, 6.3, and 6.10)

7.1.2 Syntax Parse Tables

A syntax **parse tree** is a diagram of the syntax parts which spell out a given syntax part. For example, the parse tree for C 7 + 29 = is given in Figure 7.1-1.

For better readability from the top down, the tree is shown in inverted form. The syntax parts used to make up the problem C 7 + 29 = make up the interior points (nodes) of the tree. The literal keystrokes are the leaves of the tree. In fact, the leaves are (from left to right) C, 7, +, 2, 9, =, which spell out the problem C 7 + 29 =.

Syntax parse trees are easy to understand at a glance, but become unwieldy and awkward for larger problems. Therefore, we introduce an equivalent, but more concise form for the same information, by converting trees to tables in outline form. Each node of a tree is regarded as an item of an outline. The nodes next below each node in the tree are regarded as items at the next level of outline, and are indented accordingly. The outline starts with the top node of the tree. The result is called a **parse**

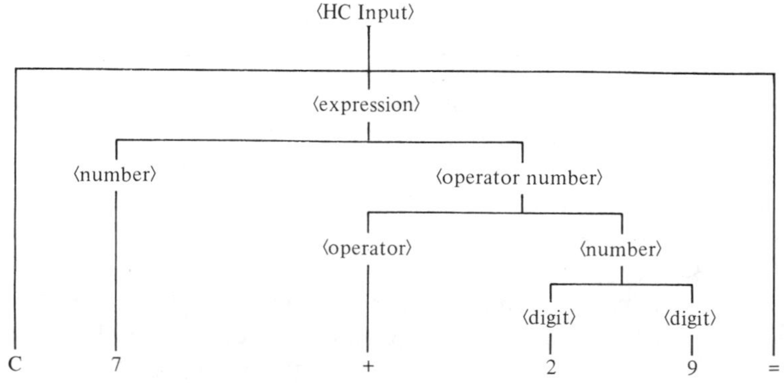

Figure 7.1-1. A Syntax Parse Tree For C 7 + 29 =.

Table 7.1-1

A Syntax Parse Table for C 7 + 29 =

<HC Input>	1.2
C	
<expression>	2
<number>	4
<digit>	6.8
7	
<operator number>	3
<operator>	5.1
+	
<number>	4
<digit>	6.3
2	
<digit>	6.10
9	
=	

table. For the problem C 7 + 29 =, whose syntax parse tree is given in Figure 7.1-1, the parse table is as given in Table 7.1-1.

The parse table also admits the explanation of each part expansion by listing the syntax rule that describes how the next level of the table is obtained.

This parse table can be constructed directly and systematically, without first inventing the parse tree. For example, the line

<HC Input> 1.2

means that <HC Input> is expanded at the next level by the rule 1.2, which is of the form

C <expression> = (rule 1.2)

Therefore, indented next below <HC Input> must be exactly (no more, no less) the items C, <expression>, and =, namely,

<HC Input> 1.2
C
<expression>
=

Now C and = are literals, so no further expansion is necessary. But <expression> is a syntax part that must be expanded. It has but one rule, but that rule, namely,

<expression> ::= <number> *<operator number>

permits any of the expansions

<number>
<number> <operator number>
<number> <operator number> <operator number>
...

For the problem at hand, C 7 + 29 =, the second expansion is needed, so the parse table becomes

<HC Input>	1.2
C	
<expression>	2
<number>	
<operator number>	
=	

The first <number> is 7 (which is also a <digit>), so, by rule 4, we obtain

<number>	4
<digit>	

and (since 7 is given by the rule 6.8 of digit),

<number>	4
<digit>	6.8
7	

The expansions for <operator> and the second <number> follow in a similar way, and will result in the original parse table shown for C 7 + 29 = in Table 7.1-1.

7.1.3 Parse Table Structures

Although the parse table above begins with the syntax part <HC Input>, every new syntax part, i.e., <expression>, <operator number>, <number>, <operator>, and <digit>, begins a new parse table of its own, which fits into the <HC Input> parse table. In each of these new parse tables, the reasoning is localized. That is, the parse table for <expression> is correct or not, independently of how it is used in <HC Input>. For example, the expression

<HC Input>
C
<expression>

(missing the final =) is incorrectly formed, but the expansion of <expression> itself can be judged correct or not independently of this error.

A syntax part is expressed correctly in a parse table only if rules exist for every line of the table. For example, the expansion

```
<number>
   7
```

is incorrect, even though "everyone knows that 7 is a number". The reason is that there is no syntax rule that states "7 is a number". Instead, <number> is one or more <digits>s and 7 is a <digit>. Therefore the correct expression is

```
<number>                  4
      <digit>             6.8
         7
```

with the rules to show for it.

In summary, as shown above, the syntax components of a syntax part are listed below and indented one level—the components are defined by the rule given at the right. The literals have no rules and spell out the syntax part being parsed. The parse table gives a compact way to prove the correctness of a syntax part, as defined in the following principle.

> **Fundamental Principle:** A syntax part is correct (correctly formed) if a parse table can be constructed in which
>
> 1. each syntax part in the table is correctly formed by components at the next level according to the rule given.
> 2. the literals of the parse table spell out the syntax part.

7.1.4 Syntax Expressions

The rules of grammar discussed above use only the metasymbols <, >, ::=, |, *. If any of these symbols were to be used in the language being defined, it would be necessary to distinguish between them as metasymbols and symbols in some way. For example, if "<", ">" were to be used as "less than", "greater than" symbols, the string "< digit >" could mean a syntax part, or a part of some expression in the language, such as (if digit is the name of a variable) "7 < digit > 9", meaning digit is greater than both 7 and 9. Ordinarily, such double uses of symbols can be distinguished by the context of their appearances. For example, it is clear that

<digit> ::= 0 | 1 | 2 | 3 | 4 | 5 | 6 | 7 | 8 | 9

represents a use of <, > as metasymbols and

<comparison> ::= < | = | >

represents a use of <, > as metasymbols on the left side and as symbols on the right side of the rule. When confusion is possible, an explanation should be given in English.

The ordinary parentheses (,) are commonly used symbols, but are also very useful in arithmetic and mathematics to group operations. For example, the expression 4 + 3 * 5 can mean

(4 + 3) * 5 = 35

or

4 + (3 * 5) = 19

depending on how the terms are grouped. This ability to group operations is so useful that parentheses are often used as metasymbols as well as ordinary symbols.

As discussed above, these distinctions must be clear in context or else explained separately. The use of parentheses permits rules of syntax to be simplified in many cases. For example, two of the rules of grammar used above for <HC Input>

2. <operator number> ::= <operator> <number>
3. <expression> ::= <number> *<operator number>

can be expressed in a single rule using parentheses as metasymbols

<expression> ::= <number> *(<operator> <number>)

and the syntax part <operator number> is no longer needed. The part rule of <HC Input> was given in the form

1. <HC Input> ::= C = | C <expression> =

and can be expressed with parentheses as

<HC Input> ::= C (| <expression>) =

That is, <HC Input> always begins and ends with C and =, but in between can consist of nothing or an <expression>. In this case the standard rules of C and = are easier to see in the rule with parentheses, than in the original rule without parentheses.

Note that parentheses can be used to show the possibility of nothing as a syntax part in a rule of syntax.

Parentheses can be used to simplify grammars originally written without them. For example, the grammar for <HC Input>

1. <HC Input> ::= C = | C <expression> =
2. <expression> ::= <number> *<operator number>
3. <operator number> ::= <operator> <number>
4. <number> ::= <digit> *<digit>
5. <operator> ::= + | − | * | /
6. <digit> ::= 0 | 1 | 2 | 3 | 4 | 5 | 6 | 7 | 8 | 9

can be transformed step by step using parenthesized expressions, as follows:

Step 1. Reform HC Input:

<HC Input> ::= C (| <expression>) =

Step 2. Expand <expression>:

<HC Input> ::= C (| <number> *(<operator> <number>)) =

Step 3. Expand <operator>:

<HC Input> ::= C (| <number> *((+ | − | * | /) <number>)) =

Step 4. Expand <number>:

<HC Input> ::= C (| (<digit> *<digit>) *((+ | − | * | /)
(<digit> *<digit>))) =

At this point, each <digit> can be expanded to express <HC Input> entirely in literals, with no other syntax parts at all. While possible, the result would not be easy to read. The best, most easily understood, grammar for a syntax part is usually a happy medium between extremes of no parentheses and no other syntax parts. In this case, perhaps the simplest grammar is defined by Step 2, with the resulting (complete) grammar

<HC Input> ::= C (| <number> *(<operator> <number>)) =
<number> ::= <digit> *<digit>
<operator> ::= + | − | * | /
<digit> ::= 0 | 1 | 2 | 3 | 4 | 5 | 6 | 7 | 8 | 9

The grammar defined by Step 3 is also very understandable,

<HC Input> ::= C (| <number> *((+ | − | * | /) <number>)) =
<number> ::= <digit> *<digit>
<digit> ::= 0 | 1 | 2 | 3 | 4 | 5 | 6 | 7 | 8 | 9

The choice between these two grammars is close. The first gives a name <operator> which helps in explanations to users; the second is one rule shorter. Either is simpler to understand than the original.

The final metasymbols we will use are brackets ([ ]) to denote an optional element of a grammar rule. For example the original rule for <HC Input> could be written as

<HC Input> ::= C [<expression>] =

The brackets mean that the enclosed syntax part may or may not be used when the rule is applied.

Recursion in Grammars

It is often natural and convenient to define language constructs recursively, i.e., in terms of themselves. A recursive production rule is one which contains the phrase being defined in its defining string of phrases and/or characters. For example, consider the following definition of a sequence control structure:

> If S1, S2,...,Sn are statements, then "**do** S1; S2;...;Sn **od**" is a sequence of statements.

To describe this structure in a grammar, we could write the following production rule:

<sequence> ::= **do** <statement>;...; <statement> **od**

but the use of ellipses (...) would create problems when we attempt to define translations based on this description. A better way to define this structure is to introduce a new syntactic entity named <statement list>, denoting any sequence of statements separated by semicolons. Then a possible set of productions is

<sequence> ::= **do** <statement list> **od**
<statement list> ::= <statement> | <statement> ; <statement list>

The rule for <statement list> is read "a <statement list> is either a <statement> or a <statement> followed by a semicolon followed by a <statement list>." From this recursive definition it follows that any sequence of statements separated by semicolons is a statement list.

In this example, we have seen how a programming language structure can be described precisely in terms of a grammar. In fact, an entire programming language can be effectively described by a set of production rules in a larger, more complex grammar. Such a grammar describes the structure of programs written in the language, as an aid to human understanding and to mechanical translation into object code. Detection of errors in a program is also simplified when a grammar is available.

7.1.5 State Machine Syntax Checkers

As surprising as it may seem, we can build a state machine which accepts one keystroke stimulus at a time, and decides if a proper <HC Input> has appeared in the immediate stimulus history. There are four kinds of literals possible in the <HC Input> syntax, namely C, =, digits, and operators. A proper input of the form <HC Input> will trace a path across the leaves of some syntax parse tree, for example, as shown in Figure 7.1-2.

By examining this tree and recalling what variations are possible (several <operator number>s, several <digit>s) it is easy to see what possibilities exist in the successive appearance of these literals. For example, C must appear first. Then following C, either an = or a digit must appear next (= from rule 1.1, digit from rule 1.2). After a digit can appear another digit, an operator, or (as on the far right side of the tree) an =. After an operator a digit must appear. We can summarize these necessary successions in the following table:

After	Can Appear
C	digit | =
digit	digit | operator | =
operator	digit

Now, consider a state machine which accepts a hand calculator stimulus history defined by this grammar and produces for each stimulus one of

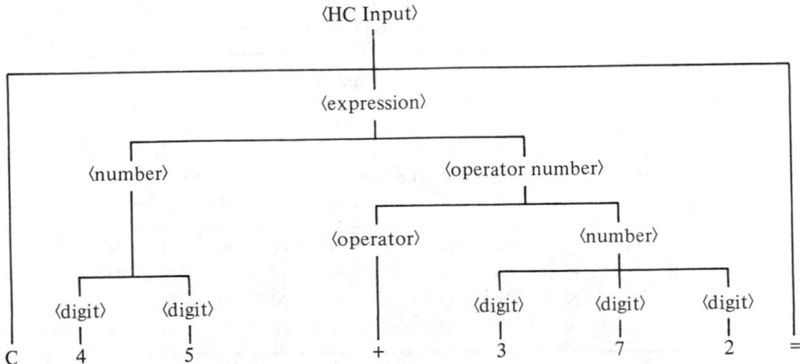

Figure 7.1-2. A Syntax Parse Tree for C 45 + 372 =.

three responses, P, Q, or N:

P: A correct problem has been entered.
Q: Part of a correct problem has been entered.
N: No correct problem is possible with the preceeding stimuli.

For a response of P, the stimulus must be =, because that is the only way to complete a correct problem. If the last response was Q, then the table of successors gives all the ways the next response is Q or P. Therefore, we can organize the required responses of the state machine by considering the last stimulus and the last response together, as shown in Table 7.1-2. The start column shows responses for the first stimulus.

With these responses known, the state is also easy to define. It is just the information needed to determine its responses, namely, the last stimulus and the last response. That is, the state can be the pair,

state = last stimulus, last response

and the machine must determine the response by Table 7.1-2 and the new state as the current stimulus and response.

While this state machine recognizes the given grammar, it also reveals shortcomings in the grammar as a specification of useful hand calculator behavior. For example, the grammar permits only a single expression to be evaluated, even if no syntax errors are present. In this case, a modified grammar beginning with the production

<HC input> ::= *(C = | C <expression> =)

permits a C stimulus following an = stimulus to begin a new expression, provided no intervening syntax errors have occurred. In addition, a hand

Table 7.1-2

State Machine Recognizer for Hand Calculator Input

		Last Response							
		Q				N or P			
Current		Last stimulus				Last stimulus			
stimulus	Start	C	d	o	=	C	d	o	=
C	Q	N	N	N	N	N	N	N	N
d	N	Q	Q	Q	N	N	N	N	N
o	N	N	Q	N	N	N	N	N	N
=	N	P	P	N	N	N	N	N	N

Key: C: Clear Key
d: digit
o: operator

Table entries are response required.

calculator should reinitialize its state on any occurrence of a C stimulus, even after a syntax error, to permit a correct expression to be entered. Such behavior would require additional modifications to the grammar, for better usability in hand calculator operation.

7.1.6 Syntax Methods in Clear Box Design

The net effect of invoking a clear box is to invoke a series of transitions of its component machines. If alternation or iteration clear box structures are contained in the overall clear box, the number and identity of machine transitions invoked will be variable, depending on the (initial) stimulus. It turns out that these machine transitions satisfy rules of syntax, which can be derived directly from the nested set of structures within the clear box. In particular, each sequence, alternation, and iteration structure corresponds to a syntax rule of concatenation, choice, or repetition.

For example, consider the clear box of Figure 7.1-3, with 3 conditions and 5 machines. Let <transition> be the series of conditions and machines invoked in an invocation of the entire clear box. Then it can be seen that

<transition> ::= C1 <upper transition> |
C1 <lower transition>

<upper transition> ::= M1 *(C2 M2) C2

<lower transition> ::= (C3 M3 M5) |
(C3 M4 M5)

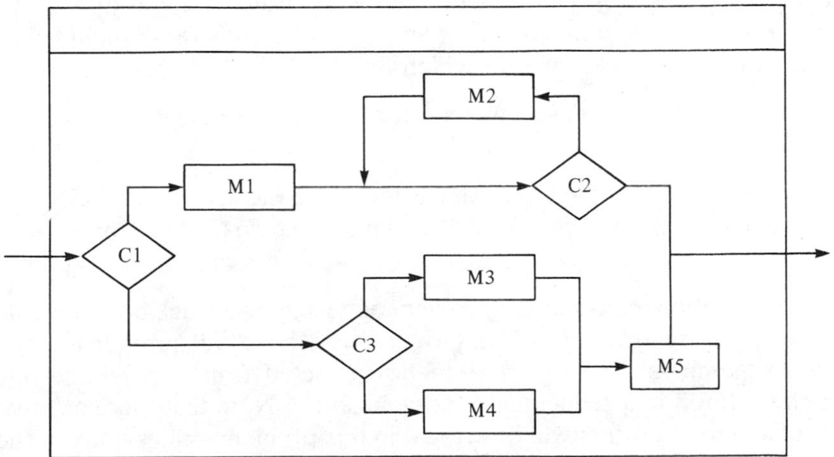

Figure 7.1-3. A Clear Box.

In fact, <transition> can be written in a single expression:

```
<transition> ::= (C1 M1 *(C2 M2) C2) |
                 (C1 C3 M3 M5) |
                 (C1 C3 M4 M5)
```

That is, <transition> describes the syntax of all possible sequences in which the conditions and machines of the clear box can be invoked. For example, the syntax of <transition> shows exactly which machines can be invoked in sequence: at most two machines can be invoked in a transition, namely, M1 before M2 (which may be repeated), M3 before M5, or M4 before M5. These relationships can also be observed in the clear box, but the syntax is much more compact, and for larger clear boxes, the syntax is more easily analyzed.

File Syntax and Clear Box Design

The property that machine transitions in clear box executions satisfy rules of syntax can be used in a decisive way for processing sequential files. The contents of sequential files also satisfy rules of syntax. Is it possible that the syntax of a sequential file is related to the syntax of a clear box which inspects or creates it? The answer is yes indeed! In fact, the clear box structure should be derived from the syntax of the file exactly as follows:

1. For every concatenation of file parts in a syntax rule, there should exist a sequence clear box structure which deals with these file parts.
2. For every choice between file parts in a syntax rule, there should exist an alternation clear box structure which deals with these file parts.
3. For every repetition of a file part in a syntax rule, there should exist an iteration clear box structure which deals with this file part.

That is, in order to design a clear box to inspect or create a sequential file, a systematic design strategy is to:

1. develop a set of syntax rules to describe the sequential file
2. derive the clear box from the syntax rules exactly as prescribed above.

In illustration, consider the problem of designing a black box to create an exception report, say, of employees who have not provided social security numbers. The report is to be extracted from a personnel file which is stored in a sequential file organization. Note that there are two sets of syntax rules that will be critical in this problem—the syntax of the personnel file and the syntax of the report. As noted before, a report may not seem to have syntax, but in fact it will.

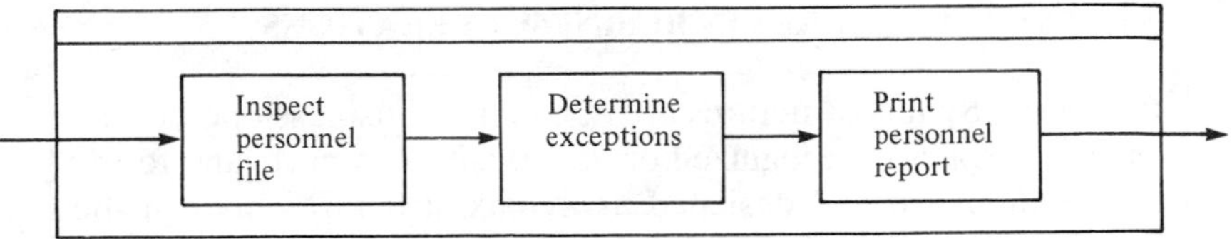

Figure 7.1-4. Exception Report Clear Box Structure.

A user will not ordinarily say "I want my report to have the following syntax." But users will frequently say "the report should be in this form," and discuss a title page, an information page, a summary page, and so on. That is, the report is to have a syntax such as

<report> ::= <title page> ∗<info page>
<summary page>

<title page> ::= <date> <department> <division>

<info page> ::= <top line> <info line> <bottom line>

<info line> ::= ...

<summary page> ::= ...

...

The black box which creates an exception report from a personnel file will treat every character of the personnel file as a stimulus, and the entire personnel file as an input. Recall that the hand calculator recognizes an input when a proper arithmetic problem has been stated in a stimulus history, and produces the answer as its output. In this case, the output of the black box is the entire exception report.

At the next level, we consider the state machine of this black box. Its state will be accumulated, stimulus by stimulus, while the personnel file is being inspected. Its machine is very complicated, and most easily described in terms of a clear box structure.

There are several possibilities for the clear box structure, depending on how much storage is available to maintain the state. In illustration, assume there is adequate storage to inspect the personnel file entirely, then determine the exceptions, and finally print the report. In this case the top level clear box structure will be as shown in Figure 7.1-4.

The first and third of these component clear boxes can now be designed exactly as prescribed above. In each case a file is to be inspected or created, and its syntax can be used to derive the clear box structure.

Summary: Syntax structures are a simple, yet powerful formalism for describing inputs, outputs, and state structures. Syntax specifications can be used in the design of system procedures that accept and validate input, organize data in storage, and prepare output reports.

7.2 SYNTAX STRUCTURES IN BUSINESS OPERATIONS

> **Preview:** Syntax structures are common in business operations, whether explicitly recognized or not. Business reports and forms can be analyzed and designed as syntax structures, as can the user languages of business information systems. User language syntax structures prescribe clear box designs for language processing.

The ability to make use of syntax structures is very valuable, because they can simplify and clarify black box, state machine, and clear box descriptions to significant degrees. In fact, practically all business communications and records have syntax structures that permit systematic analysis and description, even though they are quite different from the examples of ordinary English and arithmetic expressions that suggest syntax methods in the first place.

7.2.1 The Syntax of Files, Reports, and Business Forms

Syntax rules can be used to describe a wide variety of information structures of which proper hand calculator input is a simple case. For example, any information file is ultimately a file of characters which are organized into fields and records. A personnel file might contain a header record, a variable number of personnel records, and a summary record, according to the syntax rules

<personnel file> ::= <header> *<personnel record> <summary>

That is, the file is composed of a single syntax part <header>, zero or more syntax parts <personnel record>, and a single syntax part <summary>. If the header record contains information about the organizational unit, it might have the form

<header> ::= <organization name> <organization address> ···

the personnel record the form

<personnel record> ::= <name> *<dependents> <address> ···

(zero or more dependents), and the summary record the form

<summary> ::= <number of records> <average salary> ···

In this way the entire syntax of the personnel file can be defined in complete detail.

A generated personnel report will also have a syntax, such as

<report> ::= <front page> *<data page> <summary page>

(zero or more data pages), with front page of the form

<front page> ::= <title> <date> ···

data page of the form

<data page> ::= <top line> *<data line>

and so on.

A word processing file will have a simpler structure of blanks and words, say

<text> ::= <word> <blanks> <word>

<word> ::= <nonblank> *<nonblank>

<blanks> ::= <blank> *<blank>

and so on.

The use of lines as syntax parts permits the description of business forms of all kinds with syntax structures. A blank line (line of blank characters) can be used to take up space when space is needed. Within lines, tabs and columns can be defined by syntax parts, and so on. Syntax methods can be used to define a family of business forms, say personnel application forms, which have variations for permanent or temporary job applications.

7.2.2 The Syntax of User Languages

One of the most powerful uses of syntax is in the definition of user languages for interfacing with information systems. A grammar for a user language serves three purposes:

1. a test of completeness and consistency for the definer,
2. a basis for explaining the language to users, and
3. a basis for implementing the language by designers.

Unfortunately, user languages in information systems often grow "like Topsy", from a few commands thought up by programmers to permit the user to use a program, up through piecemeal capabilities added as afterthoughts in response to user needs and suggestions. From the point of view of a programmer trying to do some specific thing, the user language is often an afterthought. What the information systems designer must know is that the user language is as critical as the internals of the system, and must be considered from the very beginning of analysis and design. This understanding can be strengthened with an automatic recognition of the following principle.

> **Fundamental Principle:** Every user facility, no matter how simple it may seem, will require a user language, and that language should be defined with much foresight.

Defining the syntax of a user language forces a systematic analysis of what is to be required, and provides a basis for the three purposes mentioned above.

Data entry is a frequent need in information systems, and syntax methods apply to user languages needed for data entry. Consider part of a Management Information System that processes and stores data on customer calls. In addition to direct information about sales, suppose the business was also interested in customer opinions about the economy, business climate, etc., for its own forward planning of production, inventory levels, credit extensions, etc.

To be more specific, suppose the business expects a customer opinion report from each call for direct entry into the Management Information System, in order to allow nationwide and regional tabulation of customer opinions and their trends.

At first glance, there may seem to be no user language definition at all. But that is not the case. From the description above, one might surmise a language of customer reports such as given by the following grammar:

```
<customer report>  ::= <customer number> <sales trend>
                       <economy trend> <inventory plans>
                       <credit needs>
<customer number> ::= <digit> *<digit>
<sales trend>      ::= sales <trend> <amount> | none
<trend>            ::= up | down
<amount>           ::= <digit> | <digit> <digit>
<economy trend>    ::= economy <trend> <amount> | none
<inventory plans>  ::= inventory <trend> <amount> | none
<credit needs>     ::= credit <trend> <amount> | none
```

In illustration, one customer report in this language might be:

371 sales up 15 economy up 10 inventory up 5 credit up 10

Another customer report might be

421 none none none none

Still another

731 sales up 15 none none credit down 5

and so on. The language has the literals "sales", "economy", "inventory", "credit", "up", "down", "none", "0", "1", "2", ..., "9". It requires a one or two digit <amount>, but any number of digits can be a <customer number>.

Such a report could be processed automatically in a Management Information System and customer opinion reports developed from them. But before going further, the three purposes above can be examined in more detail.

1. A test of completeness and consistency for the definer. Such customer reports are the result of some business thinking. Do the reports cover the issue of interest? Do they cover all possibilities? E.g., the function of literal "none" handles a no opinion or not asked—should those two cases be distinguished in the language? The decisions are ultimately business decisions, but the language definition gives a focus to them.

2. A basis for explaining the language to users. This language is simple enough to explain, and one probably would not show users the rules of syntax directly. Instead, a tabular form such as

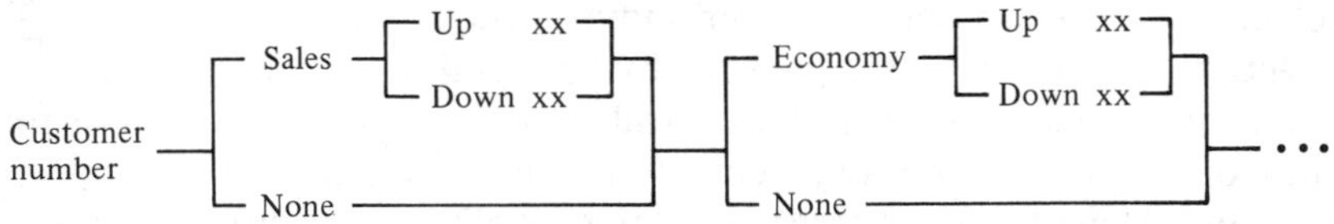

would be more easily understood. (Lesson: don't try to show users all you know in your notations; invent simpler notations for them.)

3. A basis for implementing the language by designers. The language should be checked for implementability. Is it always possible to get the meaning out of a customer report? Are there ambiguities possible? The literals "sale", "economy", "inventory", and "credit" ensure that the information following can be tabulated correctly. For example, the customer report

371 sales up 1215 economy ...

is improper (not in the language) so deciding between 12 and 15 is unnecessary.

In summary, a grammar for a user language provides many benefits in analysis and design, and in making business judgments about its use and purpose.

7.2.3 Grammars in Clear Box Design

The structure of a grammar defining the input to a clear box can often suggest a natural structure for the design of the clear box itself. For example, the input file grammar on the left side of Figure 7.2-1 leads naturally to the clear box structure on the right. The start symbol of the grammar (<input>) is defined as a three-part sequence

<header>
<body>
<trailer>

as is the corresponding clear box. The secondpart expansion of the clear box sequence is an iteration suggested by the recursive definition of <body> given in the second production rule of the grammar.

As a more complex illustration, consider the design of an on-line system to support user transactions with a banking terminal equipped with a standard keyboard and display screen. In what follows, we will (1) construct a grammar for the user input language, (2) combine elements of that grammar with elements of a grammar for the display output language to create a definition of conversational access to the terminal, and (3) create a clear box design based on the grammar structures.

A user interacts with the terminal by conducting a session, defined as a signon and zero or more requests followed by signoff. There are four types of request: withdraw, deposit, balance, and transfer, each with appropriate inputs as required. The signon allows five tries for correct user identification, and five tries for correct password. The user-language grammar is shown in Figure 7.2-1. In a specific implementation, separator characters may be required between certain syntactic entities, for exam-

```
                                              Corresponding
          Input Grammar                          Program

1.  <input>    ::= <header>               process header
                   <body>                 while
                   <trailer>                more records
2.  <body>     ::= <record> |             do
                   <record>                 process record
                   <body>                 od
3.  <record>   ::= ···                    process trailer
4.  <header>   ::= ···
5.  <trailer>  ::= ···
```

Figure 7.2-1. Corresponding Grammar and Program Structures.

1. <session> ::= <signon> *(<request>) <signoff>
2. <signon> ::= $+^5$(<id>) $+^5$(<password>)
3. <signoff> ::= OFF
4. <request> ::= <withrequest> <account> <amount> |
   <balrequest> <account> |
   <transrequest> <fromacct> <toacct> <amount> |
   <deprequest> <account> <amount> <envelope>
5. <withrequest> ::= W
6. <balrequest> ::= B
7. <transrequest> ::= T
8. <deprequest> ::= D
9. <id> ::= <letter> | +(<letter> | <digit>)
10. <password> ::= +<digit>
11. <account> ::= <number>
12. <amount> ::= <number> . <digit> <digit>
13. <toacct> ::= <number>
14. <fromacct> ::= <number>
15. <number> ::= <digit> | <digit> <number>
16. <digit> ::= 0 | 1 | 2 | ··· | 9
17. <letter> ::= A | B | C | ··· | Z
18. <envelope> ::= cash | check (physical objects)

Figure 7.2-2. Bank Terminal User Language Grammar.

ple, in production 4, but they are omitted here for the sake of simplicity. The characters W, B, T, D, and OFF denote withdraw, balance, transfer, deposit, and signoff operations, respectively. Note the use of the new metasymbol + in this grammar. Plus (+) is used in front of a syntax structure to denote the appearance of one or more instances of the structure in a valid string. The metasymbol $+^5$ means that up to five repetitions of the syntax structure can appear. Thus only five attempts are allowed to provide a valid user <id> or a <password>.

Figure 7.2-3 depicts the design of all possible conversations between the user and the system. The User Input entities are taken from the user language grammar; the Display Output entities are taken from a grammar for clear box output. The clear box output grammar is not shown; the reader may find it interesting to construct such a grammar from the information in the Display Output column of Figure 7.2-3. Note that the two grammars share many metasymbols, for example, <id> and <account>, as well as the same set of literals.

Each Display Output item is given as a set of possible responses enclosed in brackets and separated by commas, any one of which may occur depending on the current status of the conversation and the user input. This corresponds to state machine behavior, where each output depends on both input and state, and each transition produces a new state

	User Input	Display Output
1.	(no user present)	{enter id}
2.	<id>	{invalid id <id> try again, invalid id <id> session terminated, enter password}
3.	<password>	{invalid password <password> try again, invalid password <password> session terminated, enter request}
4.	<signoff>	{thank you}
5.	<withrequest><account> <amount>	{invalid request <withrequest> try again, invalid account <account> try again, overdrawn, pick up envelope}
6.	<balrequest><account>	{invalid request <balrequest> try again, invalid account <account> try again, balance is <amount>}
7.	<transrequest><fromacct> <toacct><amount>	{invalid request <transrequest> try again, invalid from account <fromacct> try again, invalid to account <toacct> try again, overdrawn from account <fromacct>, transfer completed}
8.	<deprequest><account> <amount>	{invalid request <deprequest> try again, invalid account <account> try again, insert envelope—pick up deposit slip}

Figure 7.2-3. Bank Terminal Conversational Access.

in addition to an output. Note especially that the conversation is designed to accommodate both correct and incorrect user input. That is, there is no presumption of infallibility on the part of the user; the state machine must produce an appropriate output from any state of the conversation, no matter what input the user provides.

This example illustrates the fact that grammars permit exhaustive and terse specification of all known transactions across an interface, before clear box design is begun. As a consequence, the designer is forced to address all possible user actions in developing the clear box structure. This stepwise approach can help forestall the programmer's complaint (when system operation breaks down) that "the user is doing unpredictable things", since in the definition of conversational access, nothing is unpredictable!

Figure 7.2-4 depicts the top level design of the bank terminal clear box. The structure of the conversational access has been used to help derive the clear box structure. Reading from the top down, note that the logical commentary refers to syntactic entities in the conversational ac-

```
proc terminal [process <session>]
   data
      ok: logical
      transactions: logical
      request: character
   display enter id
   [process <signon>]
   run signon (ok) [refined in Figure 7.2-5]
   [process remainder of <session>]
   if
      ok
   then [process *(<request>) <signoff>]
      do [process *(<request>)]
         transactions := true
         while
            transactions
         do
            set timer interrupt
            read request when entered
            disable timer interrupt
            if
               request = OFF
            then
               transactions := false
            else [process <request>]
               case
                  request
               part (W) [process <withrequest>]
                  run withdraw
               part (B) [process <balrequest>]
                  run balance
               part (T) [process <transrequest>]
                  run transfer
               part (D) [process <deprequest>]
                  run deposit
               part (other)
                  display invalid request <request> try again
               esac
            fi
         od
      od
      [process <signoff>]
      run signoff
   fi
corp
```

Figure 7.2-4. A Clear Box Design for Bank Terminal Conversational Access.

```
proc signon (ok) [process <signon>]
   data
      ok: logical
      count: 0..5
   do [process +^5 (<id>) +^5 (<password>)]
      count := 0
      do
         count := count + 1
         [(<id> legal → ok := true | true → ok := false)]
         run validate id (ok)
      while
         ~ok and count < 5
      do
         display invalid id <id> try again
      od
      if
         ~ok
      then
         display invalid id <id> session terminated
         wait k seconds
         display enter id
      else
         display enter password
         do [process +^5 (<password>)]
            count := 0
            do
               count := count + 1
               [(<password> legal → ok := true | true →
                  ok := false)]
               run validate password (ok)
            while
               ~ok and count < 5
            do
               display invalid password <password> try again
            od
            if
               ~ok
            then
               display invalid password <password>
                  session terminated
               wait k seconds
               display enter id
            else
               display enter request
            fi
         od
      fi
   od
corp
```

Figure 7.2-5. A Clear Box Design for the Signon Procedure.

cess in a top-down fashion as well. That is, the design of the clear box reflects the design of the conversational access.

Output from this clear box is produced by the display operation on line 37. This output corresponds to the first response in each of the Display Output sets of Figure 7.2-3 for inputs numbered 5 through 8. A time-out interrupt has been incorporated into the design to avoid tieing up the terminal (something the grammar cannot account for).

Figure 7.2-5 depicts an expansion of the signon procedure. (The procedure illustrates use of a do-while-do control structure, a variation of the while-do structure, in which procedure statements also appear before the while test.) Again, note the grammar structures named in the logical commentary, and the display output corresponding to possible conversations between a customer and the system during the signon process.

Figures 7.2-2–7.2-5 are worth some study as a miniature example of how to handle definition and control of user interfaces in a real system. Explicit interface design such as this can clarify questions and resolve issues for users and designers alike, and can help avoid surprises as systems move into operational use.

Summary: Business files, reports, and forms can naturally be defined by syntax structures for more effective analysis and design. Syntactic definition of user languages for information systems permits tests for completeness and consistency, and provides a basis for explanation and implementation. Syntax structures can define possibilities for conversational access to information systems, to help guide clear box design to support such access.

7.3 THE NEW YORK TIMES THESAURUS AND ITS GRAMMAR

Preview: A formal grammar was found for the New York Times Thesaurus of Descriptors that greatly facilitated the analysis and design required for the New York Times Information Bank. The grammar was written in terms understandable for the managers of the Information Bank.

The next two subsections are historical documents. The first of these two subsections is a New York Times document, namely the first thirteen pages of the New York Times Thesaurus of Descriptors, which serves to explain the more than thousand pages that follow it. It was an important document for the system development team, because it represented an authoritative view of the Thesaurus. It was basic to understanding the needs in automating the morgue.

The second of these two subsections is an IBM developed document, used in developing the New York Times Information Bank. It is repeated here, verbatim, just as the Thesaurus description, in order to show an actual work of information systems analysis. That is, these two sections were not made up for this book. They show a problem and a solution as they actually arose in practice.

This second document is called "A Formal Grammar for the New York Times Thesaurus." Its very title is a surprise—that the New York Times Thesaurus has a formal grammar at all. At first glance a formal grammar might seem an impossibility—for example, a formal grammar for ordinary English is beyond human capability to create or understand. It might seem of little use, even if by some miracle it could be produced. But it took no miracle and it was of much use in dealing with the Thesaurus in the system development.

In reading the first document, The New York Times Thesaurus of Descriptors, imagine facing it (with its over thousand pages of terms following) as a problem in systems analysis. It is part of the manual system which is to be automated. Can you find guiding principles or patterns in it to exploit? Instead of hand and eye scanning these pages at will, only a few lines can be shown a terminal user at a time. If you merely automate page turning, the system will be useless because there is far too much material to examine by these methods.

In reading the second document, A Formal Grammar for the New York Times Thesaurus, realize that it was written for both the system development team and for the customer. Not only is a formal grammar given for the Thesaurus, but also a tutorial explanation of what a formal grammar is and how it can be used to separate problems of specification from problems of implementation. A tutorial is frequently needed when a system analyst discovers that a new description method is applicable to a user problem.

In such a case, the tutorial and new description method should require less effort together than the old description method, or the new method shouldn't be used. It is obvious that tutorials should never be used to impress users, or because it is good for them on general grounds.

You will notice a set of metasymbols in this document that differ from those used in this book so far. In fact, the analyst discovered the formal grammar in notation very similar to that used in this book, and decided to simplify the notation in certain ways for easier understanding. It should be standard practice to simplify and adapt general notation to the specific needs of users. Usually, the analyst should think better and work faster in more compact and concise notation than the user understands, this is the right thing to do. But the second mile of the analyst is to find a simpler notation to translate results to, and to explain to the user.

7.3.1 The New York Times Thesaurus of Descriptors*

FOREWORD

The project to devise a thesaurus as an aid in processing and searching information from newspaper files was undertaken as part of an effort by The New York Times to coordinate all its information facilities. It grew out of preparations for the application of computer technology to the production of The New York Times Index. The vocabulary and structure of the Thesaurus are therefore based largely on those of the Index, but include many additional terms from the subject card file of The Times clipping "morgue" and from the vertical file catalogue of The Times Editorial Reference Library.

The following works were consulted in designing the format of the Thesaurus: The ASTIA Thesaurus of Descriptors, 2nd edition, December, 1962; the Department of Defense Manual for Building A Technical Thesaurus, Project LEX, Office of Naval Research, April, 1966; and the Engineers Joint Council's Guide for Source Indexing and Abstracting of the Engineering Literature, February, 1967. The Subject Headings Used in the Dictionary Catalog of the Library of Congress, 7th edition, 1966, was consulted in solving certain problems of terminology.

The work is a cooperative effort of the staff of The New York Times Index under the general direction of Dr. John Rothman, editor. The huge task of compiling and annotating the entries was handled by the following staff members:

Robert A. Barzilay, coordinator
Marvin M. Aledort
William F. Marshall
Robert S. Olsen
Daniel Pinzow
Susan L. Pinzow
George D. Trent

The job of final editing was shared by Dr. Rothman and Thomas R. Royston, assistant editor.

Computer programming and operations were done by Central Media Bureau, Inc., of New York.

About the Second Edition

Within a few months after publication of the Thesaurus, enough corrections and additions had accumulated to make it advisable to publish a complete revision rather than the individual pages with changes originally planned.

In all, almost a thousand changes were made by the time this Second Edition was ready for its final computer run. Many of them were based on suggestions received from Thesaurus users.

The physical format has also been improved. This edition is printed on heavier paper, which will turn more easily and be more resistant to tearing. In addition, continuation headings have been added where required.

The active interest of Thesaurus users has helped make this new edition a more useful reference tool. Your comments will always be welcome and sincerely appreciated.

Introduction

The word "thesaurus" derives from a Greek word meaning "treasure." As applied to the conventional dictionary of synonyms and antonyms, such as Roget's, it is most apt; such a thesaurus is indeed a treasure, displaying the riches, fullness, and diversity of the language.

The kind of thesaurus that has evolved in the last decade or two in the field of information processing and retrieval is not a treasure so much as the key to one. The riches lie in a file of information—a collection of books, pamphlets, reports, photographs, or newspapers—and the thesaurus is a means for their exploitation. A thesaurus of this kind is a device for ordering and controlling the file, so that new items may be added consistently to related items, and so that all relevant items are made readily and quickly accessible.

The New York Times Thesaurus of Descriptors is a structured vocabulary of terms designed to guide information specialists in processing and organizing materials from newspapers and other works dealing with current events and public affairs, and to guide users in searching collections of such materials. Because it covers the same vast variety of subject matter as the daily press, it will prove a valuable tool, we trust, not only for newspaper libraries but also for general reference libraries, educational institutions, government agencies, business and financial organizations—in short, for any organization that collects, stores, and uses information on the events of today and yesterday.

The Thesaurus consists of terms (descriptors), in a single alphabetical sequence, which denote the diverse subjects that may be found in the collection. For each descriptor, some or all of the following data are given, in the order indicated:

1. Qualifying Terms
2. Scope Notes
3. "See" or "See also" References (listed alphabetically)
4. "Refer from" References (listed alphabetically)
5. Subheadings (listed alphabetically)

These are designed to define descriptors and to correlate them with one another.

A model page appears in Figure 7.3-1. The remainder of the introduction explains the various features of the Thesaurus in detail and discusses the major principles of organizing such a file of information. It also includes some general guidelines for certain types of material (for example, foreign names and corporation names) that are not covered item by item in the Thesaurus itself. A brief index to the contents of the introduction follows.

1. Descriptors. Descriptors are primarily subject headings. Geographic names, personal names, names of companies, institutions and organizations, and other proper names are included only when they require the use of qualifying terms, scope notes, a regular pattern of cross references, or a regular pattern of subdivisions.

The Thesaurus does not include a descriptor for each individual member of a family. There would be little purpose in listing every item of furniture, every kind of weapon, or every kind of animal, vegetable, or mineral. Descriptors are given for typical items and for those requiring any special or unusual handling; and these will serve, it is hoped, as models for any similar items that are not listed.

Synonyms. Preferences between synonymous or nearly synonymous terms are indicated by "see" references (AVIATION. See Aeronautics).

Non-Standard Terms and Recent Coinages. Descriptors include terms current in the news (such as BLACK Power or BRAIN Drain) even though they are not found in standard library catalogues or dictionaries. Descriptors do not include brand names or trademarks, technical terms not normally used in newspaper articles, slang words, and terms used

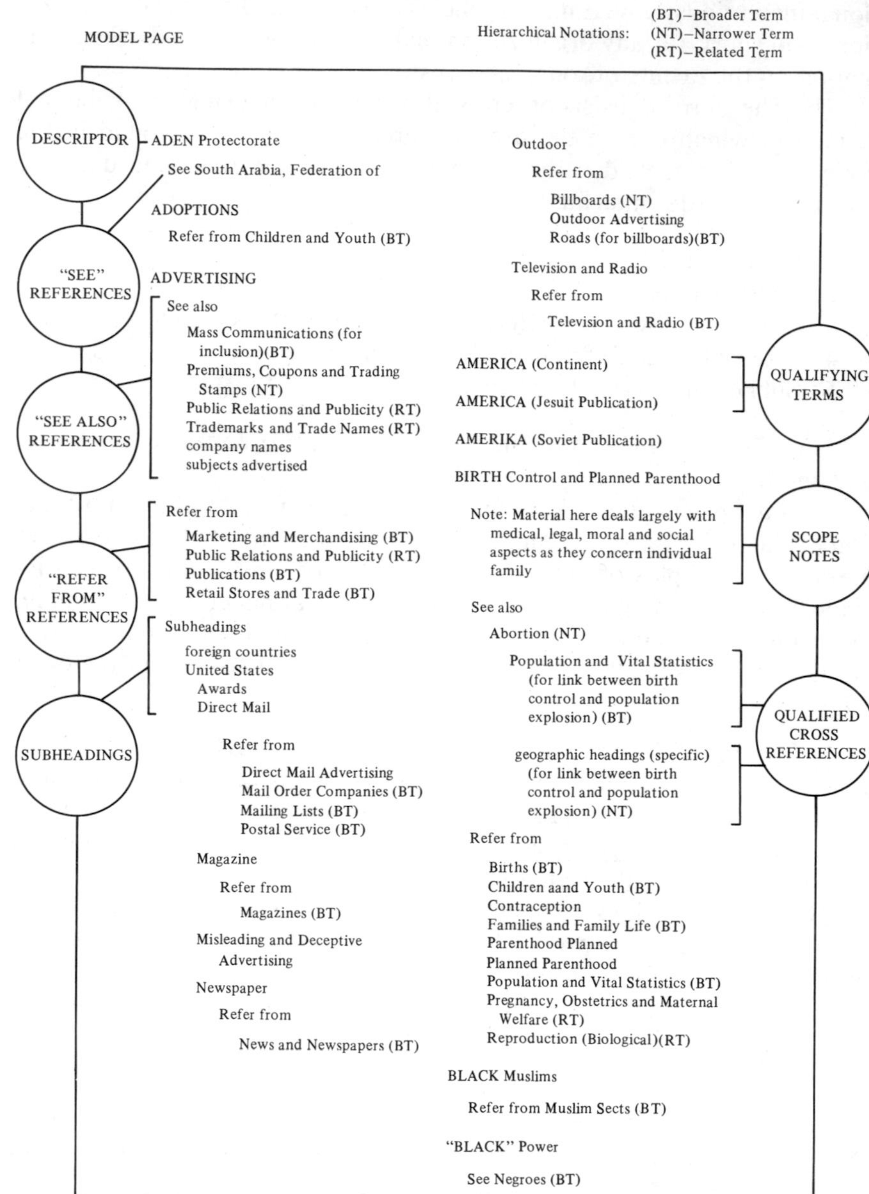

Figure 7.3-1. The Model Page.

exclusively in professional jargon. When colloquialisms, slogans, or unusual coinages are used as descriptors, they usually appear in quotation marks. Archaic or obsolete terms are included when this is considered helpful.

Abbreviations and Acronyms. Abbreviations and acronyms are used as descriptors, usually with "see" references to the name spelled out (NATO. See North Atlantic Treaty Organization). The practice may be reversed when the abbreviation is much better known and more widely used than the term it represents (DICHLORO-Diphenyl-Trichloroethane. See DDT). No attempt has been made to compile an exhaustive list of abbreviations and acronyms.

Alphabetization. To give a complete description of the alphabetization scheme followed in the Thesaurus would go far beyond the scope of this introduction; but the following are the major rules applied in alphabetizing entries here, and recommended: word-by-word order rather than letter-by-letter (AIR Pollution before AIRLINES); abbreviations filed as words (NATO between NATIONAL and NATURE); inverted headings filed before uninverted headings (NEW York, State University of, before NEW York Airways); homographs filed in the order of person, place, thing (BROOKLYN, William; BROOKLYN, NY; BROOKLYN Bridge) or in the alphabetical order of qualifying terms (MERCURY (Metal); MERCURY (Planet)); numbers filed as though spelled out (20th Century as TWENTIETH Century), except where the numerical order is clearly preferable (HENRY VII before HENRY VIII); and compound terms filed as though two words (REAL-Time before REALISM), except when the first component is a prefix (TRANS World after TRANSIT) or a term of direction (SOUTH-West) after SOUTHERN).

Specificity. In general, files of information must be so organized as to bring together all items relevant to a given inquiry and yet permit prompt access to any single, specific item. In this Thesaurus, the choice of descriptors and their degree of specificity reflect the vocabulary and scope of current journalistic writing and seek to anticipate the needs of users who consult files of newspapers, magazines, pamphlets, reports and the like for information. When the amount of material on a subject is large (for example, AERONAUTICS), separate descriptors for specific aspects are advisable (AIRLINES, AIRPLANES, AIRPORTS, etc.). When the amount of material is relatively small and should not be scattered, or when its separate aspects are not readily segregated, the use of a more comprehensive descriptor is advised. (For example, the descriptor PLASTICS is used for all kinds of plastic materials, since these are rarely differentiated in newspaper stories; obviously, such a comprehensive descriptor would be inadequate for the literature of organic chemistry.)

Generics. Because the subject fields in current events tend to overlap widely and terms are often vague and imprecise in meaning, a hierarchical or classed arrangement of descriptors was impossible to achieve. Where feasible, hierarchical relationships between descriptors are indicated by means of "broader term" (BT) and "narrower term" (NT) notations in cross references.

Geographic vs. Subject Terms. The problem of whether to organize a file by subject or by place is one of the most difficult confronting a librarian (HOUSING—New York City or NEW York City—Housing?). Except in mechanized coordinate files, the effort and expense required for complete duplication are prohibitive, and a choice between the two approaches must be made. Our preference for the subject approach is reflected in the Thesaurus. It is based on the fact that most news developments have regional rather than uniquely local significance. Much of the political and economic news deals with broad geographical areas; cities throughout the world have similar traffic, air pollution, water supply and slum housing problems; and so forth. Hence, geographic terms are used mostly for general descriptions and for general material on the economics, politics, defenses, population, history, and customs of an area; in short, for material too broad to fit under subject descriptors. Organizational material on specific government agencies (formation, budget, personnel) is covered under geographic terms; their activities are covered under appropriate subjects. Names of government agencies (except for international and American interstate agencies) are not given as descriptors. An attempt has been made to provide a list of United States (Federal) agencies (as subheadings under UNITED States), but because their names change frequently and the status of some is now in doubt, the list may not be complete and is subject to frequent revision.

Word Order in Multiple-Word Descriptors. For most subject descriptors consisting of more than one word, the natural word order is preferred and given here (AIR Pollution, not POLLUTION, Air). For personal names, the last name is always given first (JOHNSON, Lyndon Baines). For foreign personal names, determination of the correct "last name" is often troublesome; see the next section for some general rules. Geographic names usually invert from and are alphabetized under the proper-name element (PHILIPPINES, Republic of the; not REPUBLIC of the Philippines). Company names should be in natural word order (NATIONAL Broadcasting Company; not BROADCASTING Company,

National) except when inversion from a proper-name element is clearly preferable (MACY, R. H., & Co.; not R. H. Macy & Co.) (for dubious cases, the stock market tables often provide a useful guide). Names of schools, universities, and museums should generally be in natural word order (MASSACHUSETTS Institute of Technology), but there are some obvious exceptions (CHICAGO, University of; not UNIVERSITY of Chicago). Names of business, trade, civic and professional associations, labor unions, foundations, and certain other organizations should invert from an appropriate subject term or personal name (KANSAS City, Chamber of Commerce of; ADVERTISING Agencies, American Association of; CIVIL Liberties Union, American; LONGSHOREMEN'S Association, International; SLOAN, Alfred P., Foundation). It is often helpful to use inversions of word order to bring together, in the same alphabetical location, all organizations concerned with the same subject that use the descriptor for this subject as part of their names (for example, all organizations whose names contain the word EDUCATION). When the inversion is not obvious, or when there is a choice between two or more possible inversions, alternatives should be covered by "see" references to the preferred version (BROADCASTERS, National Association of Educational. See Educational Broadcasters, National Association of). Some "see" references of this type are included in the Thesaurus, especially under common words such as American, General, or International.

Foreign Names. Foreign names present problems both in determining the proper word order and in determining proper spelling for transliterations. Authoritative reference works such as Who's Who should be consulted, but even these are not always in agreement, and, of course, they cover only a limited number of names. Helpful advice can be obtained from information officers of foreign consulates, trade missions and delegations to the United Nations and other international organizations. The following rules are offered as a general guide, but they are not exhaustive, and there are many exceptions.

a. British names including two "last" names (Anthony Wedgwood Benn) usually invert from the second of these (BENN, Anthony Wedgwood).

b. Spanish names including two "last" names (Eduardo Frei Montalva) usually invert from the first of these (FREI Montalva, Eduardo).

c. European and Latin-American names containing a partitive (de, di, van, von) usually invert from the name following the partitive (GAULLE, Charles de; HASSEL, Kai-Uwe von).

d. Names containing a definite article usually invert from the article if they are French, Italian, Spanish, or Portuguese (LA Guardia, Ernesto de) and from the name following the article if they are German or Dutch (HEIDE, Gottfried von der).

e. Arabic names containing a partitive (al, el, ben, ibn) usually invert from the name following the partitive (ATTASSI, Fadhil al; BELLA, Ahmed ben).

f. Chinese, Indochinese, and Korean names invert from the last element if they have been Westernized (PARK, Chung He), but run uninverted if not (MAO Tse-tung; NGUYEN Cao Ky). (If such names become popularly known in an incorrect form, such as "Premier Ky" instead of "Premier Nguyen Cao Ky," appropriate "see" references should be run from the incorrect form to the correct form.)

g. When foreign names may be transliterated in several different ways, the preferred transliteration should be determined, if possible, and "see" references to it should be run from alternate transliterations. Among the more common instances are the following: In Arabic names, use ai instead of ei (FAISAL, not FEISAL) and use kh instead of q as the first letter (KHALIDI, not QALIDI). In Russian names, use ch instead of tch or tsch (CHERNISHEV, not TCHERNISHEV or TSCHERNISHEV) and use v instead of ff as the last letter (SUVOROV, not SUVOROFF). In Greek names, use k instead of c or ch as the initial letter (KARAMANLIS not CARAMANLIS, KRYSOSTOMOS, not CHRYSOSTOMOS). However, names for which the alternative transliteration is well established (TCHAIKOVSKY, PROKOFIEFF, CONSTANTINE) should be retained thus.

Corporation Divisions and Subsidiaries. The question of whether to establish separate descriptors for corporate divisions and subsidiaries, or to carry material about them under the name of the parent company, poses another major problem. In general, separate descriptors should be established for subsidiaries that issue their own stock, have well-known names distinct from those of the parent company, or have otherwise a separate identity (CHEVROLET Division of General Motors Corp.; IBM World Trade Corp.), and then the parent company should be linked to the subsidiary by a "see also" reference. When the subsidiaries do not have a clearly distinct identity, it is advisable to carry material about them under the name of the parent company, especially when the material does not consistently identify them by name. For example, it is virtually impossible to use separate descriptors for the overseas operating units of the major international oil companies. These are referred to sometimes by their own names (ESSO Libya Ltd.) and sometimes merely as units of the parent company (Standard Oil of New Jersey's Libyan affiliate), and there may be no way of determining whether the same unit or two different units are involved. Even when the distinction can be made, it may be better to keep material about the company together under one name than to scatter it among several names, some of which may be quite unfamiliar to the users.

Religious Denominations. When the amount of material is relatively small, material on branches, regional bodies, and other agencies of a denomination is carried under the collective name of the denomination, and not under separate descriptors. (For example, Greek Orthodox Church under ORTHODOX Churches; Southern Baptist Convention under BAPTIST Churches.) Individual congregations and parishes, if not intersectarian, should also be included under the name of the denomination, rather than given separate descriptors; but the names of well-known churches (such as St. Patrick's Cathedral in New York) should be covered by "see" references to the name of the denomination.

2. Qualifying Terms. Qualifying terms are parenthetical expressions given after certain descriptors to distinguish between homographs. For example:

MERCURY (Metal)
MERCURY (Planet)

Qualifying terms may also be used to resolve other contextual ambiguities in some descriptors. For example:

FIFTH Amendment (U.S. Constitution)

3. Scope Notes. Scope notes are notes appearing after certain descriptors to define or describe the range of subject matter encompassed by the descriptor. For example:

DRUG Addiction, Abuse and Traffic.
Note: Material here includes narcotics, stimulants, hallucinatory drugs, and others deemed socially undesirable.

Scope notes may be used at subheadings for the same purpose, and may also be used to describe the system of subdividing material under certain descriptors.

4. Cross References. Cross references serve as substitutes for multiple entries and as guides between descriptors encompassing related material. They are also used at subheadings as required.

Contrary to usual library practice, cross references have not usually been established between related descriptors that are immediately adjacent in the alphabet. (For example, there is no cross reference from

ARMORED Vehicles to ARMORED Car Services.) It was felt that the connection between such adjacent descriptors is self-evident and that cross references there would be superfluous.

See References. "See" references guide from descriptors not used for "entries" in the system to equivalent descriptors used in preference. They are used mainly between synonyms (AVIATION. See Aeronautics), and when material denoted by one descriptor is subsumed under another (ORCHESTRAS. See Music).

See also References. "See also" references guide from descriptors used for certain "entries" in the system to other descriptors where related material is entered. They may lead from more general, broader terms to more specific, narrower terms (REAL Estate. See also Housing), or vice versa (THEATER. See also Amusements). They may also lead from one descriptor to another on the same hierarchical level which may cover tangential topics or different aspects of the same topic (ROADS. See also Traffic).

Refer from References. "Refer from" references are the inverse of "see" and "see also" references. They show all the descriptors linked by "see" and "see also" references to the descriptor consulted (AERONAUTICS. Refer from Aviation).

Qualified Cross References. Numerous "see," "see also," and "refer from" references are followed by parenthetical expressions defining the particular aspect of a topic covered by the cross reference, as in DOGS. See also Blindness and the Blind (for seeing-eye dogs).

Hierarchical Notations. Many cross references are annotated to show hierarchical relationships, as follows: (NT) when the reference leads from a broader term to a narrower term (REAL Estate. See also Housing); (BT) when the reference leads from a narrower term to a broader term (THEATER. See also Amusements); and (RT) when the reference leads from one term to another on the same hierarchical level for related material (ROADS. See also Traffic). The use of these notations could not be sustained throughout the Thesaurus, however, because the subject fields covered in newspapers and other current-events publications tend to overlap widely and the vocabulary is extremely varied, complex and often imprecise; and hierarchical relationships could therefore not always be determined. (For example, CRIME and Criminals. See also Courts—which of these is the narrower descriptor, and which the broader?) In many cases, the question of hierarchy was moot, and the choice was finally governed by the descriptor from which the cross reference runs. (For example: HOUSING. See also Zoning is annotated (NT), even though zoning encompasses all kinds of land uses, because the cross reference is intended to cover a specific aspect of housing, namely, residential zoning.) Also, no attempt has been made to include cross references from all specific descriptors in a given subject field to the broader descriptor denoting the field as a whole. (For example, no broader term cross references have been made from the many specific agricultural products, such as GRAIN, to the descriptor AGRICULTURE and Agricultural Products.)

5. Subheadings. The Thesaurus lists suggested subheadings for descriptors encompassing a large amount of material. Where a category of subheadings consists of names of individual components (for example, names of countries, states, or motion pictures), only the category is given, not an inclusive list of all components.

With few exceptions, subheadings are limited to two hierarchical levels (main subheadings and sub-subheadings). Further subdivision is usually not advisable; it makes the heading structure too complex and too difficult to search. When the need for further subdivision arises, it is usually an indication that the main heading (descriptor) is too broad, and that, instead of subdividing it further, narrower descriptors should be established.

Most descriptors lend themselves to both geographic and subject subdivisions. However, it is usually not advisable to mix geographic and subject subheadings at the same level. (If under EDUCATION, for example, both Elementary and California are used as subheadings at the same level, which one would be used for material on elementary schools in California?.) The nature of the material and the interests of the users should determine whether subdivisions should be geographical or by subject.

Subheadings may appear with qualifying terms, scope notes, and cross references, just like descriptors.

6. Orientation and Format. Since the Thesaurus is based on the vocabulary used in processing information from The New York Times, it necessarily reflects the fact that The Times is published in New York. Thus, the descriptors NEW York City and NEW York State have subheadings not given for other cities and states, and New York City and New York State are used as subheadings under many descriptors that have no other city and state names as subheadings. Similarly, descriptors for local institutions (such as COLUMBIA University or NEW York Times) are shown with a detailed structure not given for similar institutions elsewhere. However, the structure outlined under NEW York City,

NEW York State, and some local institutions may be easily applied to other cities and states and their institutions in processing local newspapers and other collections there.

In this context, it should be pointed out also that the detailed structure shown under PRESIDENTIAL Election of 1968 applies to the election in any current Presidential election year. Similarly, the structure shown under JOHNSON, Lyndon Baines, applies to any President and may be applied, with any necessary modifications, to governors, mayors, heads of foreign governments, and other prominent figures.

Generally, the Thesaurus is intended, as its subtitle states, as a guide in processing and searching materials rather than as a body of firm and strict rules. Deviations from the guidelines set forth here should be made as the nature of the materials processed and the interests of their users require. In processing newspapers and other current events materials for information retrieval, flexibility is mandatory, and therefore frequent changes in the Thesaurus are envisaged. These changes may be initiated by us, or they may be made by individual users to cope with the specific problems and meet their specific needs.

It is for these reasons that the Thesaurus has been issued in looseleaf form. Even-numbered pages have been left blank to enable users to write their own notes at will opposite the appropriate Thesaurus material. Changes initiated by us will be on individual pages to be substituted or inserted. The looseleaf format permits users to insert separate sheets with their own material as desired.

7.3.2 A Formal Grammar for The New York Times Thesaurus of Descriptors

The Structure of The Times Thesaurus

An important objective of The New York Times Thesaurus of Descriptors appears on p. 13:

> Generally, the Thesaurus is intended, as its subtitle states, as a guide in processing and searching materials rather than as a body of firm and strict rules. Deviations from the guidelines set forth here should be made as the nature of the materials processed and the interests of their users require. In processing newspapers and other current events materials for information retrieval, flexibility is mandatory, and therefore frequent changes in the Thesaurus are envisaged. These changes may be initiated by us, or they may be made by individual users to cope with their specific problems and meet their specific needs.

In order to provide the kind of flexibility desired in on-line files, it is important that the computer programs not be based on a set of implicit or hidden assumptions about how the Thesaurus is handled at the present time. For this reason, a structural description (that is, a grammar) of the Thesaurus is developed here to promote future flexibility and growth through a commonly understood interface between the designers of the Thesaurus and the programmers.

The final definition for a Thesaurus, when pursued through all the intermediate definitions of the structural description, reduces to a (gigantic) natural language sequence, accessible and alphabetized on the basis of certain subsequences—Descriptors, See also References, etc. It is just that.

How this large character string is to be formatted and stored in a computing system (with auxiliary directories, pointers, counts, separator characters, etc.) is a matter of programming strategy and tactics. It is an important matter, but designers of the Thesaurus need not get tangled up with it. Rather, they need only be concerned with the Thesaurus in its external form, as a structured natural language sequence which can be queried on, and added to or deleted from, with certain automatic cross-referencing carried out thereby.

Thus, the important question for the designer is, "Is this the structure I want for the Thesaurus?" in contrast to questions of content, criteria for placing content, etc. The objective of the following description is to permit the designer to examine that question with confidence and precision. The tools may seem a little formal and formidable at first glance. But it is believed that concern will disappear with a little familiarity. The purpose is not to obscure, but to make analyses more precise and comprehensive—so that the designer can see the Thesaurus structure per se.

In this connection, the description developed below is somewhat more general than the present Thesaurus structure. It frequently happens that the simplification and unification desirable for automatic processing comes only with a certain degree of generalization. And it frequently happens that more flexibility, rather than less, accompanies such generalizations. Not all the flexibility inherent in the proposed file structure is used in present Thesaurus activities, and it is not expected that all of it will ever be used. But it is there to use, and, more important, known to be there.

The Structural Description

The structural description for the Thesaurus will be given through a series of syntactic definitions (or "syntactic equations"), each of which expands a Thesaurus term (a generic form for a part of the Thesaurus), which is being defined, into one or more patterns using simpler and more basic parts. Any term so defined is ultimately expanded thereby into natural language text, which is the unspecified primitive for the Thesaurus. As noted, the description concerns itself only with the structure of the Thesaurus and not with its contents.

The syntactic terms, or entities, used in the description are given in Table 7.3-1, first as natural language terms, and second, in a briefer symbolic form which will be used for convenience later. Notice the Thesaurus terms are in three categories. First, there is a primitive term from which the Thesaurus is ultimately constructed, which is simply natural language text. All subsequent terms are eventually decomposable into this natural language text, which is the responsibility of the designers of the Thesaurus. Second, there is a set of terms used by The New York Times which are intended to be used in the structural description exactly as The Times personnel mean them. Explanations and examples of these terms are included, based on descriptions in The New York Times Thesaurus of Descriptors. Third, there is a set of additional terms (which will be defined by syntactic equations) which serve as intermediate syntactic entities

Table 7.3-1

Thesaurus Terms

Term	Syntactic entity
Primitive Term	
Natural Language Text	<TEXT>
New York Times Terms	
Descriptor	<TERM>
Qualifying Term	<QT>
Scope Note	<SN>
Hierarchical Notation	<HN>
See Reference	<SR>
See also Reference	<SAR>
Refer from Reference	<RFR>
Subheading	<SUBH>
New York Times Thesaurus of Descriptors	<THESAURUS>
Additional Terms (defined by equations in Table 7.3-2)	
Text List	<TL>
Qualifying Terms List	<QTL>
Term Extension	<TE>
Term Extension List	<TEL>
Term Structure	<TS>
Term Structure List	<TSL>

between some of the lower and higher level terms used by The New York Times. These intermediate entities are, in fact, known in various forms to The Times personnel as well; the reason for treating them more rigorously is to improve on the precision possible over natural language descriptions.

The 16 syntactic equations of the description are given (and numbered) in Table 7.3-2, and a brief word of explanation is in order, so that the equations in Table 7.3-2 can be understood. Each equation consists of a "left-hand side" and a "right-hand side." The left-hand side consists of the syntactic entity being defined by that equation. The right-hand side is its definition.

There are two major ways a definition is made in Table 7.3-2. The first way is through an **informal definition,** given in natural language between asterisks. This kind of definition may be used when no ambiguities or misunderstandings are likely. In any case, at least one term (a primitive term such as the first one in Table 7.3-2) must be defined in some informal way or else the whole system of definitions will be circular. The second method of definition is by **syntactic formula,** which expresses one or more possible patterns of terms, using notation which we describe next.

Table 7.3-2

Thesaurus Equations

1.	<TEXT>	= *Natural Language Text*
2.	<TERM>	= <TEXT>
3.	<QT>	= (<TEXT>)
4.	<SN>	= Note: <TEXT>
5.	<HN>	= {(BT), (NT), (RT)}
6.	<SR>	= See [<TEL>] [<TL>]
7.	<SAR>	= See also [<TEL>] [<TL>]
8.	<RFR>	= Refer from <TEL>
9.	<SUBH>	= Subheadings [<TSL>] [<TL>]
10.	<THESAURUS>	= <TSL>
11.	<TL>	= *Alphabetized list of <TEXT> items*
12.	<QTL>	= *List of <QT> items*
13.	<TE>	= <TERM> [<QTL>] [<SN>] [<HN>]
14.	<TEL>	= *Alphabetized list of <TE> items*
15.	<TS>	= <TE> [{<SR> <SAR>}] [<RFR>] [<SUBH>]
16.	<TSL>	= *Alphabetized list of <TS> items*

Notation used in constructing the grammar:[a]

= read "is defined as"

, separates items in a list

[ ] enclosed item is optional

{ } precisely one item from the enclosed list be selected

[a] Note that parentheses ( ) are terminal symbols in grammar.

Note that each syntactic entity in Table 7.3-2 begins and ends with an angle bracket (<,>), which seems to enclose a meaningful acronym or word. In fact, the whole string, including the angle brackets, is to be regarded as a single symbol, and the internal sequence of characters is of mnemonic significance only. In addition to the angle brackets (which are used to construct multicharacter symbols thereby) we also use as **meta-symbols** equals (=), comma (,), square brackets ([,]), and braces ({,}). The equals has already been informally explained above, in the definition of syntactic equation. The comma is used merely to separate items in a list. The square brackets are used to enclose an item, and mean that the appearance of that item is optional, that is, it may or may not appear in the pattern given by the formula. The braces are used to enclose a list and mean that precisely one item of the list must be used in the pattern. Natural language text appearing by itself, that is, not within angle brackets or asterisks, stands for itself. For reasons that are apparent with a little reflection, such occurrences of natural language are called syntactic constants. (The expression "See also" is a frequently recurring syntactic constant in the Thesaurus, for example.) The formula of a right-hand side of a syntactic equation can thus vary, by use of brackets and braces, over several forms: the meaning of the syntactic equation is that the syntactic entity on the left is defined as any and all forms on the right side which are possible.

To illustrate these ideas, note that Equation 4 of Table 7.3-2 states that <SN> (i.e., Scope Note) consists of the five characters, "Note:", followed by the syntactic entity <TEXT>, which by Equation 1, is simply natural language text. That is, Equation 4 sets up the **syntactic constant,** "Note:", as the opening 5 characters of a Scope Note, followed by the **syntactic variable** <TEXT>, which stands for any text (sense or nonsense) desired. The first four syntactic equations can be translated back into the descriptions in the New York Times Thesaurus of Descriptors very readily. Note <QT> (Qualifying Term) is placed between parentheses in Equation 3. Equation 5 illustrates the use of braces. The entity <HN> (Hierarchical Notation) is one of the three strings of four characters, "(BT)", "(NT)", or "(RT)".

Just to check understanding, note that an equivalent form of Equation 5 is

<HN> = ({BT,NT,RT})

or even

<HN> = ({B,N,R}T)

Table 7.3-2 is a reference table rather than an exposition table. Its virtue is its conciseness and precision in defining the Thesaurus structure. But the equations leading up to Equation 10, for <THESAURUS>, take a little more examination and explanation which we go into next. The motivation for so doing is that, once understood, Table 7.3-2 is a complete and authoritative map of the structure of the Thesaurus.

More on Table 7.3-2

The idea of leading up through the higher-level entities in Table 7.3-2, to <THESAURUS>, can be illustrated by examining several instances of a <SAR>—a "See also Reference." We note a <SAR> consists of the phrase "See also", followed by one or more references to Descriptors. But, a reading of the New York Times Thesaurus of Descriptors reveals that, along with the Descriptors, may or may not come a list of <QT> (Qualifying Terms) items, a <SN> (Scope Note), and a <HN> (Hierarchical Notation). We build up these possibilities in Equation 13 (using Equation 12 first to define a list of <QT> items, in contrast to a single <QT>). Now, with each single Reference defined by Equation 13, as <TE>, we use Equation 14 to define an alphabetized list of such References, naming it <TEL>. Also, since some references may be nondescriptors ("See also foreign countries"), we build an alphabetized list of such References, naming it <TL>. Now, finally we can form <SAR> in Equation 7, as the syntactic constant "See also" followed (optionally) by a list of Descriptor References and/or a list of nondescriptor references.

We used the expansion (or synthesis) of Equation 7 to illustrate a similar process for Equations 6 and 8. Equation 9, defining Subheadings, is a little more complex, and uses what is known as a "syntactic recursion" in its definition. First, we define the structure possible under a "main heading" of the Thesaurus as <TS> (Term Structure) in Equation 15. It is a <TE>, already defined, followed (all optionally) by either or neither of <SR> or <SAR>, by <RFR>, and by <SUBH>. Next we note that a Subheading can be defined in this way, itself, if we realize two crucial points:

1. The options available include all possibilities in Subheadings, and then some—we can choose to ignore the additional possibilities if we please.

2. The relation of being a Subheading (to a Heading) can be relative rather than absolute, so that a <SUBH> under a <SUBH> (i.e., in its syntactic expansion) is an (absolute) subheading, etc.

Thus, the right-hand side of Equation 9, which defines <SUBH>, when expanded through Equations 16, 15, in turn, includes an item <SUBH>, which is the entity being defined. This is called, thereby, a recursive definition.

In more abstract topics, there are inherent theoretical difficulties with recursive definitions, but no practical ones here. What Equations 9, 15, 16 say, together, is that any number of "subhead nestings" are possible in the structural description—and this is an instance of the generality of this description. But in practice, the user will create only a given number of such nestings, the lowest subheading in the nesting will have the term <SUBH> missing on the right side of Equation 15 (the whole term [<SUBH>] is optional). Thus the full expansion of Equation 9 (or 15) in a realized file will always terminate.

It now may be somewhat of a surprise at first glance, but in defining a <TSL> in Equation 16, originally conceived to be the List of Term Structures that may be contained in a Subheading, we have, indeed, defined the Thesaurus, and Equation 10 merely records this fact. There may be a far greater number of characters and entries in the entire Thesaurus than in a typical Term Structure (the appendage to a Descriptor), but their structures are identical and that is all we are defining at this point.

Conversational Access

Access to the Thesaurus in printed form is by page turning and by eye, using the alphabetized structure inherent in its definition. The human hand and eye represent a potent search mechanism as long as the material is not voluminous and nothing further is done with the results.

In on-line conversational access, however, we must be more explicit and precise in calling for sections of the Thesaurus, at most a few lines at a time, by explicit commands rather than page turning and scanning. Therefore, we outline here a specific system for conversational access.

The basic format of the conversational access is "Request and Display." The user will enter a request at a terminal for some section of the Thesaurus and the system will display the results of that request. The results will be the section requested or else an error message, either dealing with the format of the request itself, or else stating that the section requested could not be located. The basic entry point into the Thesaurus is through Descriptors, possibly further specified by Qualifying Terms, and possibly at subheading levels in the Thesaurus. If the Descriptor entered is not a preferred term, its request will bring an automatic display of a "See Reference" list. If a Descriptor has been located which is a preferred term, it will bring a display containing Qualifying Terms, a Scope Note, and a Hierarchical Notation, to the extent present. We call this a "Base Descriptor." Now, given such a Descriptor, the user may request access to any of three lists possibly associated with it: the See also References, the Refer from References, and the Subheadings. Having requested one of these three lists, the user may then request References or Subheadings simply by asking for the "Next" item on the list, or by asking for the Descriptor itself. The display response to the "Next" request is the next Reference or Subheading, if available. A reference may be either a definite Descriptor, or an indefinite reference to a generic category of Descriptors. If no more items remain on the list (the user presumably having scanned some previously), the message "End of List" is displayed. Attention can be changed from one of the three lists to any of the others by a simple request instead of "Next" or by a Descriptor request.

The user who wants to follow out a Referenced or Subheading Descriptor (e.g., to examine its "See also References," etc.) can make a "Transfer" request, which replaces the original Base Descriptor by its Referenced or Subheading Descriptor, and access continues from the latter as indicated previously. After one or more requests for such a "Transfer," a "Return" request can be made, which replaces the current base Descriptor by the Descriptor which produced it by "Transfer." Thus, after a series of "Transfer" requests, an equal number of "Return" requests will proceed (in reverse order) through the same set of Descriptors, back to the original one.

The foregoing Requests and Displays are summarized (and numbered) in syntax form in Table 7.3-3. An examination of the table will show how each of the commands leads to a specific display. Note the only syntactic variable which can be used in a request is a <TERM> (a Descriptor)

Table 7.3-3

Conversational Access to the Thesaurus

Request	Display
1. Entry <TERM> [<QTL>]	{<TE>, <SR>, no Entry}
2. See also	{See also, no See also, no Entry}
3. Refer from	{Refer from, no Refer from, no Entry}
4. Subheading	{Subheading, no Subheading, no Entry}
5. Next	{<TE>, <TEXT>, end of list}
6. Transfer	{<TE>, no Reference/Subheading}
7. Return	{<TE>, original Entry}

followed optionally by a <QTL> (Qualifying Term List). The syntactic variables displayed are limited to <TE> (Term Extension), <SR> (See References), and <TEXT> (for generic references); but of course, just these displays permit the user to browse through any part and detail of the Thesaurus desired. The remaining requests and displays are syntactic constants. In practice, this small vocabulary of request items, all but one of which are constants, represents a simple, readily understood means for accessing any information desired in the Thesaurus.

Thesaurus Creation and Maintenance

We define Thesaurus creation and maintenance in terms of the syntactic entities of Table 7.3-1, above the level of the primitive Natural Language Text. That is, we consider only the addition and deletion of entire Thesaurus items and not portions of text. The addition and deletion of characters in text making up a file item is considered text editing, rather than Thesaurus maintenance, in this context. It is recognized that text editing is a desirable future facility in the overall process of Thesaurus maintenance, and the present emphasis reflects merely a time phasing of ultimate interests.

The process of Thesaurus creation is simply the construction of a <TSL> which is to be defined as the Thesaurus. (The problem of how such a Thesaurus is to be physically loaded into storage, with directories, etc., is a programming question not dealt with here.) For example, the New York Times Thesaurus of Descriptors, by definition, and barring typographical or logical deviations from its designer's intentions, is a <TSL>.

The process of Thesaurus maintenance is, likewise, very simple in syntactic terms. A Thesaurus addition or deletion can be defined by giving a **location** and a syntactic entity which is to be added or deleted. The location can be given in the Conversational Access requests, namely,

Entry <TERM> [<QTL>]
See also
Refer from
Subheading
Transfer

to prescribe the destination of the syntactic entity to be added or the entity to be deleted. In the case of unique items, such as a Scope Note, or a Hierarchial Notation, addition is taken to mean replacement if already present. In case of listed items, such as See or See also References, or Subheadings, addition is done automatically in alphabetized form. In the case of deletion, deleting a Descriptor automatically deletes all file items accessed by that Descriptor as well.

ILLUSTRATIONS

We use the Model Page of The New York Times Thesaurus of Descriptors, as shown in Figure 7.3-1, to illustrate the foregoing ideas concretely.

First, regard the contents of the Model Page as a miniature Thesaurus. It has the structure of the entire New York Times Thesaurus of Descriptors, only with far less text in it. It is, in fact, a <TSL> (Term Structure List) of alphabetized <TS> (Term Structure) items, which begin with Descriptors:

ADEN Protectorate
ADOPTIONS
ADVERTISING
AMERICA
AMERIKA
BIRTH Control and Planned Parenthood

(note Equations 10, 16 of Table 7.3-2 express this structural fact).

Next, any one of these <TS> consists of a <TE> (Term Extension) followed optionally by References and Subheadings (Equation 15). Some <TS> have no References or Subheadings at all, and some <TE> consist only of a <TERM> item (a Descriptor), but these are admissible possibilities in the equations. Nevertheless, in order to keep matters straight, we recognize each syntactic entity represented in the miniature (or full) Thesaurus, even though one section of natural language text may stand for several entities at once. For example, the first <TS>,

ADEN Protectorate
    See South Arabia, Federation of

represents the diagrammed entities in Figure 7.3-2. Similarly, the eighth <TS> has the structure of Figure 7.3-3.

In fact, the Model Page (by design) exhibits nearly every structure that can occur in the Thesaurus; it can be instructive to locate the syntactic equation in Table 7.3-2 that defines any given structural relationship on the Model Page.

Now, consider the miniature Thesaurus given by the Model Page to be "on-line" for conversational access. Although the eye can take in the entire page, imagine it cannot—and that only one item is available for inspection at a time. We will invoke the "Request and Display" mode of

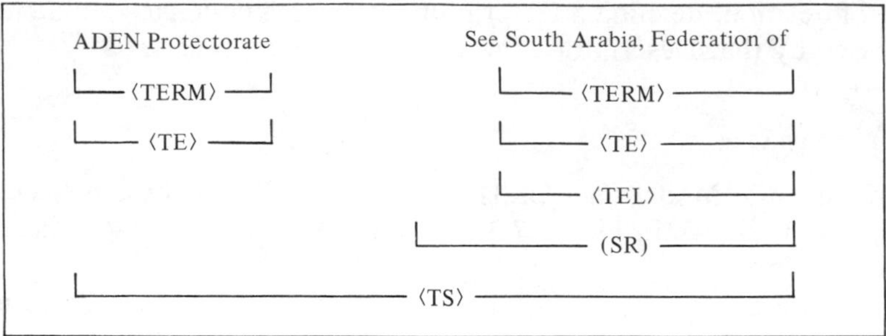

Figure 7.3-2. Diagramming the First <TS> From the Model Page of Figure 7.3-1.

conversational access to browse, in illustration, through this miniature Thesaurus. We show a conversation in Table 7.3-4.

In the conversation, the actual language itself is terse and skeletal—because we are interested only in structural aspects of the Thesaurus, and how Request and Display operations can permit a user to browse and examine the Thesaurus item by item. In practice, the Display side would be more complete, maintaining "backtrack status" information, etc., as display space permits.

Thesaurus creation is illustrated by the model page itself: natural language text with structural characteristics satisfying the equations of Table 7.3-2. For Thesaurus maintenance we consider an addition and a deletion (noting that a modification can be considered a deletion followed by an addition). Suppose we wish to add Television (NT) to the Refer from References of ADVERTISING. We form the Locator

Entry ADVERTISING, Refer from

and the item

<TE> = <TERM> <HN> = Television (NT)

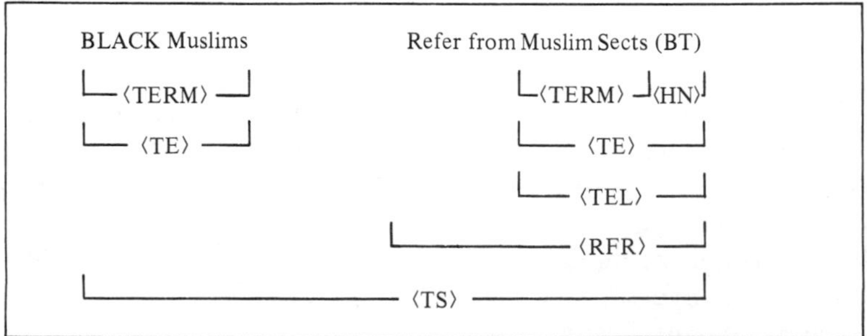

Figure 7.3-3. Diagramming the Eighth <TS> From the Model Page of Figure 7.3-1.

Table 7.3-4

A Sample Conversation

Request	Display
Entry ADEN Protectorate	ADEN Protectorate See South Arabia, Federation of
Entry ADEN	No Entry
Entry BLACK Muslims	BLACK Muslims
See also	No See also
Refer from	Refer from
Next	Muslim Sects (BT)
Next	End of List
Subheadings	No Subheadings
Entry ADVERTISING	ADVERTISING
Transfer	No Reference/Subheading
Subheadings	Subheadings
Next	Mass Communications (for inclusion) (BT)
Transfer	Mass Communications (for inclusion) (BT)
Refer from	Refer from
Next	ADVERTISING
Transfer	ADVERTISING
Return	Mass Communications (for inclusion) (BT)
Return	ADVERTISING
Return	Original Term
Subheadings	Subheadings
Next	foreign countries
Transfer	No Reference/Subheading
Next	United States
etc.	

for addition. Then Television (NT) would be automatically added (in alphabetized order) to the Refer from References of ADVERTISING. Similarly, to delete the Hierarchical Notation (BT) in the "BLACK Power" See Reference, we locate by

Entry "BLACK Power"

and delete item

<HN> = (BT)

therein.

> **Summary:** Analysis of the New York Times Information Bank led to a description of the Thesaurus in a formal grammar that was surprisingly simple yet powerful. A user language for the Thesaurus was discovered and stated in a formal grammar.

EXERCISES

1. Create parse trees and parse tables for the following hand calculator inputs:

(a) C 6 =
(b) C 31 + 42 =
(c) C 3 + 42 − 6 =

2. A grammar for decimal numbers is given as

1. <decimal number> ::= <number>.<number>
2. <number> ::= <digit> *<digit>
3. <digit> ::= 0 | 1 | 2 | 3 | 4 | 5 | 6 | 7 | 8 | 9

Determine which of the following strings are decimal numbers as defined here by writing parse tables for them, if possible:

(a) 32.45
(b) 0.456
(c) 23
(d) 23.
(e) .23

3. Modify the grammar of Exercise 2 so that all cases will be defined as decimal numbers.

4. An <identifier> in many programming languages is defined as any letter followed by zero or more letters or digits (letters and digits in any order). Define a grammar for <identifier>, first without parentheses, and second with parentheses, to make the grammar most readable.

5. Provide a grammar for dates, in three forms (three grammars):

(a) 7–4–76
(b) 4 July 1976
(c) July 4, 1976

6. In Table 7.1-2, several columns are identical. Can you express the response required in a simpler, more compact form? Can you simplify the state?

7. Design a grammar for an <identifier> in a programming language such that the first character must be a letter followed by zero or more letters, digits, or special characters (e.g., %, $, #, etc.). Define a state machine that recognizes legal identifiers for your grammar. Let the stimuli be characters in a string and the response be the last complete identifier. All possible character strings should be handled.

8. Consider defining a state machine for a spelling checker.

9. Given the grammar for HC input, design a clear box from the grammar which will determine its output.

10. Given the grammar for dates of the form

July 4, 1984

design a clear box from the grammar which will copy all dates in a character string.

11. Given a list of dates (in the form of Exercise 10) in a character string, design a clear box to determine for each adjacent pair which is earlier or if they are the same date.

12. A (modified) simple expression in Pascal is defined in syntax as:

1. <simple expression> ::= <term>|<sign><term>|
<simple expression>
<add operator><term>
2. <sign> ::= + | −
3. <add operator> ::= + | −
4. <term> ::= <factor>|<term> <multiply operator>
<factor>
5. <multiply operator> ::= * | /
6. <factor> ::= <variable>|<number>|(<simple
expression>)
7. <variable> ::= <letter> *<letter or digit>
8. <letter or digit> ::= <letter>|<digit>

9. <number> ::= <digit> *<digit>
10. <letter> ::= a | b | c | ··· | z
11. <digit> ::= 1 | 2 | 3 | 4 | 5 | 6 | 7 | 8 | 9 | 0

Build the parse table for the following legal simple expressions:

(a) abc − (20 * xyz)
(b) 80/n/65
(c) −((20 * x) + (y − z))
(d) x * y + z/5 * w

13. Based on your own imagination, invent a personnel file for a business and give its syntax.

14. Based on your own imagination, invent a personnel report based on the information in the personnel file of Exercise 13 and give its syntax.

15. Define a syntax for reserved seat tickets to a series of concerts in an auditorium with a main floor (orchestra), rows A to MM (A,...,Z,AA,...,MM), seats 1L,...,30L, 1R,...,30R, and balcony, rows A to K, seats 1L,...,25L, 1R,...,25R. The dates of the concert are August 3, 4, 5, 1987.

16. Design an input grammar for set expressions. Elements of a set are single letters and sets are represented by brackets (e.g., [A, P, C, X]). Null sets, [ ], are legal. Three set operations are used in the expressions

∪ Union,
∩ Intersection, and
− Difference

Parentheses are used to group operations.

Using the designed grammar, build a parse tree (or table) for the expression

([A, B] ∪ [A, C, D]) − [A]

Chapter 8 Data Structures in Information Systems

8.1 DATA STRUCTURES

> **Preview:** The analysis and design of state data representations are critical tasks in information systems development. Data structures, file systems, and database systems are three levels of data design that provide increasing data independence to the user. Data design requirements include techniques and information to support data access, retention, and control. The basic data structures are data items, arrays, records, lists, trees, and graphs.

8.1.1 Data Analysis and Design

Effective representation of state data is a crucial aspect of box structure design. A large part of information system implementation is devoted to the mechanics of accessing, updating, and managing data. Efficient system implementations depend on good data representations, to simplify and systematize these operations. Data representation possibilities can be organized into three levels of design, namely, data structures, file systems, and database systems, as illustrated in Figure 8.1-1. Each level of data design is built upon the levels below it in the hierarchy. The use of data at one level is **independent** of implementation details of the lower levels. More precisely:

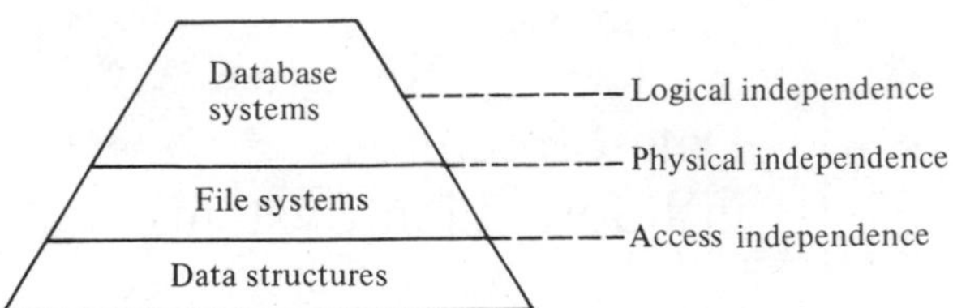

Figure 8.1-1. Hierarchy of Data Design.

Data Structures. Data is represented and stored in terms of the storage media of an information system. Data structures allow similar, and dissimilar but related, data items to be accessed and retained as coherent units. **Access independence** relieves the user from storage details.

File Systems. Files group collections of similar data structures for common access and retention. File systems provide **physical independence** to users. Physical independence hides details of underlying file organization. Details of data structure representation and storage in a file are managed by the file system and need not be visible to the user.

Database Systems. A database is a grouping of data files and a representation of the relationships among the files. A database system manages details of access, retention, and control to present the user with a greater degree of data independence, in a high-level conceptual view of the data environment. In addition to physical independence, database systems also provide **logical independence,** through multiple user controls, that allow users to view only the data needed for their applications.

The objective of state data analysis and design is to provide effective and efficient use of data in an information system. However, the data design problem must not be separated from the overall system design. Otherwise, unforeseen consequences can arise, as these examples illustrate:

A new information system is to be built based upon the functions of one or more existing systems, both manual and automated. To avoid data restructuring, the designer adopts the data designs in the current systems, and attempts to integrate them into a new data design.

A designer has a preconceived idea that a database system should be used for data management. Thus, the data design will influence the system design to justify use of the database system.

A designer has considerable expertise and experience with one type of data design, for example, file systems, or one particular data management package. The system design will be tailored to fit a data design compatible with the designer's experience.

Hardware and software for an information system is selected before systems analysis and design is completed. The system design, and thus, the data design, is severely constrained by hardware and software capabilities.

In short, data design must be based on analysis of state requirements in the box structure hierarchy, and not necessarily on preconceived or pre-existing designs, as summarized in the following fundamental principle.

Fundamental Principle: Data design is an integral part of box structure design and must not be considered separately.

State definitions play a major role in box structure methodology. A black box is a state-free description of behavior in terms of stimulus history. However, at the state machine level, state definitions replace stimulus histories in describing behavior. State and machine definitions are separate yet interdependent, and must be analyzed and designed together. At the clear box level, state definitions exist (although always subject to better ideas) and procedures are created to carry out machine transitions.

Data design is the process of selecting data representations for all of the state information in a box structure design. Two fundamental principles guide data design in the box structure methodology.

Fundamental Principles: Data Design.

1. Data design proceeds top-down in a box structure hierarchy. The top level state should be designed first, followed by all other states in top-down fashion.
2. The functionality of all machines within the clear box expansion under a state should guide the data design for that state.

Top-down data design is intrinsic to the box structure methodology. Common states at higher levels in a box structure contain the most widely shared states in the system. This implies the need for strong controls on data integrity, consistency, reliability, and access efficiency. Effective data representation of these states is crucial for good system design. A corollary to the top-down design principle is that higher level data designs can influence lower level data designs, but not vice versa.

The state transitions carried out by the machine of a state machine guide data design to support those transitions. That is, data design is predicated upon requirements for state operations defined by the state machine and carried out by its clear box expansion. These **data requirements** can be categorized in three levels, as shown in Figure 8.1-2:

1. Data Access. Box structure data access requirements include the ability to store and retrieve state data. Any data representation must support these functions.

2. Data Retention. Box structures retain state data between transitions. Retention must be supported by a data representation that maintains data integrity while providing for data access. The effective organization of data in storage is also a goal of data retention support.

3. Data Control. An information system may require special controls for access to information. In such a case, box structures must support control functions, including data sharing among multiple users without compromising data integrity, providing data security, and providing mechanisms for data recovery in the event of system failure.

Data access requirements are the most fundamental in an information system. Data retention requirements imply data access. (What good is information that cannot be accessed?) Data control requirements imply data retention. (Why control information that is not saved?) A data representation that meets the outer data requirements in Figure 8.1-2 must also meet the inner requirements as well. Thus, data design involves more than selection of data structures for state items. It also requires design of data structures to support machine processing for access, retention, and control of those state items. This observation is generalized in the following fundamental principle:

> **Fundamental Principle:** Data to support data access, retention, and control is itself part of the state in box structure design. The design of a data representation for a state must include consideration of these data requirements.

8.1.2 Data Representation

Data in an information system is represented in a variety of forms. All data representation is based upon a standard building block, namely, a **data item.** Just as a brick can be used to construct many different forms of

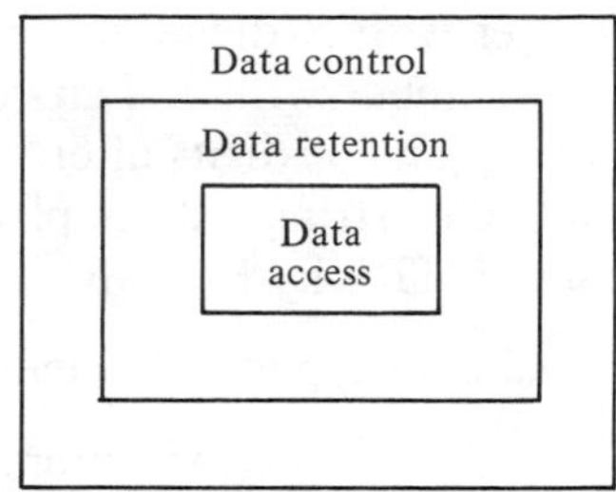

Figure 8.1-2. Levels of Data Requirements.

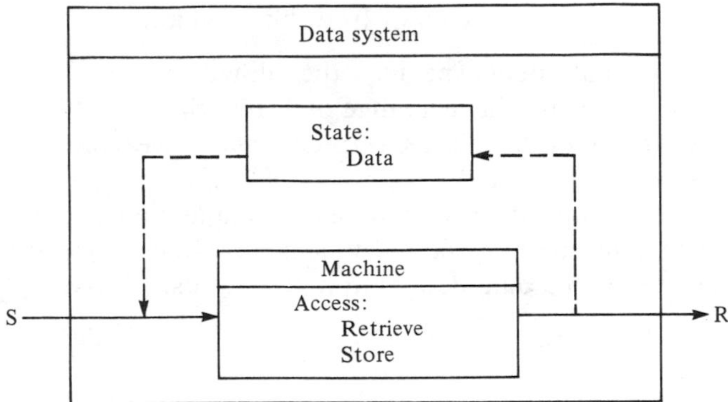

Figure 8.1-3. Data State Machine.

buildings, from a simple house to a complex mansion, data items can be used to construct many different forms of data support, from simple data structures to complex database systems.

Figure 8.1-3 illustrates a state machine representation of a system that accesses state data. The principal access capabilities are storage and retrieval. The remainder of this section discusses state data representation and corresponding storage and retrieval techniques.

Data Items

Data items are atomic units of information to be represented in a system. Their physical representations depend upon the medium on which it is recorded. An employee number, for example, may be represented as typed numbers on a payroll check, as a magnetic strip on an identification card, or as a sequence of bits in a computer system.

Data items are defined on a **domain.** A domain definition has two parts; a **data type** and a **range.** The data type specifies the form of the data item. Standard data types include:

INTEGER (X):	Integer with X digits
REAL (X.Y):	Real number with X digits before the decimal point and Y digits after
CHAR (X):	Character string with X places
BOOLEAN:	One of {TRUE, FALSE}

Many systems allow special data types to be defined; for example,

DATE (XX/YY/ZZ):	Date in the form month (XX), day (YY), and year (ZZ)
TIME (XX:YY:ZZ):	Time in the form hour (XX), minute (YY), and second (ZZ)
COLOR:	One of {red, blue, white,...}

The range of data items specifies the allowed values that data items may hold. AGE data may have a range of 0–100, SALARY data may have a range of $100–$100,000, and GRADE data may have a range of A, B, C, D, F, I, W.

Data items are specified in BDL as a combination of a name and a domain. Data items are referred to by name in a BDL procedure, and are defined separately in a data declaration prior to use. Thus,

```
a, b, c:  INTEGER
d:        CHAR
e:        hand calculator key
```

are data item declarations that associate a data type with each name (a, b, c as integers, d as a character, e as a hand calculator key). Simple data types are part of BDL syntax, to be defined according to problem needs. The general criteria for simple data types is that they refer to units of data not ordinarily decomposed further. In addition to identifying data types by a term (integer) or phrase (hand calculator key), a data type can be described by an enumeration of all possible values, for example,

```
BR:       (B, C)
weekday:  (Mon, Tue, Wed, Thu, Fri, Sat, Sun)
```

or by describing a range of items in a known list, by stating the first and last members separated by a double period, for example,

```
workday:       (Mon..Fri)
digit:         (0..9)
nonzerodigit:  (1..9)
```

A data type and a range can be specified together in BDL, for example,

SSN:	CHAR (11)
	Range = (000-00-0000..999-99-9999)
DEPT#:	INTEGER (2)
	Range = (1..99)
SIZE:	REAL

A range specification (e.g., SIZE) in a domain is optional. The default range would be only limited by the ability of the system to handle data values. Thus, computer systems can only accommodate numbers to a certain precision, and business forms typically place constraints on the number of characters that can be entered on each line.

Systems should provide a means of ensuring that a data item only contains values in its domain. Operations that insert a new value or update an old value should be monitored to guarantee the validity of the new value, otherwise the operation should not be allowed. These checks maintain the domain integrity of data items.

The Box Structure of Data Items. As surprising as it may seem, even simple data storage and retrieval operations on data items can be represented by a black box, specifically, a common service black box. The black box of data items has a complex stimulus defined in two parts, namely, the operation, or **service,** to be performed on the variable, and the variable value itself. The services are storage (ST for short) and retrieval of a value (RT for short). In illustration, the sequence of assignment statements

```
do
   c := 2;
   b := 1;
   a := b + c;
   b := a;
   c := c + b;
   d := b
od
```

defines the following stimulus/response history for the black box of b:

S		R
service	value	
ST	1	
RT		1
ST	3	
RT		3
RT		3

where blanks in the table signify a null value. That is, the black box stores and retains a value for b through successive retrievals until a new value is stored. The black box specification for b is as follows:

```
define BB b
  stimulus
    S:(ST,RT)
    V:number
  response
    R:number
  transition
    if S = ST then
      R := null
    fi;
    if S = RT then
      R := V(k), where V(k) represents the last value
              stimulus, such that S(k) = ST.
    fi
```

Note that the storage service produces no response, and depends on the definition of a black box transition to append the stimulus to the stimulus history for use by the next retrieval service.

A data item is thus a pure storage and retrieval black box, the simplest one possible. The corresponding state machine description is

```
define SM b
  stimulus
    S:(ST,RT)
    V:number
  response
    R:number
  state
    E:number
  machine
    if S = ST then
      E := V
    fi;
    if S = RT then
      R := E
    fi
```

where b is the black box name and E is the state variable.

Thus, when a data item appears on either side of a data assignment as, for b,

a := b
b := 2

we can interpret the appearances as shorthand for transition invocations of a common service state machine, either to retrieve its state value or to store a new state value. Also, proper use of the state machine requires a retrieval transition to be preceded by a storage transition. Failure to do so results in the common mistake known as an "uninitialized value". Note too, that successive storage transitions without an intervening retrieval represent a somewhat subtler mistake known as an "unreferenced value," where a stored state value is not used in some transition before it is overwritten.

ARRAYS

An **array** is an ordered set of identical data elements of finite length. Arrays can be organized into 1-, 2-, 3-,... dimensional structures. A list of 5 numbers

3 6 1 9 6

can be stored in an array

stimulus:**array**(1...5) of number

such that

stimulus(4) = 9, stimulus(1) = 3

and so on. The expression stimulus(6) is undefined because of the **array** bounds 1...5 given in the declaration.

A magic square

1	14	7	12
15	4	9	6
10	5	16	3
8	11	2	13

(what's magic about it?) can be stored in an array

magic square:**array**(1...4, 1...4) of number

such that

magic square(2, 3) = 9
magic square(4, 4) = 13
magic square(3, 1) = 10

Records

A **record** is a grouping of data elements that, together, describe objects in an information system. The objects are termed **entities,** and are described by **attributes.** Entities and attributes describe information at a conceptual level. At the data design level, data items represent attributes and records represent instances of entities.

BDL records are data structures that permit collections of data elements to be named and organized into a hierarchy. For example, the hierarchical structure

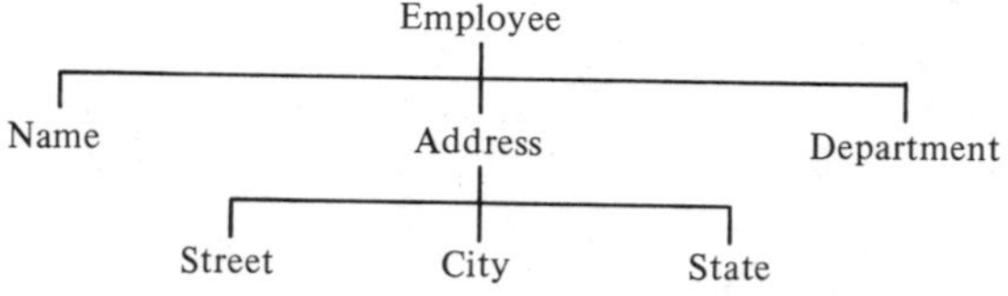

can be declared by keyword **record** and a tabular typographic form,

```
employee: record
  name: text
  address: record
    street: text
    city: text
    state: text
  end record
  department: number
end record
```

where employee names a collection of three items (name, address, and city), as does address (street, city, state). Elements in a record can be referred to in a procedure by a dot notation, for example

employee.name := 'Jones, John'

or simply

name := 'Jones, John'

where no ambiguity results (that is, name is not used as a variable or as another record qualifier in the procedure).

Variant Records

Variant records generalize the idea of a record to include multiple possibilities. In BDL, variant records permit records of varying structure, depending on data found in the record itself. A variant record contains a **tag field,** which defines the specific variant of the record. Variant records are especially useful in dealing with generic information that can occur in different forms. For example, an employee record may be quite different for permanent and temporary employees. A variant record could be defined in this situation with tag field named employee type:

```
employee: variant record
              case employee type: (perm, temp) of
                perm:record
                       employee number:number
                       ...
                     end record
                temp:record
                       ...
                     end record
              esac
            end record
```

FILES

Files group records of the same entity-type. For example, all employee records of business constitute an employee file. Records can be organized into a file in several different ways as described in the next section.

In BDL, files can be described as, for example,

emp: **file** of employee **records**

The concept of a **key** is of great importance for file organization. A **candidate key** (or simply a key) is an attribute or group of attributes of an entity that has a unique value for each instance of the entity. An entity may have many candidate keys. Consider an employee entity with attributes employee number (EMP#), social security number (SSN), name (NAME), address (ADDR), and department number (DEPT#). Based upon standard semantic assumptions, we could state that EMP#, SSN, or even NAME are each candidate keys of the employee entity. In fact, any grouping of attributes, as long as the group contained one of {EMP#, SSN, NAME}, would meet the definition of a candidate key. However, candidate keys are normally considered to be **minimal,** such that no candidate key contains another candidate key as a subset of its attributes.

A primary key is a candidate key selected to serve as the unique access key of the entity. Only one primary key exists for an entity. Most file organizations store and retrieve data based upon the value of the primary key. In the employee entity, defined above, the logical choice for a primary key would be EMP#, since values of employee numbers are controlled by the organization and can be guaranteed to be unique.

A **secondary key** is any attribute in an entity that is not, by itself, a candidate key. Secondary keys provide the potential for secondary indexes to be constructed for a file. Such indexes allow access to groups of records that contain the same value in the secondary key. In the employee entity the attributes ADDR and DEPT# are considered secondary keys.

8.1.3 Linear Data Structures

Sequential access of data requires that each data item be available to a process in a well-defined order. This order can be determined in a number of ways. The value of the data item may determine the order, for example, a sorted list of test scores. Data items may also be ordered by time of entry into the data structure, for example, a first-in-first-out (FIFO) queue at a bank. When a well-defined order of data items is required for sequential processing, a linear data structure is appropriate. This section describes the principal forms of linear data structures, namely, lists, queues, and stacks.

Lists

Lists are a flexible data structure in which data are maintained in a particular order; however, individual elements can be accessed, inserted, deleted, or updated at any position. Lists can grow and shrink to arbitrary sizes. List applications include lists of tasks to perform, lists of past months' sales and purchases, lists of customers, etc. A list is defined in BDL as:

l:**list** of number
x, y:number

A list data design includes the necessary functions to satisfy the data requirements of the system. The following functions for a list state machine represent list processing possibilities:

insert(l,x): Insert element x into the list l at its correct position based upon list ordering.
delete(l,x): Delete element x from list l.
locate(l,x): Locate the element x in the list l.
first(l,x): Retrieve the first element from list l and place it in element x.

next(l,x): Based upon the last transition, retrieve the next element in order from list l and place it in element x.

prior(l,x): Based upon the last transition, retrieve the previous element in order from list 1 and place it in element x.

These functions provide a range of processing capabilities on lists. The most effective use of lists comes in organizing and retrieving data in a linear order. While retrieval of arbitrary data elements can be done, for example with the locate function, the search requires retrieval of all data elements up to the desired element in order.

The ordering of a list is defined by the data designer, and must be maintained by insertion, deletion, and update transitions designed into the system. **Queues** and **stacks** are two special types of lists. In these structures, special ordering and access disciplines are enforced.

QUEUES

A queue is a dynamic data structure with a FIFO (First-In-First-Out) access discipline. A value stored in a queue goes to the **back** of the queue, a value is retrieved from the **front** of the queue. In BDL, after declarations

q:**queue** of number
x, y:number

the assignment

back(q) := x

stores the value of x at the back of the queue q; the assignment

y := **front**(q)

moves the value at the front of q, if any, to y (removing that value from q). If q is empty the assignment has no effect. The condition

isempty(q)

is true or false depending on whether q is empty.

In illustration, representing a queue as a list with front at the left, back at the right,

if q = (3 6 1 9), x = 6

after **back**(q) := x,

q = (3 6 1 9 6), x = 6,

and after x := **front**(q),

q = (6 1 9 6), x = 3

and

isempty(q) = false.

The Box Structure of a Queue. Data storage and retrieval operations for a queue can be represented by a black box. In illustration, consider the following two histories for a given black box, with ST for store, RT for retrieve:

S		R	S		R
service	value		service	value	
ST	3		ST	3	
ST	6		RT		3
ST	1		ST	6	
RT		3	RT		6
RT		6	ST	1	
ST	9		ST	9	
ST	6		RT		1
RT		1	RT		9
RT		9	ST	6	
RT		6	RT		6

In both cases, the order of values retrieved (3 6 1 9 6) corresponds to the order of values stored (3 6 1 9 6), although the two sequences of storage and retrieval stimuli are different. With a little thought the black box specification can be written as

```
define BB queue
  stimulus
    S:(ST, RT)
    V:number
  response
    R:number
  transition
    if S = ST then
      R := null
    fi;
    if S = RT then
      R := V(j), where V(j) indicates the first stimulus history mem-
            ber not yet retrieved such that S(j)=ST.
    fi
```

Any black box that exhibits this behavior is called a queue.

The state machine for a queue black box can be written as,

```
define SM queue
  stimulus
    S:(ST, RT)
    V:number
  response
    R:number
  state
    Q:queue of number
  machine
    if S = ST then
      back(Q) := V
    fi;
    if S = RT and not isempty (Q) then
      R := front(Q)
    fi
```

with behavior identical to the black box above.

Many uses of queues exist in information systems. For example, customer service satisfaction is dependent upon perception of fair treatment. The fairest scheduling discipline for customer service is the FIFO discipline of a queue.

An important variation on a queue is a **priority queue,** where the element at the front of the queue depends on a priority formula. For example, the next order to be processed at a warehouse may depend upon the total cost of the order, the number of items in the order, the size and weight of the items, and the location to which the order must be sent. A formula based upon these factors would assign a priority to each order as it is entered into the queue. The arrangement of elements in the priority queue is based upon the priority value.

The BDL description of a priority queue requires a different function for queue assignment. If Q is a queue, then

priority(Q) := E

places element E into its correct priority position in Q. The **front**(Q) function operates as before, by accessing the first queue element (i.e., with highest priority) and deleting the element from the queue.

Stacks

A stack is a dynamic data structure with a LIFO (Last-In-First-Out) access discipline. Values are stored at, and retrieved from, the top of a stack. In BDL, after declarations

```
s:stack of number
x, y:number
```

the assignment

top(s) := x

stores the value of x at the top of the stack; the assignment

y := **top**(s)

moves the value at the top of s, if any, to y (removing that value from s). If s is empty the assignment has no effect. The condition

isempty(s)

is true or false depending on whether s is empty. In illustration, representing a stack as a list with top at the left,

if s = (6 1 9 6), x = 3, y = 6

after **top**(s) := x,

s = (3 6 1 9 6), x = 3, y = 6

and after y := **top**(s)

s = (6 1 9 6), x = 3, y = 3

and

isempty(s) = false.

The black box and state machine representations for a stack of numbers are as follows:

```
define BB stack
  stimulus
      S:(ST,RT)
      V:number
  response
      R:number
  transition
      if S = ST then
        R = null
      fi;
      if S = RT then
        R := V(j), where V(j) is the last stimulus not yet retrieved
               such that S(j)=ST.
      fi
```

```
define SM stack
  stimulus
      S:(ST,RT)
      V:number
  response
      R:number
  state
      T:stack of number
  machine
      if S = ST then
        top(T) := V
      fi;
      if S = RT and not isempty(T) then
        R := top(T)
      fi
```

8.1.4 Nonlinear Data Structures

While linear data structures provide access efficiencies for sequential processing of ordered data, many processing applications require random processing of ordered data, or deal with unordered data. Three important classes of nonlinear data structures are useful in these cases, namely, sets, trees, and graphs.

SETS

A set is a structure in which the data items have no distinguished order. It is a grouping of similar data items, for example, a set of tasks to be accomplished in no particular order, or a group of newly received orders to be assigned order numbers and placed into a database. In BDL, a set is declared as, for example,

```
v, s, t:set of number
  x, y:number
```

The following functions for a set state machine represent set processing possibilities.

null(s):	Makes set s empty.
insert(s, x):	Insert element x into set s.
delete(s, x):	Delete element x from set s.

member(s, x): Logical function returns true if element x is in set s, otherwise returns false.

isempty(s): Logical function returns true if set s is empty, otherwise returns false.

union(v,s,t): Set v becomes the union of elements in sets s and t.

intersection(v,s,t): Set v becomes the intersection of elements in sets s and t.

difference(v,s,t): Set v becomes the result of eliminating all elements of set t from set s.

select(s,x): If set s is not empty an arbitrary element is selected from s and placed into x (the element is not removed from set s).

Sets provide clear advantages for access and storage of unordered data. The set operations of union, intersection, and difference provide powerful data manipulation capabilities. Processing all elements of a set can be accomplished by repeatedly selecting (**select**(s,x)) an arbitrary element of the set, processing it, and then deleting the element (**delete**(s,x)). Retrieving a specific element in a set is less efficient, however. Since a set is maintained in no particular order, a search over all set elements may be required, in the worst case.

The implementation of a set data structure depends upon the storage medium and the size of the set. A **bit-vector** can be used for sets in a limited value range. For example, if set elements are known to have values in the range 1..N, the set can be implemented as an N bit vector. The ith bit will be 1 if that element is in the set. More general sets will require some form of **linked-list** implementation, where each element is defined and linked in a list in no particular order.

TREES

A tree data structure places data elements in a hierarchical arrangement. Tree structures provide improved access efficiency on ordered data and the ability to represent hierarchical relationships among data.

Each data element is termed a **node** in a trcc. A trcc has a single **root** node. A tree is defined recursively as (Figure 8.1-4):

1. A single node is a tree.
2. Given a root node, all remaining nodes are partitioned into disjoint sets, $T_1, \ldots, T_n$, and each of these sets is in turn a tree.

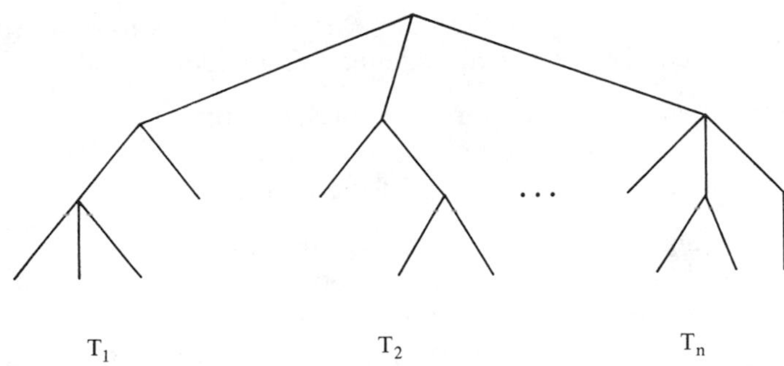

Figure 8.1-4. A Tree.

Tree data structures are common in information systems. For example, trees can be used to represent organization charts or parts explosion diagrams. The outline of this book can itself be represented by a tree. The book is divided into eight chapters (e.g., C3), each chapter into major sections (e.g., C 3.1), and each major section into minor sections (e.g., C 3.1.2).

A tree data structure can also be used with ordered data to improve access efficiency on random retrievals. By structuring ordered data into a tree, sequential search is no longer required to find a particular element. The requested element is compared to an element in a tree node (a search would begin at the root). Based upon the comparison, either the requested element is found or one of the subtrees of the node is identified as containing it; that subtree is then searched, continuing in this fashion until the element is located.

This retrieval principle is illustrated by the **binary tree** in Figure 8.1-5. A binary tree contains one element per node and each node has at most two subtrees. Data elements are ordered such that elements with value less than the current node are placed in the left subtree, elements with value greater are placed in the right subtree.

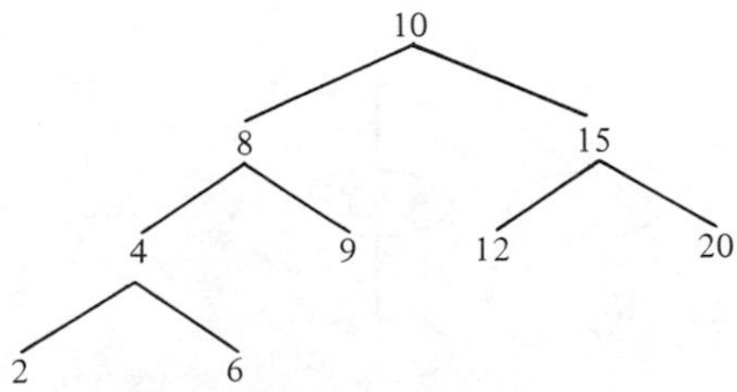

Figure 8.1-5. A Binary Tree.

Sequential processing of the data elements requires an **inorder traversal** of the tree. The traversal is defined recursively at each node as

1. Perform an inorder traversal of the left subtree.
2. Process the current node.
3. Perform an inorder traversal of the right subtree.

Using the algorithm on the example binary tree in Figure 8.1-5, the processing order of the nodes is (as expected): 2, 4, 6, 8, 9, 10, 12, 15, 20.

Random retrieval on the example binary tree requires, at most, four comparisons. Whereas a random retrieval on a linear structure of this data would require, at most, nine comparisons. Tree structures provide significant performance advantages for applications that include random processing requirements.

The following variations of tree data structures are widely used:

Balanced Binary Trees. These variations of the binary tree follow insertion and deletion rules that maintain approximately the same number of data elements in both subtrees of all tree nodes. Tree balancing provides close to optimal random retrieval performance.

B-Trees and 2–3 Trees. By allowing more than one data element in each tree node, multiple branching from each node occurs. For structures with a large number of elements, this grouping is important to decrease the number of levels in the tree. However, several comparisons are now required at each node to identify the proper subtree to search next (if necessary). The simplest form of B-Tree is a 2–3 Tree wherein each node contains two data elements and three subtrees per node. Figure 8.1-6 shows the example data from Figure 8.1-5 in a 2–3 tree structure.

Tries

Tries (from re*trie*val) are used primarily for searches on character strings. Each element in the string defines the path in the trie to follow.

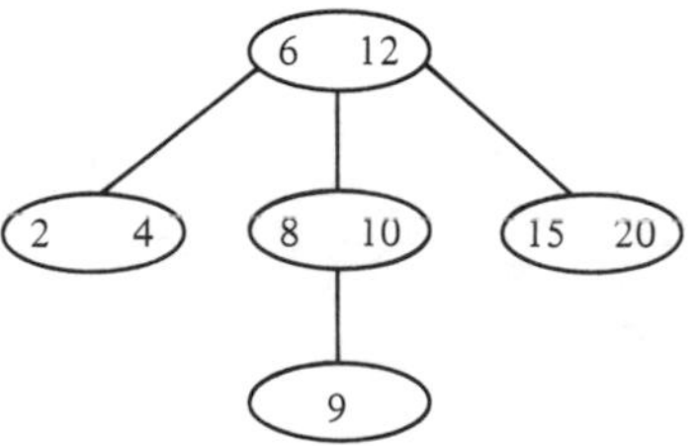

Figure 8.1-6. A 2–3 Tree.

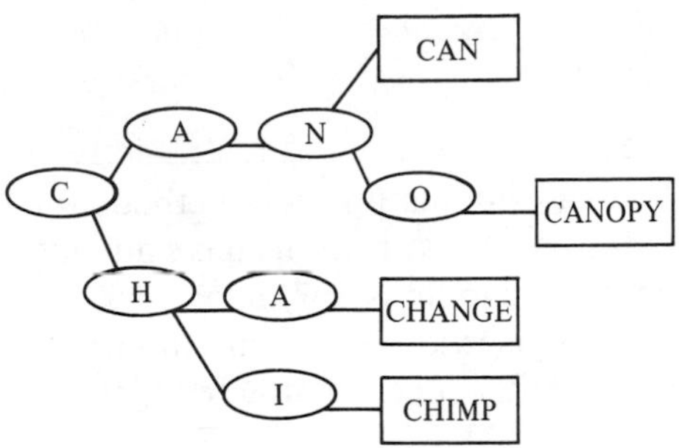

Figure 8.1-7. A Trie.

The search continues until the data element required is uniquely identified. Figure 8.1-7 illustrates a trie structure.

Graphs

Graphs or **network** data structures do not restrict data to a linear or hierarchical structure. Graphs are used to represent more complex relationships among data elements. The data elements in a graph may be unordered or ordered.

Unordered graphs can be represented with data elements as nodes, with undirected arcs representing direct relationships between nodes. For example, a route map with pickup and delivery nodes and arcs representing (two-way) routes between nodes is an unordered graph. The shortest paths in the graph would determine a routing schedule.

If the data elements have a defined ordering or dependency relationship, then a directed graph can be used. In a directed graph, each arc is one-way, denoted by a directed arrow. An example of an ordered graph is a **project network,** in which the nodes are events and the dependency ordering among events is represented by the directed arcs.

Summary: Data structures provide a basis for storing and retrieving data in information systems. Linear data structures facilitate sequential retrieval by ordering data in lists, queues, or stacks. Nonlinear data structures provide faster retrieval capabilities on ordered data or handle unordered data. Nonlinear data structures include sets, trees, and graphs.

8.2 FILE SYSTEMS

> **Preview:** File organizations provide methods for data retention and access. Various file organizations have advantages and disadvantages based on capabilities for data storage, sequential and random access, and insertion, deletion, and update. Inverted indexes provide additional access paths for files.

Files are groupings of similar records. These records are maintained together as a unit in a system over time. The long-term state of a system is normally made up of a number of files. Files are stored in a file organization. The primary distinction between various file organizations is the ability to support **sequential access** of records versus **random access** of records. Sequential processing of a file requires that each record be accessed in a defined logical order. This order is normally defined by the value of the primary key in the file. All records are accessed and processed in this order. Random processing requires the individual retrieval of records given their primary keys. Certain file organizations are designed to provide rapid sequential access of records; while some organizations are designed for rapid random access of records. Other file organizations provide both sequential and random access capabilities.

Businesses have varying requirements for file retention and access. In addition to sequential and random access, requirements for record insertion, modification, and deletion must be handled. The amount of storage in memory and on secondary storage devices must also be considered. This section describes advantages and disadvantages of popular file organizations.

Figure 8.2-1 shows a state machine representation of a generic file system. File systems provide access and retention functionality.

If F is declared as a file in the state, X is declared as a record variable, and key(X) is the primary key of the record type, then the following operations represent file processing possibilities in a file system state machine:

Sequential Retrieval:

X := **getfirst**(F)	Retrieve the first record of file F and assign it to X
X := **getnext**(F)	Based on the last record retrieved, retrieve the next record in primary key order and assign it to X

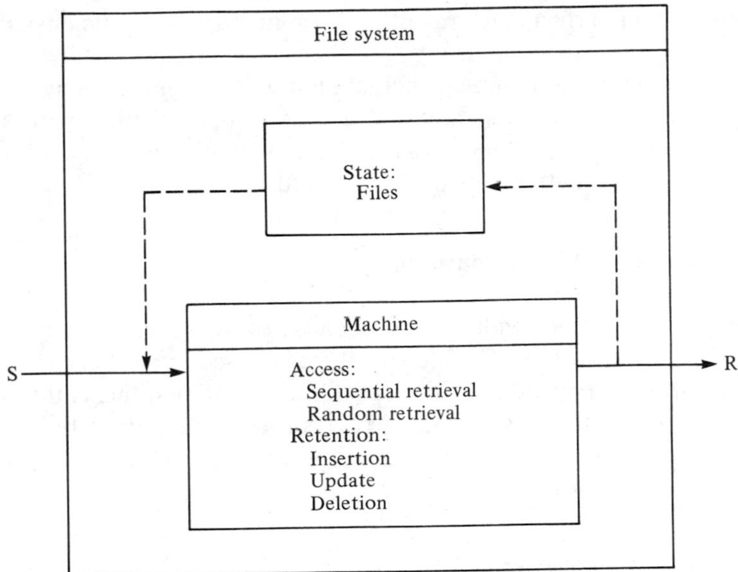

Figure 8.2-1. File System State Machine.

X := **getprior**(F) — Based on the last record retrieved, retrieve the prior record in primary key order and assign it to X

Random Retrieval:

X := **getrandom**(key(X)) — Given the primary key of record, retrieve that record and assign it to X

Insertion:

insert(F, X) — Insert record X into file F

Update:

replace(F, X) — Locate the record with key(X) in file F and replace it with record X

Deletion:

delete(F, key (X)) — Delete record with key(X) from file F

These operations could be provided by a common service box structure, whose users would remain independent of the file organizations and access methods employed. The user would not be independent, however, of poor system performance resulting from inappropriate file design decisions.

In what follows, the three principal file organizations, namely, sequential, direct, and indexed sequential, are described together with advantages and disadvantages of each. The section concludes with a discussion of access paths on attributes using inverted indexes.

8.2.1 Sequential File Organizations

Sequential files are widely used because of their simple structure and easy use in applications. The file structure is based upon record access in order of primary key value. For many file applications, this is the natural and preferred method of accessing records. The characteristics of a sequential file organization vary depending upon its implementation as a **table** or as a **linked list.**

Table Implementation

In a table implementation, records are stored in order by physical contiguity. A record physically follows the previous record in the file in logical order. A sequential file stored in this manner can be placed on all types of storage media, including tapes, disks, and random access memory.

The table implementation of a sequential file has a simple storage structure. Sequential record access is inherent in the structure. The major disadvantages are poor random access capability and the cost of maintaining physical record contiguity in the presence of insertions, deletions, and updates.

1. **Storage.** An advantage of the sequential file in a table implementation is that no storage space is wasted. No pointers, indexes, or free space are used in the file structure.
2. **Sequential Access.** Table implementations of sequential file provide rapid sequential record access.
3. **Random Access.** Sequential files have poor random record access performance. Locating a particular record, given a primary key value, requires a sequential search through the file until the key is matched. On average, such a search requires accessing one half of the records in the file.
4. **Insertion.** Inserting a new record requires first locating the position in the file to place the record. This is similar to a single record random

access. Next, all records after the new record must be physically moved in order to open space for the new record to be inserted. This can be a difficult and costly operation.

5. **Update.** Once a record is accessed, either sequentially or randomly, the values of the nonprimary key attributes may be updated. Note that updating a primary key is considered to be a deletion of the old record and an insertion of a new record. For **fixed-length** records, the updated record can then be written back into the same storage location. **Variable-length** records may require shifting other records to accommodate the larger or smaller record size. Since such shifting may be costly, variable-length records are rarely used in a sequential file using a table implementation.

6. **Deletion.** For deletion, first the record must be located. Next, the records after the deleted record must be moved forward into the newly freed space. Again, moving records can be a costly operation. A more efficient technique that can be applied to all file organizations is to **mark** deleted records. A marked record is treated as nonexistent in the file. At periodic intervals a procedure known as **garbage collection** is performed on the files to physically delete the marked records. Periodic garbage collection is essential, since marked records degrade performance on all operations.

Linked-List Implementation

In this form of sequential file implementation, records are maintained in the logical order of their primary keys by means of pointers (or links). This implementation need not maintain physical contiguity of records in logical order. Each record contains one or more pointer fields to provide record ordering. A **forward pointer** links a record to the next logical record in the file. A **backward pointer** links a record to the previous logical record in the file.

A storage area dedicated to a linked-list sequential file would be divided into two lists: a list of file records and a list of free space. The **free-space list** provides access to areas of storage that are not used. The operations of insertion, update, and deletion are handled efficiently through use of the free-space list. Only forward pointers are needed in a linked-list; however, backward pointers provide efficiencies during insertion and deletion of records.

Maintenance of a linked-list file structure places major importance on pointer integrity. An incorrect pointer value can destroy the file organization and render the data inaccessible to the user. The file system must maintain pointers; users are rarely given access to the pointer fields in the records. Linked-list implementations offer advantages in dealing with variable-length records and in file modification operations (insertion, update, and deletion).

1. **Storage.** Pointer fields require additional space in each record. For efficient operation, a certain amount of storage must be dedicated to the free-space list.

2. **Sequential Access.** Records are accessed sequentially from one record to the next by means of the forward pointers. This may be less efficient than the table implementation if linked records reside in different areas of storage.

3. **Random Access.** As in the table implementation, random access of a record would require a sequential search of the file. On average, one half of the records would be accessed.

4. **Insertion.** First the position of the new record must be located. Storage from the free-space list would be used to hold the data of the new record, which would be linked into the file by updating the forward pointer of the previous record, the backward pointer of the next record, and the pointers of the new record. Both fixed and variable-length records can be handled in this manner.

5. **Update.** First, the record must be accessed and the appropriate attribute values changed. The record is then written back into the same location. If the size of the record is increased or decreased substantially, adjustments must be made to the free-space list.

6. **Deletion.** A record to be deleted, once located, must be removed from the file and the storage added to the free-space list. The forward pointer of the previous record must be set to the address of the next record, and the backward pointer of the next record set to the address of the previous record.

8.2.2 Direct File Organizations

Many applications require random access of records based upon the primary key value (e.g., retrieval of a specific employee's information given an employee number). For an information system that handles a significant number of random access requests, a file structure that supports rapid, direct access to individual records is needed.

In this section, two methods of random access are discussed: **directory access** and **hashing.** A third method of random access, **indexing,** is described in the context of indexed sequential files.

Directory Access

For directory access, records are physically stored in secondary storage without regard to logical ordering. A **directory** is constructed in main memory that records the primary key of each record and the address at which it is stored. The directory is maintained in logical record order for rapid retrieval of a required record based on primary kcy valuc.

This file organization is effective for smaller files. However, as the size of the directory grows, large amounts of memory must be used and searching the directory becomes costly.

1. **Storage.** Efficient use is made of secondary storage. The directory, however, requires allocation of more costly main memory.
2. **Sequential Access.** Sequential access of records in the directory is fast since the directory is sorted on primary keys. The access of records from secondary storage, though, is done one at a time from, perhaps, different storage areas.
3. **Random Access.** Random access is efficient since the search for a required record is carried out in main memory. Fast searching techniques (e.g., binary search) can be used to locate the key. The address is then used to directly access the record. Directory search time becomes significant when the file size is large.
4. **Insertion.** Insertion is performed by adding the record in a free location in secondary storage and placing the (key, address) values in the directory in sorted order of the key.
5. **Update.** The record is randomly accessed, updated, and returned to its position in storage. The directory need not be modified.
6. **Deletion.** Deletion returns the record storage to free space and the (key, address) entry is removed from the directory.

Hashing

The hashing access method achieves the goal of retrieving a single record of a file in time independent of the size of the file. Note that this is not the case in directory access. Searching the directory requires time proportional to the number of entries (i.e., number of records in the file) in the directory.

Hashing achieves this file size independence by directly incorporating the primary key value of a record into its physical storage address. Thus, if x represents the key of a record, then a function f(x) calculates the address of the record; f is called the **hash function.** The secondary storage area in which the records are placed is called a **hash table.** A hash table, as shown in Figure 8.2-2, contains b buckets, numbered 0 to b − 1, with each bucket capable of holding s records. The parameters b and s are chosen based upon the size of storage available for the file and the optimal blocking size for the system. The hash function f is chosen to be easily computable and to minimize **collisions.** A collision occurs when the hash function maps two different keys, $x \neq y$, into the same bucket, $f(x) = f(y)$. The goal is to find a hash function such that $f(x) = i$, $0 \leq i \leq b - 1$, with probability 1/b, for all values in the primary key domain. In this way, an equal number of records would be placed in each bucket, and collisions minimized.

A number of hash functions have been proposed and analyzed. The technique that has proven to be the most useful is **division hashing.** In division hashing, the primary key value is divided by the number of buckets. The remainder of the division is taken as the bucket address in which the record is stored. Thus, $f(x) = x \text{ MOD } b$. Within the limits of available storage, the parameter b must be chosen to enhance the goal of giving f(x) a uniform distribution of values. Analysis results indicate that b should be a prime number that is not close in value to a power of 2.

The hashing file organization must have a technique for handling collisions. Records can hash into the same bucket until the bucket becomes full (i.e., it contains s records). Further records in the same bucket are handled by one of two **overflow techniques: open addressing** or **overflow linking.**

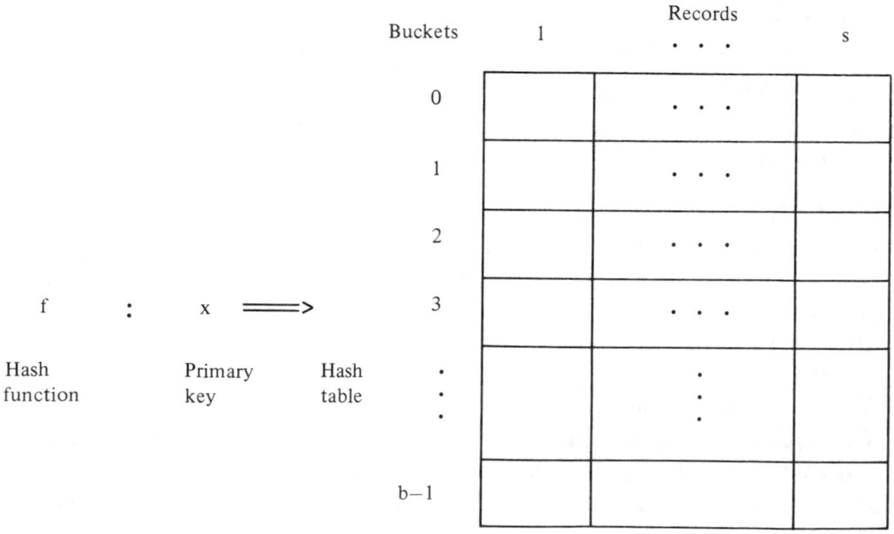

Figure 8.2-2. Hashing.

In linear open addressing, the overflow record would be placed into the next bucket in the table. If that bucket is full, then it is placed in the next bucket, and so on. Note that buckets are considered a cycle, such that the next bucket after b − 1 is 0.

The more popular overflow linking technique provides better record access performance, but requires additional storage for overflow records. An overflow area is designated to hold file records. When a record overflows its hash bucket, it is placed in the overflow area and linked to the bucket in the hash table.

The disadvantages of overflow linking are the additional storage needed for overflow and pointers, and the need to maintain pointer integrity in the file structure.

Hashing is an excellent file organization for random access requirements.

1. **Storage.** Hashing requires space for a hash table and an overflow area if overflow linking is used. This storage must be allocated at the beginning of file design. In general, the loading factor of a hash table runs between 50% and 80% full. Thus, a significant amount of space is unused at any time.

2. **Sequential Access.** A hash file is not designed for sequential record access. Sequential access requires each primary key, in turn, to be hashed and the record retrieved individually. Another method to achieve sequential access is to retrieve the entire hash table and overflow area and to sort all records in memory. This is also a costly operation.

3. **Random Access.** Hashing is extremely effective for random access. Random record access requires only the primary key to be hashed and the designated bucket to be retrieved. Based upon the overflow technique used, additional record accesses may be required in the overflow area or in further buckets in the hash table.

4. **Insertion.** Insertion is performed by locating the hash table bucket in which the new record is to be placed. If that bucket is full, the overflow technique is used to place the record in storage. As more and more records are inserted, the hash table may approach a loading factor of 80% to 90%. At this point, the performance of the file organization will be severely degraded because of the additional overflow accesses required for an arbitrary record. In this situation, the file can be reorganized by allocating more storage and altering the file parameters of b, s, and the hash function f.

5. **Update.** The record is retrieved, updated, and replaced into the same storage location.

6. **Deletion.** If open addressing is used for overflow, then deletion of a record from a hash table requires marking the new free space as previ-

ously used. This is essential since retrieval of overflow records continues until a nonfull bucket is found. The marked space can be reused by inserting new records. For overflow linking, a deleted record in the overflow area requires returning the space to the free list and modifying the chain from the appropriate bucket. If a record is deleted from the hash table bucket, then an overflow record (if any) can be moved into the bucket to improve performance.

8.2.3 Indexed Sequential File Organizations

The indexed sequential file organization is widely used to provide both efficient sequential and random access to records. Records in secondary storage are organized in logical, primary key order. In addition, a multilevel index is constructed based upon the primary key values. A random access uses the index, usually placed in memory, to isolate the block of storage in which the required record is located. Thus, both sequential processing and individual record processing can be performed efficiently at the expense of an index structure.

There are two methods of implementing an indexed sequential file organization: **Indexed Sequential Access Method** (ISAM) and **Virtual Sequential Access Method** (VSAM).

INDEXED SEQUENTIAL ACCESS METHOD (ISAM)

The ISAM architecture is presented in Figure 8.2-3. The ISAM terminology corresponds closely with the physical description of a disk. While the terms are suggestive, however, there need not be a direct mapping between the ISAM architecture and its physical representation on a disk.

The design of the ISAM file organization takes into account the size and volatility of the file to be stored. In each **cylinder,** the storage area is divided among **data tracks, overflow tracks,** and a small area that holds the **track index.** For files that have a higher rate of insertions and deletions, a proportionally larger overflow area is needed. The larger the file, the more levels of indexes are needed to provide fast access to individual records.

Loading an ISAM file requires the records to be in logical order of the primary key. Records are placed on data tracks in order, completely filling one track and then going to the next. The entire file is loaded across as many cylinders as necessary. No data are placed into the overflow tracks during loading. The index structure is built as the records are loaded. The track index contains one entry for each data track in the cylinder and an overflow pointer to the next free record position in overflow. Each data track entry has four fields:

High Key	Track High Key	Track Pointer	Overflow Pointer

High Key. All records with a primary key value greater than the high key of the previous track and less than or equal to this high key are placed on this track or in its overflow. This value does not change over time.

Track High Key. As records from this track are placed in overflow, the high key value physically on the data track will decrease.

Track Pointer. This pointer defines the address of the beginning of the data track. This value does not change over time.

Overflow Pointer. This pointer defines the address of the first record, in logical order, in overflow controlled by this data track. As records move into overflow, each data track maintains its records in logical order by linking.

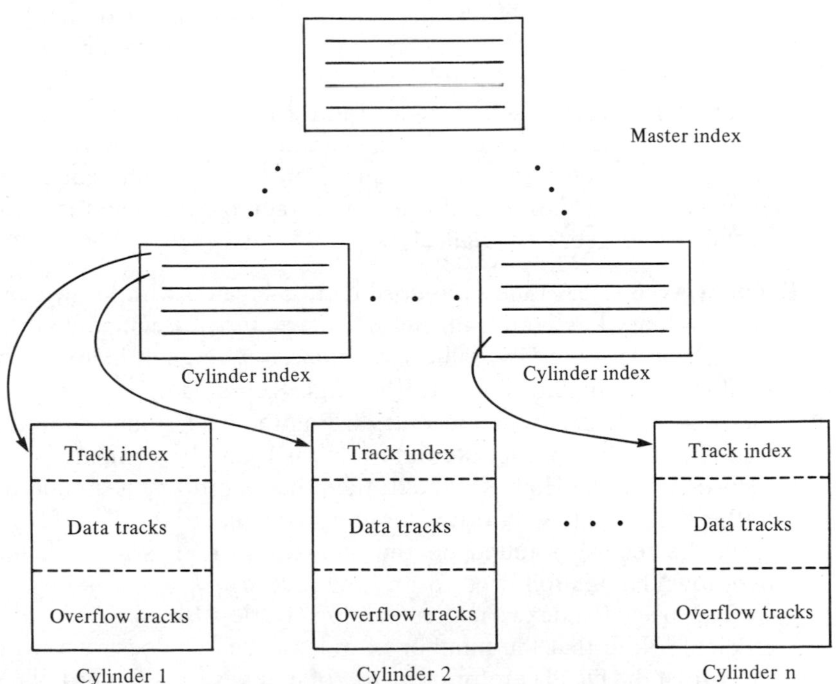

Figure 8.2-3. ISAM File Organization.

The higher level indexes, Cylinder Index and Master Index, are built by recording the high key and address for each index at the next lower level.

ISAM organizations require periodic reorganizations. A reorganization is triggered by running out of overflow space or by recognizing a performance degradation caused by lengthy overflow chains. A reorganization is performed by copying the file in logical order onto another storage device (e.g., tape storage). Then the ISAM organization is redesigned, the file is reloaded in logical order, and the new index structures are built.

ISAM provides efficient sequential and random access at the cost of index overhead and pointer chains through secondary storage.

Storage. In addition to data tracks, ISAM allocates space for overflow tracks and index structures. The index structures above the track index are normally maintained in memory.

Sequential Access. A sequential pass of the file does not use the high level index structure; only the track indexes are used. Each data track is accessed in order. For each track, the track index entry is used. The Track Pointer points to the beginning of the data track, and these records are accessed. The Overflow Pointer points to the beginning of the overflow chain for the track. This chain is followed by accessing each record in turn until the chain is complete. Thus, sequential processing can be done efficiently. ISAM is slightly less efficient for sequential processing than sequential file organizations because of the use of track indexes and the need to follow the overflow chains.

Random Access. A random record can be located with knowledge of its primary key by use of the index structures. Beginning at the topmost index, the key value is compared with each index entry until it is less than or equal to the High Key field. The next level index is accessed from the entry's Pointer field. Finally, at the track index, the appropriate data track is found by comparing the record key to the High Key field; then the record key is compared to the Track High Key. If it is less than or equal to the Track High Key, the record is found on the data track. If it is greater, the overflow chain is followed to find the record.

The use of indexes makes random retrievals in ISAM quite efficient. Note that the number of index accesses does vary with the size of the file. Therefore, the use of indexes does not provide a direct access cost independent of file size as does a hashing file organization. However, by placing the higher index levels in main memory, the index access time can be small.

Insertion. Insertion is performed by locating the correct position of the record through the indexes. The record is then placed on the data track or the overflow chain as previously discussed. The movement of data on thc data track can be costly.

Update. The record to be updated is retrieved through the indexes, and updated and replaced in the same storage location.

Deletion. Marking is an efficient means of deletion. The disadvantage is the growth of unused space throughout the file organization. Reorganization performs the garbage collection of this unused space.

Virtual Sequential Access Method (VSAM)

VSAM is an indexed sequential file organization that does not require periodic reorganization. The techniques of **distributed free space** and **cellular splitting** allow the file and index structures to expand and contract with the file. The VSAM architecture and terminology is given in Figure 8.2-4.

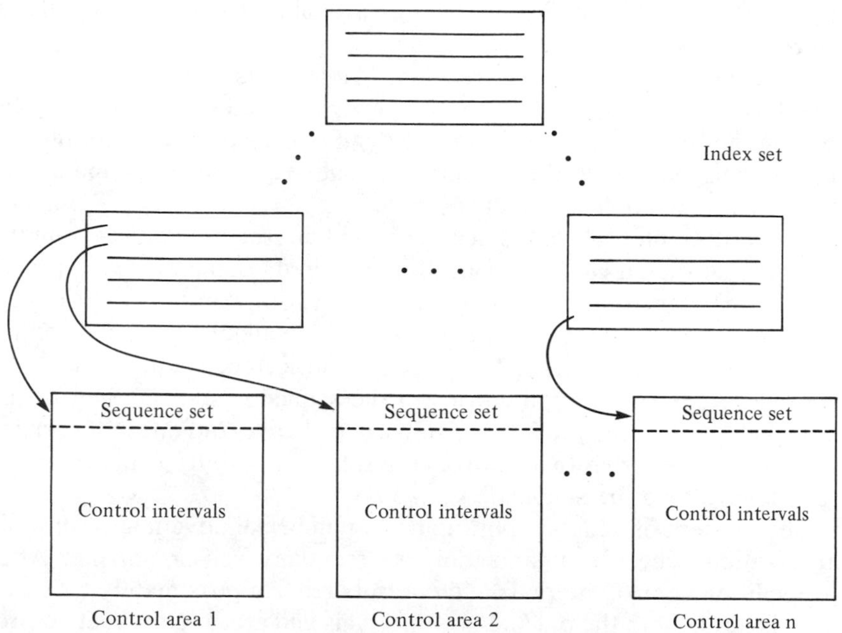

Figure 8.2-4. VSAM File Organization.

The design of VSAM includes the distribution of free space throughout the file organization. Based upon the volatility of the file, the initial VSAM design would allocate a certain percentage of free control areas. Within each control area, a percentage of control intervals would be unused. Within each control interval, a percentage of storage would be unused. Normally 10% to 40% of the space in control intervals and control areas would be initially allocated to free space.

Once the free space percentages are established, the file is loaded by inserting records in logical order into control intervals and control areas. VSAM indexes are built as the load is performed. The sequence set is the lowest level of index. The sequence set for each control area is built with an entry for each control interval. Each entry contains the High Key in that control interval and a Pointer to the beginning of the interval. The control interval entries are maintained in logical order of the High Key values. The higher level indexes in the index set contain the High Keys of and Pointers to the indexes at the next lower level.

The procedure for new record insertion uses **cellular splitting.** The control interval for the new record is found by comparing the primary key of the record to the high keys in the indexes. The new record is placed in the control interval in its logical position. If sufficient space is free on the interval, the records are moved to accommodate the new record in its correct position. If insufficient space is available in the interval, then a **control interval split** occurs.

In a control interval split, a free control interval is used. The records in the full control interval, including the newly inserted record in its logical position, are divided in two parts. One half is placed in the free interval and one half remains in the old interval. The sequence set is updated to include the new High Key and is sorted.

If a control interval split is necessary and no free intervals exist in the control area, then a **control area split** is required. Using the same procedure, a free control area is used. First, the control interval split is performed as if a virtual free control interval were available. A virtual control interval entry is placed in the sequence set and sorted. Then, based on the ordering of control interval entries in the sequence set, one half of the control intervals is moved into the new control area, and one half remains in the old control area. In a control area split, a new entry is placed in the index level above the sequence sets.

The concept of cellular splitting has a number of advantages for a file organization. The file organization is essentially self-organizing. After each split on control intervals and control areas, approximately 50% free space is left in both the old and new intervals and areas. Note that control intervals within control areas and control areas do not need to be maintained in logical order. Indexes serve to maintain logical order among and within control areas. The principles behind cellular splitting can also apply to cellular merging when adjacent intervals or areas are both less than 50% full.

VSAM provides the advantages of dynamic self-organization in addition to efficient sequential and random access.

Storage. The use of distributed free space requires a considerable storage allocation in secondary storage. Space for high-level indexes is also required in memory.

Sequential Access. Different from ISAM, the VSAM organization requires using the index set to locate each control area in logical order. Within each control area, the sequence set must be used to access each control interval in logical order. Records in the intervals are maintained in sequential order. Overall, however, the cost of sequential access is similar between ISAM and VSAM, since VSAM does not require the access of overflow chains as does ISAM.

Random Access. Given the primary key of a record, it can be retrieved by following through the index levels to the control interval containing the record. VSAM has more efficient random access characteristics than ISAM. At most, only the records within a control interval need to be searched, as overflow chains are avoided.

Insertion. Insertion involves placing new records in logical order in control intervals. Cellular splitting dynamically reorganizes the file organization and provides distributed free space.

Update. The record is randomly accessed, updated, and replaced in the file organization.

Deletion. A record to be deleted is located in a control interval. The storage used by the record is denoted as free space in the interval and reused as subsequent records are inserted in the interval. Under conditions when adjacent control intervals have a high percentage of free space, a cellular merging procedure can be used to create a free control interval. Likewise, control areas can be merged to form a free control area.

8.2.4 Multiple Key Access

The retrieval of specified records from a file can be formally stated as a **query** on the file. A query is made up of one or more **attribute value clauses** connected by the boolean operators AND, OR, and NOT. Each clause specifies a value or range of values that an attribute must contain. The clause is evaluated on each file record to be TRUE or FALSE.

To illustrate, consider the following attribute value clauses for a softball team file; TEAM (Player Number, Player Name, Position, Games Played, Batting Average, Runs-Batted-In):

C1: (Position = IF)

C2: (Batting Average $\geq$ .300)

C3: (Games Played < 10)

Each of these clauses will be TRUE or FALSE for each of the records in the file. For a record (Player Number = 17, Player Name = Ellis, Position = OF, Games Played = 8, Batting Average = .350, Runs-Batted-In = 9), clause C1 is FALSE, clause C2 is TRUE, and clause C3 is TRUE.

A query forms clauses into a boolean formula that evaluates into a TRUE or FALSE value for each file record. The boolean operators and their respective truth tables for clauses P and Q are found in Table 8.2-1.

A query on the softball team file could be

Q1: (Position = IF) AND (Batting Average $\geq$ .300)
OR (Runs-Batted-In > 10).

To evaluate this query on each record, the precedence of the boolean operators must be determined. In most systems, the precedence order is NOT, then AND, then OR. Parentheses may also be used to establish the order of operator evaluation. The use of parentheses is highly recommended to clarify the semantic meaning of a query. With operators of equal precedence, the order of operation is from left to right. Query Q1,

Table 8.2-1

Boolean Truth Tables

P	Q	P AND Q	P OR Q	NOT P
T	T	T	T	F
T	F	F	T	F
F	T	F	T	T
F	F	F	F	T

then, can be stated, "Retrieve players who are infielders with batting averages greater than or equal to .300 or players who have batted in more than 10 runs."

The semantic meaning of the query can be changed by adding parentheses as

Q2: (Position = IF) AND
((Batting Average ≥ .300) OR (Runs-Batted-In > 10)).

Query Q2 is "Retrieve players who are infielders and whose batting averages are greater than or equal to .300 or whose runs-batted-in are greater than 10."

Queries can contain clauses involving any attribute in a file. Clauses based upon primary keys (e.g., (Player Number ≥ 30)) will require the use of the file organization structure to retrieve records. Clauses on secondary keys (e.g., (Position = IF)), however, cannot make use of the file structure since it is based solely on the primary key. The secondary key access structures of **inverted indexes** provide an efficient capability to retrieve records based upon secondary key values. This structure can be implemented in conjunction with any of the previously described file organizations.

An essential part of file design is to decide which of the secondary attributes should have an access path on its values. The decision is usually made based upon the characteristics of the attribute and the frequency of its use in file queries. The decision trade-off is the improvement in retrieval time for queries that use the secondary access structures versus the cost of maintaining and updating the structures during insertions, updates, and deletions on the file. This analysis will be discussed for inverted indexes.

An inverted index uses a directory of record addresses to provide direct access to all records having a specific attribute value. In main storage, the directory is maintained with one entry for each attribute value. Each entry is of the form:

Attribute Value	Record Addresses

Entries are variable length since the number of records having the attribute value varies among values. Insertions, deletions, and updates of records in the file will cause the directory entries to be updated. Therefore, significant directory storage space is required for an inverted index and directory maintenance is needed during insertions, deletions, and updates.

Position Inverted Index Directory

Value	Record Addresses
IF	A(1), A(23), A(32), A(34), A(49)
OF	A(5), A(7), A(27), A(45)
P	A(16), A(41)
C	A(38)

Batting Average Inverted Index Directory

Value	Record Addresses
0.000 TO 0.250	A(16), A(41)
0.251 TO 0.300	A(23), A(34), A(49)
0.301 TO 0.350	A(1), A(27), A(32)
0.351 TO 0.400	A(38)
0.401 TO 1.000	A(5), A(7), A(45)

Figure 8.2-5. Inverted Index Example.

Figure 8.2-5 shows sample inverted indexes for Position and Batting Average on the softball team file with 12 players. The notation A(1) is the address of the record for player number 1. The following queries demonstrate the use of inverted indexes.

Query 1 (Position = OF):

The Position directory is searched for the outfielder entry. The desired records are directly retrieved via the entry address list.

Query 2 (Batting Average ≥ .325):

The player records are retrieved for the final three entries in the Batting Average directory. The records in the .301 to .350 range are checked for averages less than .325.

Query 3 (Runs-Batted-In = 10):

Since Runs-Batted-In is not indexed, all records must be retrieved and checked.

Query 4 (Position = OF) AND (Batting Average > .400):

A principal advantage of inverted indexes is that addresses for all records are in the directory. Address lists can be compared among indexes on different attributes. To solve AND conditions, the record addresses that satisfy each clause can be **intersected** to find those records that satisfy *both* clauses. Then these records are retrieved. For this query, the required players would be found by:

$$\underbrace{(A(5), A(7), A(27), A(45))}_{\text{Position = OF}} \cap \underbrace{(A(5), A(7), A(45))}_{\text{Batting Average} > .400}$$

$$= (A(5), A(7), A(45))$$

Query 5 (Position = P) OR (Batting Average $\leq$ .300):

The OR condition uses a **union** of record addresses satisfying each clause to find the query result. Thus, the result of this query is:

$$\underbrace{(A(16), A(41))}_{\text{Position = P}} \cup \underbrace{(A(16), A(23), A(34), A(41), A(49))}_{\text{Batting Average} \leq .300}$$

$$= (A(16), A(23), A(34), A(41), A(49))$$

Query 6 (Position = IF) AND (Games Played $<$ 10):

Since Games Played is not indexed, the infielder records would be retrieved and checked for the Games Played value.

Query 7 (Batting Average $>$ .350) OR (Runs-Batted-In $\geq$ 10):

All records must be retrieved in an OR operation if one or both of the attributes in the clauses are not indexed.

The above query optimization guidelines are also used in more complex queries. Once the precedence order of the boolean operations is established, then an overall strategy for the query can be formulated. For example, given the inverted indexes on Position and Batting Average, consider the following two queries.

Query 8: (Position = IF) OR
((Games Played = 10) AND (Position = OF))

The AND operator is performed first. The Position index is used to retrieve outfielders whose records are checked for Games Played value. This result is then combined (unioned) with all infielder records for the result.

Query 9: (Runs-Batted-In $\geq$ 10) OR
((Position = P) AND (Batting Average $>$ .250))

The AND operation can use the indexes on Position and Batting Average. However, retrieving the records that satisfy the AND operation is wasted effort since executing the OR operation requires that all player records be retrieved anyway. Thus, query optimization would find that the best way to perform this query is to retrieve all records and check the appropriate attribute values.

Inverted indexes provide efficient access capabilities at the expense of directory storage in memory.

Storage. The directory can require considerable memory for large files. No extra secondary storage space is needed in the file organization.

Record Access. Records are directly accessed via their pointers in the directory. The ability to perform intersect (for AND), union (for OR), and complement (for NOT) on address sets provides efficient query optimization on files.

Record Retention. The insertion, deletion, and update of records will require maintenance of the inverted index directory. This expense may be large for volatile files. The inverted index structure is relatively independent of file reorganization. The only changes occur in the directories since the file structure does not include index information.

Summary: Files should be organized to best satisfy system requirements for storage, access, and maintenance. Sequential files are designed for storage efficiency and sequential access. Direct files provide excellent random access. Indexed sequential files allow efficient sequential and random access. Inverted indexes provide additional access paths on files for query optimization.

8.3 DATABASE SYSTEMS

Preview: Database systems manage information about entities and relationships in an application environment. Provisions for data access, retention, and control are included in database systems.

8.3.1 Database System Architecture

A database forms information from an application environment into an integrated representation. The information is described in terms of the entities involved and the relationships among the entities. A database can be defined in a three level database architecture as shown in Figure 8.3-1.

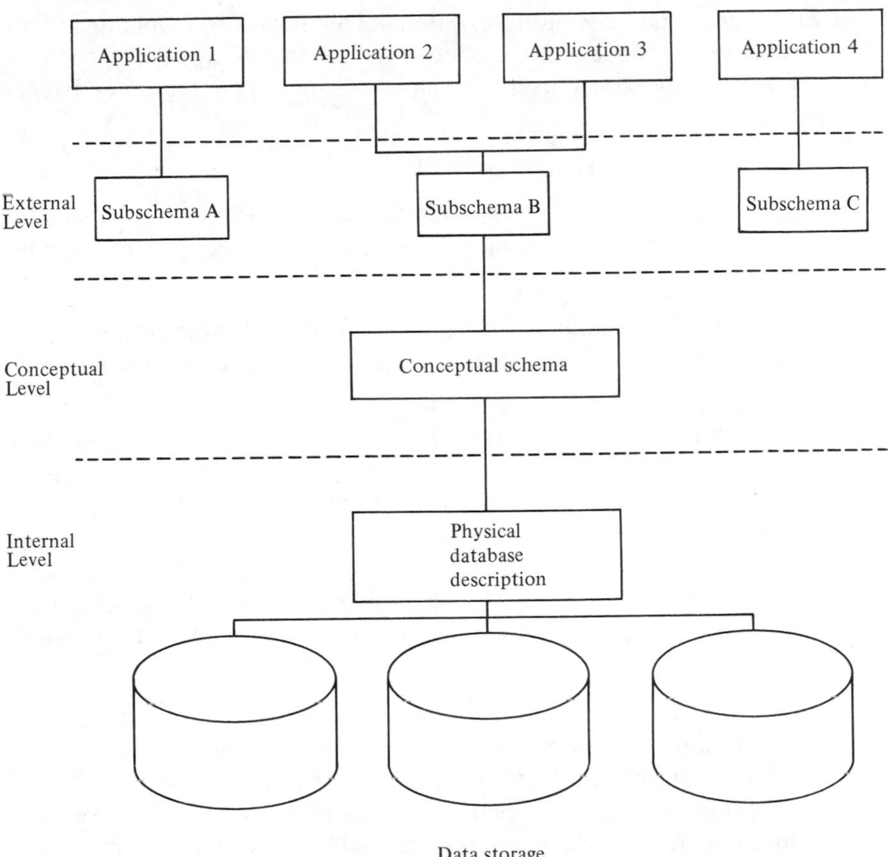

Figure 8.3-1. A Three Level Database System Architecture.

The **Internal Level** describes the organization of data in secondary storage. The file organizations and access paths (e.g., indexes) are recorded in the physical database description. The **Conceptual Level** represents the database as an integrated schema, or picture, of the entities and relationships. The schema provides a semantic understanding of the information that is independent of the actual data storage details of the physical level. This concept is called **physical data independence.** The database system provides this independence by the automatic mapping from the conceptual schema into the physical database description. The **External Level** supports subschema declarations for application programs. Subschemas are specialized views of the schema designed for different applications. A mapping that provides **logical data independence** is performed between the subschema and schema descriptions. This form of independence allows a user to be independent of conceptual schema details that are irrelevant or unavailable to that user because of security and integrity considerations.

A database system is composed of

Hardware. A general purpose computer system or a specially designed database machine is used for database processing and storage.

Software. The Database Management System (DBMS) is a sophisticated software system that manages and controls all database activities.

Data. The information in the database is of two types. The data values of the application environment are stored at the physical level on secondary storage devices. Other information, or meta-data, is needed to allow the management and control of the database. For example, descriptions of each level in the database architecture are required for the mappings between the levels. This meta-data is usually stored and maintained in a **Data Dictionary/ Directory System.**

Users. A database system is used by different groups that have different user requirements. A **superuser** group includes a database administrator and system operators. They are responsible for describing the data at all levels of the database system. They require intimate access to the complete system. **Application users** need a programming level of access to the database system. They need facilities for declaring subschemas and accessing the database through high-level programming languages such as COBOL and PL/I. Newer capabilities for application generation, such as form interfaces and report writers, are becoming more available. **Casual users** make use of query language interfaces to the database. These interfaces include easy-to-learn, easy-to-use nonprocedural languages, such as SQL and QBE, for manipulation of the database.

As a box structure, a database system can be viewed in a state machine diagram as shown in Figure 8.3-2. In addition to the access and retention capabilities provided to users through host languages and query languages, a major responsibility of database systems is to provide extensive control facilities. The control processing includes automatic schema mapping and integrity provisions. The database user may not be aware of this control, but these control functions are what support the implementa-

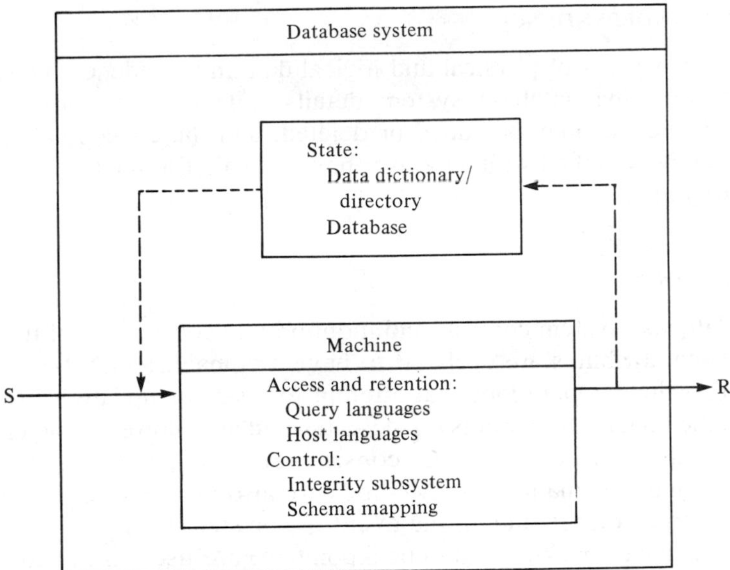

Figure 8.3-2. Database System State Machine.

tion of the database concept. The database concept, as embodied in database systems, provides substantially more capability and control over data than do file systems. The most important objectives of a database system are as follows.

Data Integration

A database is viewed by the user as a centralized repository of all organizational information. Access paths provide for the efficient integration of data from multiple files. Harmful redundancy of data can be eliminated by only recording a unit of information once and allowing all applications access to the data. Boundaries that inhibit information flow, such as departments or geographical areas, are removed in the database concept.

Data Sharing

Database systems of reasonable size provide multiple user capabilities. Multiple retrieval applications are allowed. However, control must be provided for update operations on the same data. Database systems provide methods of mutual exclusion (e.g., locking) so that data integrity is not corrupted by losing valid updates or reading invalid data.

Data Independence

The provisions of physical and logical data independence relieve the user of knowing database system details. File organizations may be changed, indexes may be added or deleted, and the conceptual schema may even be modified without a user having to alter a functional application program.

Data Integrity

A database system controls and monitors the correctness of the database. Data attributes are declared to have a domain, a data type, and a range of values. Operations that attempt to insert or update values that violate the domain constraints are detected and not allowed. When redundant data is used in a database, consistency of values among the data instances must be maintained. Methods are also provided to recover the database to a correct state in the event of a system failure. Techniques, such as logging, backups, and checkpointing, are used to minimize the database damage and lost work when a failure occurs.

Data Security

The centralization of information presents a threat for the compromise of all information during a breach of security. Therefore a database system must have strong security measures. Such measures include password authentication on subschemas, data encryption, and user access lists that limit the capabilities (e.g., read, write, insert, delete, update) a user has on parts of the database. The use of subschemas allows the user to only view the portion of the database in the subschema. These software techniques are used with physical security measures (e.g., locking doors and terminals, shielding transmission lines) to provide a comprehensive data security plan.

Application Generation

The important concept of data independence supports a wide range of application generation capabilities. Nonprocedural query languages have become very popular because of their English-like syntax (e.g., SQL) and use of two-dimensional screen presentation (e.g., QBE). A new generation of application generators, so-called Fourth Generation Languages (4GL), include sophisticated forms systems, report writers, and icon-based office information systems. The goals of such database interfaces are to improve productivity in developing applications and to allow casual users to access the database with more powerful capabilities.

System Performance

The extensive capabilities of a database system require considerable processing and storage resources. System performance is an important measure of effectiveness in any organization. The database algorithms for query processing, integrity control, concurrency control, recovery, and security must be efficient. The selection and tuning of file organizations and indexes is a critical component of performance. Database systems normally maintain hardware and software monitors to measure system performance.

System Administration

Responsibility for the administration and control of a database system lies in the role of the **Database Administrator (DBA).** The DBA may be one or more individuals depending upon the size of the organization. The responsibilities of this role include day-to-day system operation, user liaison and support, and system maintenance and control. The DBA must ensure that the system delivers adequate levels of integrity, security, reliability, and performance to the users. Provisions for auditing the system may also be required in many environments.

Data Models

Database systems differ in the type of **data model** used to describe the conceptual schema of the database. A database designer uses a schema to represent the required semantics in an application environment. A data model supports this conceptual modeling task, while leaving the physical database details for the next step of database design. The three prevalent data models are the **relational, hierarchical,** and **network** data models.

To illustrate these models, a simple application environment will be considered. An organization has a number of departments, defined by department number (D#), name (DNAME), a manager (MGR#), and location (LOC). A department has employees, defined by employee number (E#), name (ENAME), position (POS), and salary (SAL). An employee works in one department. Each department is responsible for an equipment inventory. Each piece of equipment has attributes inventory number (INV#), description (DESC), and cost (COST). Employees work on projects, defined by project number (P#), project name (PNAME), and work site (SITE). An employee may work on many projects. The hours (HRS) worked on each project are recorded. Each project is administered by one department, and a department can administer many projects.

This description allows the designer to recognize the entities and relationships in this environment. With each relationship the functionality of the relationship must also be discovered. Three types of relationships exist:

One-to-One Relationship. An instance of one entity is related to at most one instance of the other entity and vice versa. For example, an employee can manage at most one department and a department is managed by at most one employee.

One-to-Many or Many-to-One Relationship. An instance of the one entity is related to any number of instances of the many entity. However, an instance of the many entity is related to at most one instance of the one entity. For example, a department has many employees but an employee is in at most one department.

Many-to-Many Relationships. An instance of an entity is related to any number of instances of the other entity and vice versa. For example, a project is worked on by many employees and an employee can work on many projects.

Table 8.3-1 lists the entities and relationships found in the example application environment. As we describe the manner in which the different data models handle this example, note that each represents relationships differently.

8.3.2 Relational Databases

In the relational data model the mathematical concept of a **relation** is used to conceptually structure data. A relation is a two-dimensional, tablelike structure as shown in Figure 8.3-3. The number of columns of the relation is known as the **degree** of the relation, and the number of rows (called tuples) is known as the **cardinality** of the relation. The similarity of a relation and a flat file can easily be observed where columns represent attributes and tuples represent records in the file.

For relational data modeling, each entity forms a separate relation. Relationships are represented by adding attributes to entity relations for one-to-one and one-to-many relationships or building new relations for

Table 8.3-1

Entities and Relationships

ENTITIES

Name	Attributes
Department	D#, DNAME, MGR#, LOC
Employee	E#, ENAME, POS, SAL
Equipment	INV#, DESC, COST
Project	P#, PNAME, SITE

RELATIONSHIPS

Name	Type	Attributes
Employees manage departments	One-to-One	
Employees work in departments	Many-to-One	
Equipment is inventoried in departments	Many-to-One	
Employees work on projects	Many-to-Many	HRS
Departments administer projects	One-to-Many	

many-to-many relationships. By matching the values of attributes among relations the relationships in the database are modeled. Relational data modeling is illustrated for the example database.

Each entity, as shown in Table 8.3-1, becomes a relation with appropriate attributes.

Relation	Column 1	Column 2	. . .	Column n
Tuple 1			. . .	
Tuple 2			. . .	
Tuple 3			. . .	
Tuple 4			. . .	
Tuple 5			. . .	
Tuple 6			. . .	
	⋮	⋮	⋮	⋮
Tuple m			. . .	

Figure 8.3-3. A Relation.

DEPARTMENT (D#, DNAME, MGR#, LOC)
EMPLOYEE (E#, ENAME, POS, SAL)
EQUIPMENT (INV#, DESC, COST)
PROJECT (P#, PNAME, SITE)

The one-to-one "manages" relationship is already represented in DEPARTMENT by including the attribute MGR#. MGR# and E# in EMPLOYEE are defined on the same domain allowing values to be matched across these two attributes. Thus, a department can be related to the employee who manages it, and vice versa.

One-to-many relationships are represented by placing a new attribute in the "many" entity that uniquely identifies the instance of the "one" entity to which it is related. For the "works-in" relationship between EMPLOYEE and DEPARTMENT, and attribute D#, the unique primary key of DEPARTMENT, would be added to EMPLOYEE.

EMPLOYEE (E#, ENAME, POS, SAL, D#)

This new attribute is termed a **foreign key,** since it relates to a primary key in another relation. Now an employee is related to a unique department and a department is related to a set of employees. In the same way the "inventory" and "administers" relationships are represented as follows:

EQUIPMENT (INV#, DESC, COST, D#)
PROJECT (P#, PNAME, SITE, D#)

A many-to-many relationship requires the construction of a new relation that contains an attribute for the primary key of both involved entities. Data that is recorded based upon the relationship of specific instances of each entity is stored as an attribute in this relation. For example, the "works-on" relationship between EMPLOYEES and PROJECTS generates a new relation with foreign keys E# and P# and intersection data of the hours that a specific employee works on a specific project.

WORKS-ON (E#, P#, HRS)

By integrating data across the three relations EMPLOYEE, WORKS-ON, and PROJECT, information on what project an employee works on and information on what employees work on a project can be obtained in a symmetric manner. The complete relational schema for the example database is shown in Figure 8.3-4.

The relational data model has many attractive features for conceptual modeling. Its basis in mathematical theory provides several advantages. A relational algebra and a relational calculus have been modified into user-friendly query languages such as SQL, QUEL, and QBE. Relational

DEPARTMENT	(D#, DNAME, MGR#, LOC)
EMPLOYEE	(E#, ENAME, POS, SAL, D#)
EQUIPMENT	(INV#, DESC, COST, D#)
PROJECT	(P#, PNAME, SITE, D#)
WORKS-ON	(E#, P#, HRS)

Figure 8.3-4. Relational Schema.

theory also is used to develop and analyze methods to achieve query optimization, data integrity, concurrency control, and recovery, among others. Finally the data model uses only one consistent data structure at the conceptual level, the relation. The ability of this simple structure to represent all forms of semantic meaning is an extremely powerful concept.

8.3.3 Navigational Databases

The hierarchical and network data models are called **navigational** models because they exhibit explicit access paths through the conceptual data representation. Users, thus, navigate the database along these paths in applications. Relationships among entities in the navigational data models are represented by the existence of an access path from one entity to the other. The explicit representation of relationships as access paths, while providing navigational guidance to application programmers and physical database designers, hampers the goal of data independence by requiring that physical data structures support the access paths.

The hierarchical data model represents all information in tree-like, parent-child structures. A parent entity instance is related to any number of child entity instances. A child entity instance is related to one and only one parent entity instance. Thus, the parent-child relationship is one-to-many or one-to-one. Many-to-many relationships cannot be directly represented in a tree structure. A hierarchical database system, such as IMS, gets around this problem by describing a database as a set of tree structures with logical access paths between trees to allow many-to-many relationships. Each tree in the physical database can be clustered in hierarchical sequence for access efficiency. Logical access paths are implemented as pointers between records in secondary storage.

Figure 8.3-5 shows the example database represented in the IMS hierarchical data model. The example database can be represented by two trees with two logical access paths between them. The many-to-many relationship requires that the employee entity and the project entity be in separate trees. The "works-on" information then is stored in TREE 2 as a

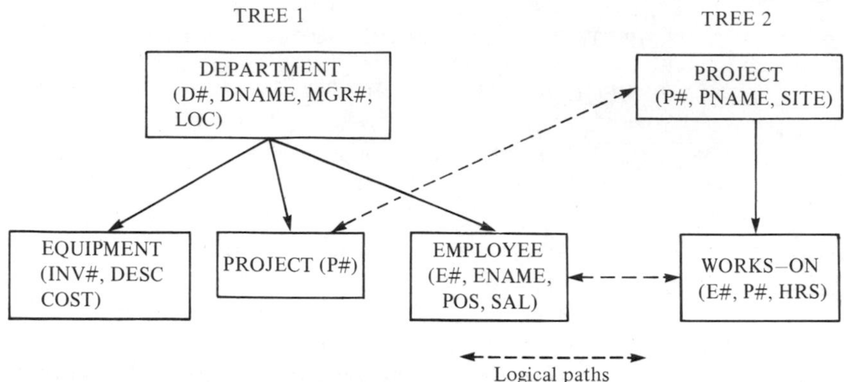

Figure 8.3-5. Hierarchical Schema.

child of the appropriate project with a pointer from each "works-on" record to the appropriate employee record in TREE 1. Pointers in the other direction allow an employee to be related to his or her appropriate "works-on" records. Since the complete project information is stored in TREE 2 there is no reason to duplicate this data in TREE 1. However, project records must appear in TREE 1 to show the one-to-many relationship between departments and projects. This is done by placing skeleton project records in TREE 1 that have pointers to the corresponding complete project record in TREE 2.

The network data model uses access paths between entities in a general network structure. Each access path, called a **set,** links an owner entity with a member entity in a one-to-many relationship. As in the hierarchical data model, a many-to-many relationship cannot be directly implemented. However, in a network this relationship can be easily represented with two access paths. A group of intersection records is formed into an entity. This entity then serves as the member entity through access paths from both owner entities.

The structures in the network data model can be observed in the example database as modeled in Figure 8.3-6. Each set is implemented in the physical database as follows. In the owner entity, each instance is the head of a linked list that connects it to each of its member instances in the member entity. Access efficiencies can be gained by physically clustering an owner entity with one of the member entities to which it is related.

We note that both of the navigational data models rely on some form of pointer implementation in the physical database design. Careful maintenance and control of these pointers and the resulting access paths are essential.

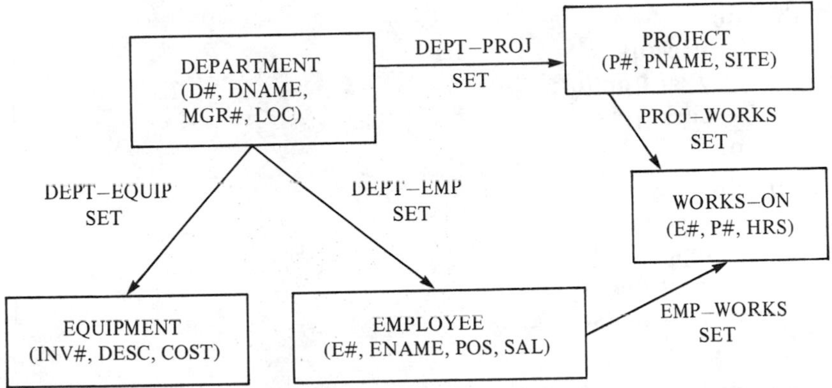

Figure 8.3-6. Network Schema.

Summary: Database systems implement the database concept. Principal features of the database concept are data sharing, data independence, and control over all information in an application environment. The relational data model presents a conceptually clean (e.g., no pointers) method of viewing information. Navigational data models (i.e., hierarchical and network) represent relationships as access paths in the conceptual schema.

EXERCISES

1. Construct record declarations for
   (a) dates
   (b) baseball teams
   (c) coefficients of quadratic expressions
2. Given the declarations

   q:queue of number
   s:stack of number
   x, y:number

   and q = (3 6 1), s = (3 6), x = 9, y = 6, describe the effects of each of the following sequences

   (a) **do**

   ```
       back(q) := x;
       back(q) := y;
       y := front(q)
   od
   ```

(b) **do**

```
    back(q) := x;
    y := front(q);
    back(q) := y
od
```

(c) **do**

```
    top(s) := x;
    top(s) := y;
    x := top(s);
    y := top(s)
od
```

(d) **do**

```
    x := back(q);
    y := top(s);
    top(s) := x;
    front(q) := y
od
```

3. Specify the set operations of Union, Intersection, and Difference as black boxes and state machines for the set data structure. Consider the stimulus to be the appropriate operator, the response to be the resulting set, and the state to include the sets participating in the operation.

4. You are asked to develop an effective file organization for your local softball team data. The file has the following attributes:

TEAM (PLAYER#, PNAME, POS, GAMES, BA, RBI)

You decide to compare the methods of sequential files, ISAM files, and hash files. The structure of each file is as follows:

(i) Sequential File. Records are stored and maintained in a physically contiguous manner. There is sufficient space to store all records.

(ii) ISAM File. There are four tracks per cylinder (1 track is overflow) and each track holds four fixed length records. The track index is stored in the first record of the first track.

(iii) Hash File. There are 13 buckets in the hash table, each holding one record. There are additional record locations in an overflow area. Use the hash function f(x) = x mod 13, and overflow chaining for collisions.

(a) Load the following example data into **each** file organization. Use diagrams with addresses and pointers to best convey the file organization. Discuss briefly the advantages and disadvantages of each organization in terms of storage space required.

SOFTBALL TEAM

PLAYER#	PNAME	POS	GAMES	BA	RBI
5	Abel	IF	8	.290	6
12	Baker	OF	9	.320	10
19	Jones	C	9	.275	5
25	Mills	IF	7	.310	11
29	Smith	P	5	.150	2
32	Hill	OF	9	.333	15
33	Mars	IF	9	.295	10
38	North	OF	8	.380	12
49	Lopes	IF	9	.265	8
50	Toms	P	4	.190	2

(b) Perform the following operations on **each** file organization. Discuss briefly the advantages and disadvantages of each organization in terms of ease of insertion and deletion.

- Insert record (23, Giles, OF, 7, .325, 5).
- Insert record (3, Rowe, IF, 8, .288, 8).
- Delete record with PLAYER# = 38.
- Insert record (6, Wicks, P, 3, .200, 4)

(c) Describe how each file organization supports i) sequential processing of the entire file and ii) random retrieval of a single record. Discuss the advantages and disadvantages of each organization on these operations.

5. Consider the file in problem 4. as initially loaded (i.e., before any insertions or deletions) in the ISAM file organization.

(a) Construct inverted indexes on attributes position (POS), games played (GAMES), and batting average (BA).

(b) Given the indexes in (a) design the best access method (i.e., minimize the number of records accessed) to find the PLAYER records that satisfy the following queries:

(i) List infielders (IF) who have a batting average less than .300.

(ii) List players who have played in 9 games and who are batting over .300 or have 10 or more RBI's.

6. Consider the EMPLOYEE file given below:

EMPLOYEE

ADDRESS	E#	NAME	D#	SKILL	SAL (xK)
101	731	Smith	10	PL	32
102	745	Jones	10	AD	28
103	752	Wilson	20	PL	35
104	780	Bell	20	SE	40
105	802	Walker	10	AD	26
106	810	Hill	20	AD	24
107	820	Jones	20	PL	36
108	831	Thomas	20	FI	31
109	850	Heck	40	AD	22
110	853	Miller	20	AD	42

(a) Construct inverted indexes on attributes D# and SKILL.

(b) Given the index structures that you constructed in a. design the best access method (i.e., minimize the records accessed) to find the EMPLOYEE records that satisfy the following queries:

- List all employees who work in department 10 or who earn between 30K and 40K salary.
- List all plumbers (PL) in department 20.
- List all adjusters (AD) who do not work in department 20.

7. Briefly discuss the differences between database systems and file systems.

8. Consider the following description of a database for a small college.

(i) The college is divided into departments defined by a department name [DNAME] and the chairman [CHAIR] who is a faculty member.

(ii) Faculty members are assigned to a department. Faculty are identified by a faculty number [F#], a name [FNAME], and a phone number [PH#].

(iii) Students are defined by their student number [S#], name [SNAME], address [ADDR], and grade point average [GPA]. A student is assigned to a department as a major.

(iv) Courses are given within departments. A course has a course number [C#] and a title [TITLE].

(v) A class is defined as a course taught by a faculty member during a specific semester [SEM]. Students attend classes and receive a grade [GRADE] in the class.

Design the small college database in the following data models:

(a) Relational Data Model

(b) Hierarchical Data Model

(c) Network Data Model

Index

E

F

P

Q

R

S

T

U

V

W